Lecture Notes in Physics

Springer
Berlin
Heidelberg
New York
Barcelona
Hong Kong
London
Milan
Paris
Tokyo

Physics and Astronomy

ONLINE LIBRARY

http://www.springer.de/phys/

The Editorial Policy for Edited Volumes

The series *Lecture Notes in Physics* (LNP), founded in 1969, reports new developments in physics research and teaching - quickly, informally but with a high degree of quality. Manuscripts to be considered for publication are topical volumes consisting of a limited number of contributions, carefully edited and closely related to each other. Each contribution should contain at least partly original and previously unpublished material, be written in a clear, pedagogical style and aimed at a broader readership, especially graduate students and nonspecialist researchers wishing to familiarize themselves with the topic concerned. For this reason, traditional proceedings cannot be considered for this series though volumes to appear in this series are often based on material presented at conferences, workshops and schools.

Acceptance

A project can only be accepted tentatively for publication, by both the editorial board and the publisher, following thorough examination of the material submitted. The book proposal sent to the publisher should consist at least of a preliminary table of contents outlining the structure of the book together with abstracts of all contributions to be included. Final acceptance is issued by the series editor in charge, in consultation with the publisher, only after receiving the complete manuscript. Final acceptance, possibly requiring minor corrections, usually follows the tentative acceptance unless the final manuscript differs significantly from expectations (project outline). In particular, the series editors are entitled to reject individual contributions if they do not meet the high quality standards of this series. The final manuscript must be ready to print, and should include both an informative introduction and a sufficiently detailed subject index.

Contractual Aspects

Publication in LNP is free of charge. There is no formal contract, no royalties are paid, and no bulk orders are required, although special discounts are offered in this case. The volume editors receive jointly 30 free copies for their personal use and are entitled, as are the contributing authors, to purchase Springer books at a reduced rate. The publisher secures the copyright for each volume. As a rule, no reprints of individual contributions can be supplied.

Manuscript Submission

The manuscript in its final and approved version must be submitted in ready to print form. The corresponding electronic source files are also required for the production process, in particular the online version. Technical assistance in compiling the final manuscript can be provided by the publisher's production editor(s), especially with regard to the publisher's own LaTeX macro package which has been specially designed for this series.

LNP Homepage (http://link.springer.de/series/lnp/)

On the LNP homepage you will find:
−The LNP online archive. It contains the full texts (PDF) of all volumes published since 2000. Abstracts, table of contents and prefaces are accessible free of charge to everyone. Information about the availability of printed volumes can be obtained.
−The subscription information. The online archive is free of charge to all subscribers of the printed volumes.
−The editorial contacts, with respect to both scientific and technical matters.
−The author's / editor's instructions.

C. Fiolhais F. Nogueira M. Marques (Eds.)

A Primer in Density Functional Theory

 Springer

Editors

Carlos Fiolhais
Departamento de Física,
Universidade de Coimbra,
Rua Larga,
3004 - 516, Coimbra, Portugal

Miguel A. L. Marques
Donostia International Physics Center
(DIPC),
P. Manuel Lardizábal 4,
20080 San Sebastián, Spain

Fernando Nogueira
Departamento de Física,
Universidade de Coimbra,
Rua Larga,
3004 - 516, Coimbra, Portugal

Cataloging-in-Publication Data applied for

A catalog record for this book is available from the Library of Congress.

Bibliographic information published by Die Deutsche Bibliothek

Die Deutsche Bibliothek lists this publication in the Deutsche Nationalbibliografie;
detailed bibliographic data is available in the Internet at http://dnb.ddb.de

ISSN 0075-8450

ISBN 978-3-642-05704-5   e-ISBN 978-3-540-37072-7

Springer-Verlag Berlin Heidelberg New York
a part of Springer Science+Business Media

http://www.springer.de

© Springer-Verlag Berlin Heidelberg 2010
Printed in Germany

Camera-data conversion by Steingraeber Satztechnik GmbH Heidelberg
Cover design: *design & production*, Heidelberg

Printed on acid-free paper
54/3111- 5 4 3 2 1

Preface

Density functional theory is a clever way to solve the Schrödinger equation for a many-body system. In the formulation given by Kohn, Hohenberg, and Sham in the 1960's the real system is described by an effective one-body system. To achieve that goal, the complex many-body wave function, which is the solution of the Schrödinger equation, is abandoned in favour of the density which only depends on the three spatial coordinates. The energy is just a function of this function, i.e., a density functional.

This book, which intends to be an introduction to density functional theory, collects the lectures presented in the second Coimbra School on Computational Physics. In a way, it is a sequel to the sold-out Lecture Notes in Physics vol. 500 (ed. D. Joubert). This Summer School took place in late August of 2001 in the nice scenery of the Caramulo mountains, in central Portugal, some 50 km away from the old University of Coimbra. It was organized by the recently established (1998) Center for Computational Physics of the University of Coimbra, and was the second of a series which started, in 1999, with a school on "Monte Carlo Methods in Physics".

Like the summer school in South-Africa which originated the volume 500, the Coimbra School on Computational Physics devoted to density functional methods was a good opportunity for graduate students to enter the realm of density functionals, or to enlarge their previous knowledge in this fast expanding branch of physics and chemistry. About 50 students from different countries attended the School. Some teachers, who were also present at the South-African School (John Perdew, Reiner Dreizler and Eberhard Gross), were joined by new ones (Eberhard Engel, Rex Godby, Fernando Nogueira and Miguel Marques). The school was possible due to the support of Fundação para a Ciência e Tecnologia, Fundação Calouste Gulbenkian and the University of Coimbra, whom we would like to acknowledge here.

The contents of this volume are as follows. The theoretical foundations of the theory are reviewed by Stefan Kurth and John Perdew, in a chapter which is essentially an updated version of the article published in the above mentioned volume 500. The recent orbital dependent functionals are presented by Eberhard Engel. Two important extensions to the standard theory follow: relativistic systems, by Reiner Dreizler; and time-dependent non-relativistic problems by Miguel Marques and Eberhard Gross. In the next chapter Rex

Godby and Pablo García-González discuss some of the shortcomings of density functional theory and contrast it with conventional many-body theory. A tutorial, by Fernando Nogueira, Alberto Castro, and Miguel Marques, on practical applications of the formalism to atoms, molecules, and solids closes this book.

From the school and from this book emerges the view that, even though the "divine functional" – the energy functional with exact exchange and exact correlation – is yet a vision far on the horizon, extraordinary progress has been made since the seminal works of Kohn, Hohenberg, and Sham (not to speak about the early work in the thirties by Thomas and Fermi). The local density approximation to exchange and correlation from the sixties has been surpassed by the now standard generalized-gradient approximations. In principle more precise approaches like the meta-generalized gradient approximation or hybrid functionals are now being developed and applied, climbing what John Perdew called picturesquely "Jacob's Ladder" towards the "divine functional". The Chemistry Nobel prize awarded in 1998 to John Pople and Walter Kohn indeed gave a major impulse to the dissemination of density functional theory in physics and chemistry (several applications in biology and geology have also appeared!), but in order to have "chemical accuracy" further steps have to be taken.

It is the task of the new generation to continue the past and present efforts in this exciting field. We hope with this "primer" in density functional theory to provide students, and even established researchers, an overview of the present state and prospects of density functional methods.

Last but not least, the Coimbra school was also an opportunity to recognize the work of an active player in the field – Reiner Dreizler – on the occasion of his retirement, which took place in September 2001. The organizers would like to dedicate the present book to him. Although they know that he is not keen of homages and that his activity in physics is not over, we think that it is fully justified to summarize here his curriculum, emphasizing some of his achievements.

Reiner Dreizler was born 1936 in Stuttgart, Germany. In 1961, he received his "Diploma" in theoretical physics at the Albert Ludwigs Universität, in Freiburg, and in 1964, the title of Doctor of Philosophy in theoretical physics at the Australian National University, Canberra. From 1964 to 1966 he was Research Associate at the University of Pennsylvania, Philadelphia, USA and thereafter, until 1972, Assistant Professor of Physics at the same University. From 1972 to his retirement, he was Full Professor of Theoretical Physics at the Johann Wolfgang Goethe Universität, Frankfurt am Main, Germany. He has been guest lecturer all around the world, namely in Romania, Australia, Portugal, Russia, Ukraine, Japan, the USA and Brazil. Regarding positions and honours: He was Dean in 1981/1982 of the Faculty of Physics, Universität Frankfurt, and became Fellow of the American Physical Society in 1995; In 1999 he received the endowed chair "S. Lyson Professor der Physik".

His research interests have been very diverse. Besides the development and application of density functional methods, he studied the many-body problem in nuclear, atomic and molecular physics, and the theory of atomic scattering processes. He also investigated variational, iterative and projective techniques in handling quantum-mechanical problems and made contributions to the quantum-field description of many-body systems. Over the years he accumulated more than 230 contributions to refereed journals, 27 conference contributions, and four books. These include two Plenum Press Proceedings volumes, that stemmed from schools on density functinal theory (one of them in Alcabideche, Lisbon), and the famous Springer texbook on density functional theory co-authored by his ex-student and friend Eberhard Gross. He was supervisor of many PhDs. (including one of the school organizers and two of the school speakers) and Diploma theses.

In a world where science is more and more specialized, it is more and more difficult to meet someone like Reiner Dreizler, who covered with his work the whole spectrum of quantum mechanics from Particle to Solid State Physics, through Atomic, Molecular and Cluster Physics. May his example be followed by others!

Coimbra,   *Carlos Fiolhais*
December 2002   *Fernando Nogueira*
                                                       Miguel Marques

Table of Contents

List of Contributors

Alberto Castro
Departamento de Física Teórica,
Universidad de Valladolid,
47011 Valladolid, Spain
alberto.castro@tddft.org

Reiner Dreizler
Institut für Theoretische Physik,
J. W. Goethe - Universität Frankfurt
Robert-Mayer-Straße 6-8,
60054 Frankfurt/Main, Germany
dreizler@th.physik.
uni-frankfurt.de

Eberhard Engel
Institut für Theoretische Physik,
J. W.Goethe - Universität Frankfurt
Robert-Mayer-Straße 6-8,
60054 Frankfurt/Main, Germany
engel@th.physik.uni-frankfurt.de

Carlos Fiolhais
Departamento de Física,
Universidade de Coimbra,
Rua Larga,
3004 – 516, Coimbra, Portugal
tcarlos@teor.fis.uc.pt

Pablo García-González
Departamento de Física Fundamental, Universidad Nacional
de
Educación a Distancia, Apto. 60141,
28080 Madrid, Spain
pgarcia@fisfun.uned.es

Rex W. Godby
Department of Physics,
University of York,
Heslington, York YO10 5DD,
United Kingdom
rwg3@york.ac.uk

Eberhard K. U. Gross
Institut für Theoretische Physik,
Freie Universität Berlin,
Arnimallee 14,
14195 Berlin, Germany
hardy@physik.fu-berlin.de

Stefan Kurth
Institut für Theoretische Physik,
Freie Universität Berlin,
Arnimallee 14,
14195 Berlin, Germany
kurth@physik.fu-berlin.de

Miguel A. L. Marques
Donostia International Physics
Center (DIPC),
P. Manuel Lardizábal 4,
20080 San Sebastián, Spain
marques@tddft.org

Fernando Nogueira
Departamento de Física,
Universidade de Coimbra,
Rua Larga,
3004 – 516, Coimbra, Portugal
fnog@teor.fis.uc.pt

John P. Perdew
Department of Physics and Quantum
Theory Group,
Tulane University, New Orleans
LA 70118, USA
perdew@frigg.phy.tulane.edu

1 Density Functionals for Non-relativistic Coulomb Systems in the New Century

John P. Perdew* and Stefan Kurth[†]

* Department of Physics and
Quantum Theory Group, Tulane University,
New Orleans LA 70118, USA
perdew@frigg.phy.tulane.edu

[†] Institut für Theoretische Physik,
Freie Universität
Berlin, Arnimallee 14, 14195 Berlin, Germany
kurth@physik.fu-berlin.de

John Perdew

1.1 Introduction

1.1.1 Quantum Mechanical Many-Electron Problem

The material world of everyday experience, as studied by chemistry and condensed-matter physics, is built up from electrons and a few (or at most a few hundred) kinds of nuclei . The basic interaction is electrostatic or Coulombic: An electron at position $\mathbf{r}$ is attracted to a nucleus of charge Z at $\mathbf{R}$ by the potential energy $-Z/|\mathbf{r} - \mathbf{R}|$, a pair of electrons at $\mathbf{r}$ and $\mathbf{r}'$ repel one another by the potential energy $1/|\mathbf{r} - \mathbf{r}'|$, and two nuclei at $\mathbf{R}$ and $\mathbf{R}'$ repel one another as $Z'Z/|\mathbf{R} - \mathbf{R}'|$. The electrons must be described by quantum mechanics, while the more massive nuclei can sometimes be regarded as classical particles. All of the electrons in the lighter elements, and the chemically important valence electrons in most elements, move at speeds much less than the speed of light, and so are non-relativistic.

In essence, that is the simple story of practically everything. But there is still a long path from these general principles to theoretical prediction of the structures and properties of atoms, molecules, and solids, and eventually to the design of new chemicals or materials. If we restrict our focus to the important class of ground-state properties, we can take a shortcut through density functional theory.

These lectures present an introduction to density functionals for non-relativistic Coulomb systems. The reader is assumed to have a working knowledge of quantum mechanics at the level of one-particle wavefunctions $\psi(\mathbf{r})$ [1]. The many-electron wavefunction $\Psi(\mathbf{r}_1, \mathbf{r}_2, \ldots, \mathbf{r}_N)$ [2] is briefly introduced here, and then replaced as basic variable by the electron density $n(\mathbf{r})$. Various terms of the total energy are defined as functionals of the electron density, and some formal properties of these functionals are discussed. The most widely-used density functionals – the local spin density and generalized gradient approximations – are then introduced and discussed. At the end, the reader should be prepared to approach the broad literature of quantum chemistry and condensed-matter physics in which these density functionals are applied to predict diverse properties: the shapes and sizes of molecules, the crystal structures of solids, binding or atomization energies, ionization energies and electron affinities, the heights of energy barriers to various processes, static response functions, vibrational frequencies of nuclei, etc. Moreover, the reader's approach will be an informed and discerning one, based upon an understanding of where these functionals come from, why they work, and how they work.

These lectures are intended to teach at the introductory level, and not to serve as a comprehensive treatise. The reader who wants more can go to several excellent general sources [3–5] or to the original literature. Atomic units (in which all electromagnetic equations are written in cgs form, and the fundamental constants $\hbar$, e^2, and m are set to unity) have been used throughout.

1.1.2   Summary of Kohn–Sham Spin-Density Functional Theory

This introduction closes with a brief presentation of the Kohn-Sham [6] spin-density functional method, the most widely-used method of electronic-structure calculation in condensed-matter physics and one of the most widely-used methods in quantum chemistry. We seek the ground-state total energy E and spin densities $n_\uparrow(\mathbf{r})$, $n_\downarrow(\mathbf{r})$ for a collection of N electrons interacting with one another and with an external potential $v(\mathbf{r})$ (due to the nuclei in most practical cases). These are found by the selfconsistent solution of an auxiliary (fictitious) one-electron Schrödinger equation:

$$\left(-\frac{1}{2}\nabla^2 + v(\mathbf{r}) + u([n];\mathbf{r}) + v_{\mathrm{xc}}^\sigma([n_\uparrow, n_\downarrow];\mathbf{r}) \right) \psi_{\alpha\sigma}(\mathbf{r}) = \varepsilon_{\alpha\sigma}\psi_{\alpha\sigma}(\mathbf{r}) \,, \quad (1.1)$$

$$n_\sigma(\mathbf{r}) = \sum_\alpha \theta(\mu - \varepsilon_{\alpha\sigma})|\psi_{\alpha\sigma}(\mathbf{r})|^2 \,. \quad (1.2)$$

Here $\sigma = \uparrow$ or $\downarrow$ is the z-component of spin, and α stands for the set of remaining one-electron quantum numbers. The effective potential includes a classical Hartree potential

$$u([n];\mathbf{r}) = \int \mathrm{d}^3 r' \, \frac{n(\mathbf{r}')}{|\mathbf{r} - \mathbf{r}'|} \,, \quad (1.3)$$

$$n(\mathbf{r}) = n_\uparrow(\mathbf{r}) + n_\downarrow(\mathbf{r}) \,, \quad (1.4)$$

and $v_{\mathrm{xc}}^\sigma([n_\uparrow, n_\downarrow];\mathbf{r})$, a multiplicative spin-dependent exchange-correlation potential which is a functional of the spin densities. The step function $\theta(\mu - \varepsilon_{\alpha\sigma})$ in (1.2) ensures that all Kohn-Sham spin orbitals with $\varepsilon_{\alpha\sigma} < \mu$ are singly occupied, and those with $\varepsilon_{\alpha\sigma} > \mu$ are empty. The chemical potential μ is chosen to satisfy

$$\int d^3r\, n(\mathbf{r}) = N \ . \tag{1.5}$$

Because (1.1) and (1.2) are interlinked, they can only be solved by iteration to selfconsistency.

The total energy is

$$E = T_\mathrm{s}[n_\uparrow, n_\downarrow] + \int d^3r\, n(\mathbf{r})v(\mathbf{r}) + U[n] + E_\mathrm{xc}[n_\uparrow, n_\downarrow] \ , \tag{1.6}$$

where

$$T_\mathrm{s}[n_\uparrow, n_\downarrow] = \sum_\sigma \sum_\alpha \theta(\mu - \varepsilon_{\alpha\sigma})\langle\psi_{\alpha\sigma}| -\frac{1}{2}\nabla^2|\psi_{\alpha\sigma}\rangle \tag{1.7}$$

is the non-interacting kinetic energy, a functional of the spin densities because (as we shall see) the external potential $v(\mathbf{r})$ and hence the Kohn-Sham orbitals are functionals of the spin densities. In our notation,

$$\langle\psi_{\alpha\sigma}|\hat{O}|\psi_{\alpha\sigma}\rangle = \int d^3r\, \psi_{\alpha\sigma}^*(\mathbf{r})\hat{O}\psi_{\alpha\sigma}(\mathbf{r}) \ . \tag{1.8}$$

The second term of (1.6) is the interaction of the electrons with the external potential. The third term of (1.6) is the Hartree electrostatic self-repulsion of the electron density

$$U[n] = \frac{1}{2}\int d^3r\int d^3r'\, \frac{n(\mathbf{r})n(\mathbf{r}')}{|\mathbf{r} - \mathbf{r}'|} \ . \tag{1.9}$$

The last term of (1.6) is the exchange-correlation energy, whose functional derivative (as explained later) yields the exchange-correlation potential

$$v_\mathrm{xc}^\sigma([n_\uparrow, n_\downarrow]; \mathbf{r}) = \frac{\delta E_\mathrm{xc}}{\delta n_\sigma(\mathbf{r})} \ . \tag{1.10}$$

Not displayed in (1.6), but needed for a system of electrons and nuclei, is the electrostatic repulsion among the nuclei. E_xc is defined to include everything else omitted from the first three terms of (1.6).

If the exact dependence of E_xc upon $n_\uparrow$ and $n_\downarrow$ were known, these equations would predict the exact ground-state energy and spin-densities of a many-electron system. The forces on the nuclei, and their equilibrium positions, could then be found from $-\frac{\partial E}{\partial \mathbf{R}}$.

In practice, the exchange-correlation energy functional must be approximated. The local spin density [6,7] (LSD) approximation has long been popular in solid state physics:

$$E_\mathrm{xc}^\mathrm{LSD}[n_\uparrow, n_\downarrow] = \int d^3r\, n(\mathbf{r})e_\mathrm{xc}(n_\uparrow(\mathbf{r}), n_\downarrow(\mathbf{r})) \ , \tag{1.11}$$

where $e_{xc}(n_\uparrow, n_\downarrow)$ is the known [8–10] exchange-correlation energy per particle for an electron gas of uniform spin densities $n_\uparrow$, $n_\downarrow$. More recently, generalized gradient approximations (GGA's) [11–21] have become popular in quantum chemistry:

$$E_{xc}^{GGA}[n_\uparrow, n_\downarrow] = \int d^3r \, f(n_\uparrow, n_\downarrow, \nabla n_\uparrow, \nabla n_\downarrow) \,. \qquad (1.12)$$

The input $e_{xc}(n_\uparrow, n_\downarrow)$ to LSD is in principle unique, since there is a possible system in which $n_\uparrow$ and $n_\downarrow$ are constant and for which LSD is exact. At least in this sense, there is no unique input $f(n_\uparrow, n_\downarrow, \nabla n_\uparrow, \nabla n_\downarrow)$ to GGA. These lectures will stress a conservative "philosophy of approximation" [20,21], in which we construct a nearly-unique GGA with all the known correct formal features of LSD, plus others. We will also discuss how to go beyond GGA.

The equations presented here are really all that we need to do a practical calculation for a many-electron system. They allow us to draw upon the intuition and experience we have developed for one-particle systems. The many-body effects are in $U[n]$ (trivially) and $E_{xc}[n_\uparrow, n_\downarrow]$ (less trivially), but we shall also develop an intuitive appreciation for E_{xc}.

While E_{xc} is often a relatively small fraction of the total energy of an atom, molecule, or solid (minus the work needed to break up the system into separated electrons and nuclei), the contribution from E_{xc} is typically about 100% or more of the chemical bonding or atomization energy (the work needed to break up the system into separated neutral atoms). E_{xc} is a kind of "glue", without which atoms would bond weakly if at all. Thus, accurate approximations to E_{xc} are essential to the whole enterprise of density functional theory. Table 1.1 shows the typical relative errors we find from selfconsistent calculations within the LSD or GGA approximations of (1.11) and (1.12). Table 1.2 shows the mean absolute errors in the atomization energies of 20 molecules when calculated by LSD, by GGA, and in the Hartree-Fock approximation. Hartree-Fock treats exchange exactly, but neglects correlation completely. While the Hartree-Fock total energy is an upper bound to the true ground-state total energy, the LSD and GGA energies are not.

In most cases we are only interested in small total-energy changes associated with re-arrangements of the outer or valence electrons, to which the inner or core electrons of the atoms do not contribute. In these cases, we can replace each core by the pseudopotential [22] it presents to the valence electrons, and then expand the valence-electron orbitals in an economical and convenient basis of plane waves. Pseudopotentials are routinely combined with density functionals. Although the most realistic pseudopotentials are nonlocal operators and not simply local or multiplication operators, and although density functional theory in principle requires a local external potential, this inconsistency does not seem to cause any practical difficulties.

There are empirical versions of LSD and GGA, but these lectures will only discuss non-empirical versions. If every electronic-structure calculation

Table 1.1. Typical errors for atoms, molecules, and solids from selfconsistent Kohn-Sham calculations within the LSD and GGA approximations of (1.11) and (1.12). Note that there is typically some cancellation of errors between the exchange (E_x) and correlation (E_c) contributions to E_{xc}. The "energy barrier" is the barrier to a chemical reaction that arises at a highly-bonded intermediate state

Property	LSD	GGA
E_x	5% (not negative enough)	0.5%
E_c	100% (too negative)	5%
bond length	1% (too short)	1% (too long)
structure	overly favors close packing	more correct
energy barrier	100% (too low)	30% (too low)

Table 1.2. Mean absolute error of the atomization energies for 20 molecules, evaluated by various approximations. (1 hartree $= 27.21\,\text{eV}$) (From [20])

Approximation	Mean absolute error (eV)
Unrestricted Hartree-Fock	3.1 (underbinding)
LSD	1.3 (overbinding)
GGA	0.3 (mostly overbinding)
Desired "chemical accuracy"	0.05

were done at least twice, once with nonempirical LSD and once with nonempirical GGA, the results would be useful not only to those interested in the systems under consideration but also to those interested in the development and understanding of density functionals.

1.2   Wavefunction Theory

1.2.1   Wavefunctions and Their Interpretation

We begin with a brief review of one-particle quantum mechanics [1]. An electron has spin $s = \frac{1}{2}$ and z-component of spin $\sigma = +\frac{1}{2}$ ($\uparrow$) or $-\frac{1}{2}$ ($\downarrow$). The Hamiltonian or energy operator for one electron in the presence of an external potential $v(\mathbf{r})$ is

$$\hat{h} = -\frac{1}{2}\nabla^2 + v(\mathbf{r}) \,. \tag{1.13}$$

The energy eigenstates $\psi_\alpha(\mathbf{r}, \sigma)$ and eigenvalues ε_α are solutions of the time-independent Schrödinger equation

$$\hat{h}\psi_\alpha(\mathbf{r}, \sigma) = \varepsilon_\alpha\psi_\alpha(\mathbf{r}, \sigma) \,, \tag{1.14}$$

and $|\psi_\alpha(\mathbf{r}, \sigma)|^2 d^3r$ is the probability to find the electron with spin σ in volume element d^3r at $\mathbf{r}$, given that it is in energy eigenstate ψ_α. Thus

$$\sum_\sigma \int d^3r \, |\psi_\alpha(\mathbf{r}, \sigma)|^2 = \langle \psi | \psi \rangle = 1 \,. \tag{1.15}$$

Since $\hat{h}$ commutes with $\hat{s}_z$, we can choose the ψ_α to be eigenstates of $\hat{s}_z$, i.e., we can choose $\sigma = \uparrow$ or $\downarrow$ as a one-electron quantum number.

The Hamiltonian for N electrons in the presence of an external potential $v(\mathbf{r})$ is [2]

$$\hat{H} = -\frac{1}{2} \sum_{i=1}^N \nabla_i^2 + \sum_{i=1}^N v(\mathbf{r}_i) + \frac{1}{2} \sum_i \sum_{j \neq i} \frac{1}{|\mathbf{r}_i - \mathbf{r}_j|}$$

$$= \hat{T} + \hat{V}_{\text{ext}} + \hat{V}_{\text{ee}} \,. \tag{1.16}$$

The electron-electron repulsion $\hat{V}_{\text{ee}}$ sums over distinct pairs of different electrons. The states of well-defined energy are the eigenstates of $\hat{H}$:

$$\hat{H} \Psi_k(\mathbf{r}_1 \sigma_1, \ldots, \mathbf{r}_N \sigma_N) = E_k \Psi_k(\mathbf{r}_1 \sigma_1, \ldots, \mathbf{r}_N \sigma_N) \,, \tag{1.17}$$

where k is a complete set of many-electron quantum numbers; we shall be interested mainly in the ground state or state of lowest energy, the zero-temperature equilibrium state for the electrons.

Because electrons are fermions, the only physical solutions of (1.17) are those wavefunctions that are antisymmetric [2] under exchange of two electron labels i and j:

$$\Psi(\mathbf{r}_1 \sigma_1, \ldots, \mathbf{r}_i \sigma_i, \ldots, \mathbf{r}_j \sigma_j, \ldots, \mathbf{r}_N \sigma_N) =$$
$$- \Psi(\mathbf{r}_1 \sigma_1, \ldots, \mathbf{r}_j \sigma_j, \ldots, \mathbf{r}_i \sigma_i, \ldots, \mathbf{r}_N \sigma_N) \,. \tag{1.18}$$

There are $N!$ distinct permutations of the labels $1, 2, \ldots, N$, which by (1.18) all have the same $|\Psi|^2$. Thus $N! \, |\Psi(\mathbf{r}_1 \sigma_1, \ldots, \mathbf{r}_N \sigma_N)|^2 d^3 r_1 \ldots d^3 r_N$ is the probability to find *any* electron with spin σ_1 in volume element $d^3 r_1$, etc., and

$$\frac{1}{N!} \sum_{\sigma_1 \ldots \sigma_N} \int d^3 r_1 \ldots \int d^3 r_N \, N! \, |\Psi(\mathbf{r}_1 \sigma_1, \ldots, \mathbf{r}_N \sigma_N)|^2 = \int |\Psi|^2 = \langle \Psi | \Psi \rangle = 1 \,. \tag{1.19}$$

We define the electron spin density $n_\sigma(\mathbf{r})$ so that $n_\sigma(\mathbf{r}) d^3 r$ is the probability to find an electron with spin σ in volume element $d^3 r$ at $\mathbf{r}$. We find $n_\sigma(\mathbf{r})$ by integrating over the coordinates and spins of the $(N-1)$ other electrons, i.e.,

$$n_\sigma(\mathbf{r}) = \frac{1}{(N-1)!} \sum_{\sigma_2 \ldots \sigma_N} \int d^3 r_2 \ldots \int d^3 r_N \, N! |\Psi(\mathbf{r}\sigma, \mathbf{r}_2 \sigma_2, \ldots, \mathbf{r}_N \sigma_N)|^2$$

$$= N \sum_{\sigma_2 \ldots \sigma_N} \int d^3 r_2 \ldots \int d^3 r_N \, |\Psi(\mathbf{r}\sigma, \mathbf{r}_2 \sigma_2, \ldots, \mathbf{r}_N \sigma_N)|^2 \,. \tag{1.20}$$

Equations (1.19) and (1.20) yield

$$\sum_\sigma \int d^3r\, n_\sigma(\mathbf{r}) = N \ . \tag{1.21}$$

Based on the probability interpretation of $n_\sigma(\mathbf{r})$, we might have expected the right hand side of (1.21) to be 1, but that is wrong; the sum of probabilities of all mutually-exclusive events equals 1, but finding an electron at $\mathbf{r}$ does *not* exclude the possibility of finding one at $\mathbf{r}'$, except in a one-electron system. Equation (1.21) shows that $n_\sigma(\mathbf{r})d^3r$ is the average number of electrons of spin σ in volume element d^3r. Moreover, the expectation value of the external potential is

$$\langle \hat{V}_{\mathrm{ext}} \rangle = \langle \Psi | \sum_{i=1}^N v(\mathbf{r}_i) | \Psi \rangle = \int d^3r\, n(\mathbf{r})v(\mathbf{r}) \ , \tag{1.22}$$

with the electron density $n(\mathbf{r})$ given by (1.4).

1.2.2   Wavefunctions for Non-interacting Electrons

As an important special case, consider the Hamiltonian for N non-interacting electrons:

$$\hat{H}_{\mathrm{non}} = \sum_{i=1}^N \left[-\frac{1}{2}\nabla_i^2 + v(\mathbf{r}_i) \right] \ . \tag{1.23}$$

The eigenfunctions of the one-electron problem of (1.13) and (1.14) are spin orbitals which can be used to construct the antisymmetric eigenfunctions Φ of $\hat{H}_{\mathrm{non}}$:

$$\hat{H}_{\mathrm{non}}\Phi = E_{\mathrm{non}}\Phi \ . \tag{1.24}$$

Let i stand for $\mathbf{r}_i, \sigma_i$ and construct the Slater determinant or antisymmetrized product [2]

$$\Phi = \frac{1}{\sqrt{N!}} \sum_P (-1)^P \psi_{\alpha_1}(P1)\psi_{\alpha_2}(P2)\dots\psi_{\alpha_N}(PN) \ , \tag{1.25}$$

where the quantum label α_i now includes the spin quantum number σ. Here P is any permutation of the labels $1, 2, \dots, N$, and $(-1)^P$ equals $+1$ for an even permutation and -1 for an odd permutation. The total energy is

$$E_{\mathrm{non}} = \varepsilon_{\alpha_1} + \varepsilon_{\alpha_2} + \dots + \varepsilon_{\alpha_N} \ , \tag{1.26}$$

and the density is given by the sum of $|\psi_{\alpha_i}(\mathbf{r})|^2$. If any α_i equals any α_j in (1.25), we find $\Phi = 0$, which is not a normalizable wavefunction. This is the Pauli exclusion principle: two or more non-interacting electrons may not occupy the same spin orbital.

As an example, consider the ground state for the non-interacting helium atom ($N = 2$). The occupied spin orbitals are

$$\psi_1(\mathbf{r}, \sigma) = \psi_{1s}(\mathbf{r})\delta_{\sigma,\uparrow} , \tag{1.27}$$

$$\psi_2(\mathbf{r}, \sigma) = \psi_{1s}(\mathbf{r})\delta_{\sigma,\downarrow} , \tag{1.28}$$

and the 2-electron Slater determinant is

$$
\begin{aligned}
\Phi(1,2) &= \frac{1}{\sqrt{2}} \begin{vmatrix} \psi_1(\mathbf{r}_1, \sigma_1) & \psi_2(\mathbf{r}_1, \sigma_1) \\ \psi_1(\mathbf{r}_2, \sigma_2) & \psi_2(\mathbf{r}_2, \sigma_2) \end{vmatrix} \\
&= \psi_{1s}(\mathbf{r}_1)\psi_{1s}(\mathbf{r}_2)\frac{1}{\sqrt{2}} \left(\delta_{\sigma_1,\uparrow}\delta_{\sigma_2,\downarrow} - \delta_{\sigma_2,\uparrow}\delta_{\sigma_1,\downarrow} \right) ,
\end{aligned} \tag{1.29}
$$

which is symmetric in space but antisymmetric in spin (whence the total spin is $S = 0$).

If several different Slater determinants yield the same non-interacting energy E_{non}, then a linear combination of them will be another antisymmetric eigenstate of $\hat{H}_{non}$. More generally, the Slater-determinant eigenstates of $\hat{H}_{non}$ define a complete orthonormal basis for expansion of the antisymmetric eigenstates of $\hat{H}$, the interacting Hamiltonian of (1.16).

1.2.3   Wavefunction Variational Principle

The Schrödinger equation (1.17) is equivalent to a wavefunction variational principle [2]: Extremize $\langle \Psi | \hat{H} | \Psi \rangle$ subject to the constraint $\langle \Psi | \Psi \rangle = 1$, i.e., set the following first variation to zero:

$$\delta \left\{ \langle \Psi | \hat{H} | \Psi \rangle / \langle \Psi | \Psi \rangle \right\} = 0 . \tag{1.30}$$

The ground state energy and wavefunction are found by minimizing the expression in curly brackets.

The Rayleigh-Ritz method finds the extrema or the minimum in a *restricted* space of wavefunctions. For example, the Hartree-Fock approximation to the ground-state wavefunction is the single Slater determinant Φ that minimizes $\langle \Phi | \hat{H} | \Phi \rangle / \langle \Phi | \Phi \rangle$. The configuration-interaction ground-state wavefunction [23] is an energy-minimizing linear combination of Slater determinants, restricted to certain kinds of excitations out of a reference determinant. The Quantum Monte Carlo method typically employs a trial wavefunction which is a single Slater determinant times a Jastrow pair-correlation factor [24]. Those widely-used many-electron wavefunction methods are both approximate and computationally demanding, especially for large systems where density functional methods are distinctly more efficient.

The unrestricted solution of (1.30) is equivalent by the method of Lagrange multipliers to the unconstrained solution of

$$\delta \left\{ \langle \Psi | \hat{H} | \Psi \rangle - E \langle \Psi | \Psi \rangle \right\} = 0 , \tag{1.31}$$

i.e.,

$$\langle \delta\Psi|(\hat{H} - E)|\Psi\rangle = 0 . \tag{1.32}$$

Since $\delta\Psi$ is an arbitrary variation, we recover the Schrödinger equation (1.17). Every eigenstate of $\hat{H}$ is an extremum of $\langle\Psi|\hat{H}|\Psi\rangle/\langle\Psi|\Psi\rangle$ and vice versa.

The wavefunction variational principle implies the Hellmann-Feynman and virial theorems below and also implies the Hohenberg-Kohn [25] density functional variational principle to be presented later.

1.2.4 Hellmann–Feynman Theorem

Often the Hamiltonian $\hat{H}_\lambda$ depends upon a parameter λ, and we want to know how the energy E_λ depends upon this parameter. For any normalized variational solution Ψ_λ (including in particular any eigenstate of $\hat{H}_\lambda$), we define

$$E_\lambda = \langle\Psi_\lambda|\hat{H}_\lambda|\Psi_\lambda\rangle . \tag{1.33}$$

Then

$$\frac{dE_\lambda}{d\lambda} = \frac{d}{d\lambda'}\langle\Psi_{\lambda'}|\hat{H}_\lambda|\Psi_{\lambda'}\rangle\Big|_{\lambda'=\lambda} + \langle\Psi_\lambda|\frac{\partial\hat{H}_\lambda}{\partial\lambda}|\Psi_\lambda\rangle . \tag{1.34}$$

The first term of (1.34) vanishes by the variational principle, and we find the Hellmann-Feynman theorem [26]

$$\frac{dE_\lambda}{d\lambda} = \langle\Psi_\lambda|\frac{\partial\hat{H}_\lambda}{\partial\lambda}|\Psi_\lambda\rangle . \tag{1.35}$$

Equation (1.35) will be useful later for our understanding of E_{xc}. For now, we shall use (1.35) to derive the electrostatic force theorem [26]. Let $\mathbf{r}_i$ be the position of the i-th electron, and $\mathbf{R}_I$ the position of the (static) nucleus I with atomic number Z_I. The Hamiltonian

$$\hat{H} = \sum_{i=1}^{N} -\frac{1}{2}\nabla_i^2 + \sum_i\sum_I\frac{-Z_I}{|\mathbf{r}_i - \mathbf{R}_I|} + \frac{1}{2}\sum_i\sum_{j\neq i}\frac{1}{|\mathbf{r}_i - \mathbf{r}_j|} + \frac{1}{2}\sum_I\sum_{J\neq I}\frac{Z_I Z_J}{|\mathbf{R}_I - \mathbf{R}_J|} \tag{1.36}$$

depends parametrically upon the position $\mathbf{R}_I$, so the force on nucleus I is

$$-\frac{\partial E}{\partial\mathbf{R}_I} = \left\langle\Psi\left|-\frac{\partial\hat{H}}{\partial\mathbf{R}_I}\right|\Psi\right\rangle$$

$$= \int d^3r\, n(\mathbf{r})\frac{Z_I(\mathbf{r} - \mathbf{R}_I)}{|\mathbf{r} - \mathbf{R}_I|^3} + \sum_{J\neq I}\frac{Z_I Z_J(\mathbf{R}_I - \mathbf{R}_J)}{|\mathbf{R}_I - \mathbf{R}_J|^3} , \tag{1.37}$$

just as classical electrostatics would predict. Equation (1.37) can be used to find the equilibrium geometries of a molecule or solid by varying all the $\mathbf{R}_I$ until the energy is a minimum and $-\partial E/\partial\mathbf{R}_I = 0$. Equation (1.37) also forms the basis for a possible density functional molecular dynamics, in which the nuclei move under these forces by Newton's second law. In principle, all we need for either application is an accurate electron density for each set of nuclear positions.

1.2.5 Virial Theorem

The density scaling relations to be presented in Sect. 1.4, which constitute important constraints on the density functionals, are rooted in the same wavefunction scaling that will be used here to derive the virial theorem [26].

Let $\Psi(\mathbf{r}_1, \ldots, \mathbf{r}_N)$ be any extremum of $\langle \Psi | \hat{H} | \Psi \rangle$ over normalized wavefunctions, i.e., any eigenstate or optimized restricted trial wavefunction (where irrelevant spin variables have been suppressed). For any scale parameter $\gamma > 0$, define the uniformly-scaled wavefunction

$$\Psi_\gamma(\mathbf{r}_1, \ldots, \mathbf{r}_N) = \gamma^{3N/2} \Psi(\gamma \mathbf{r}_1, \ldots, \gamma \mathbf{r}_N) \tag{1.38}$$

and observe that

$$\langle \Psi_\gamma | \Psi_\gamma \rangle = \langle \Psi | \Psi \rangle = 1 . \tag{1.39}$$

The density corresponding to the scaled wavefunction is the scaled density

$$n_\gamma(\mathbf{r}) = \gamma^3 \, n(\gamma \mathbf{r}) , \tag{1.40}$$

which clearly conserves the electron number:

$$\int \mathrm{d}^3 r \, n_\gamma(\mathbf{r}) = \int \mathrm{d}^3 r \, n(\mathbf{r}) = N . \tag{1.41}$$

$\gamma > 1$ leads to densities $n_\gamma(\mathbf{r})$ that are higher (on average) and more contracted than $n(\mathbf{r})$, while $\gamma < 1$ produces densities that are lower and more expanded.

Now consider what happens to $\langle \hat{H} \rangle = \langle \hat{T} + \hat{V} \rangle$ under scaling. By definition of Ψ,

$$\left. \frac{\mathrm{d}}{\mathrm{d}\gamma} \langle \Psi_\gamma | \hat{T} + \hat{V} | \Psi_\gamma \rangle \right|_{\gamma=1} = 0 . \tag{1.42}$$

But $\hat{T}$ is homogeneous of degree -2 in $\mathbf{r}$, so

$$\langle \Psi_\gamma | \hat{T} | \Psi_\gamma \rangle = \gamma^2 \, \langle \Psi | \hat{T} | \Psi \rangle , \tag{1.43}$$

and (1.42) becomes

$$2 \langle \Psi | \hat{T} | \Psi \rangle + \left. \frac{\mathrm{d}}{\mathrm{d}\gamma} \langle \Psi_\gamma | \hat{V} | \Psi_\gamma \rangle \right|_{\gamma=1} = 0 , \tag{1.44}$$

or

$$2 \langle \hat{T} \rangle - \langle \sum_{i=1}^{N} \mathbf{r}_i \cdot \frac{\partial \hat{V}}{\partial \mathbf{r}_i} \rangle = 0 . \tag{1.45}$$

If the potential energy $\hat{V}$ is homogeneous of degree n, i.e., if

$$V(\gamma \mathbf{r}_i, \ldots, \gamma \mathbf{r}_N) = \gamma^n V(\mathbf{r}_i, \ldots, \mathbf{r}_N) , \qquad (1.46)$$

then

$$\langle \Psi_\gamma | \hat{V} | \Psi_\gamma \rangle = \gamma^{-n} \langle \Psi | \hat{V} | \Psi \rangle , \qquad (1.47)$$

and (1.44) becomes simply

$$2 \langle \Psi | \hat{T} | \Psi \rangle - n \langle \Psi | \hat{V} | \Psi \rangle = 0 . \qquad (1.48)$$

For example, $n = -1$ for the Hamiltonian of (1.36) in the presence of a single nucleus, or more generally when the Hellmann-Feynman forces of (1.37) vanish for the state Ψ.

1.3 Definitions of Density Functionals

1.3.1 Introduction to Density Functionals

The many-electron wavefunction $\Psi(\mathbf{r}_1 \sigma_1, \ldots, \mathbf{r}_N \sigma_N)$ contains a great deal of information – all we could ever have, but more than we usually want. Because it is a function of many variables, it is not easy to calculate, store, apply or even think about. Often we want no more than the total energy E (and its changes), or perhaps also the spin densities $n_\uparrow(\mathbf{r})$ and $n_\downarrow(\mathbf{r})$, for the ground state. As we shall see, we can formally replace Ψ by the observables $n_\uparrow$ and $n_\downarrow$ as the basic variational objects.

While a *function* is a rule which assigns a number $f(x)$ to a number x, a *functional* is a rule which assigns a number $F[f]$ to a function f. For example, $h[\Psi] = \langle \Psi | \hat{H} | \Psi \rangle$ is a functional of the trial wavefunction Ψ, given the Hamiltonian $\hat{H}$. $U[n]$ of (1.9) is a functional of the density $n(\mathbf{r})$, as is the local density approximation for the exchange energy:

$$E_\mathrm{x}^{\mathrm{LDA}}[n] = A_\mathrm{x} \int \mathrm{d}^3 r \, n(\mathbf{r})^{4/3} . \qquad (1.49)$$

The *functional derivative* $\delta F / \delta n(\mathbf{r})$ tells us how the functional $F[n]$ changes under a small variation $\delta n(\mathbf{r})$:

$$\delta F = \int \mathrm{d}^3 r \left(\frac{\delta F}{\delta n(\mathbf{r})} \right) \delta n(\mathbf{r}) . \qquad (1.50)$$

For example,

$$\delta E_\mathrm{x}^{\mathrm{LDA}} = A_\mathrm{x} \int \mathrm{d}^3 r \left\{ [n(\mathbf{r}) + \delta n(\mathbf{r})]^{4/3} - n(\mathbf{r})^{4/3} \right\}$$
$$= A_\mathrm{x} \int \mathrm{d}^3 r \, \frac{4}{3} n(\mathbf{r})^{1/3} \delta n(\mathbf{r}) ,$$

so

$$\frac{\delta E_{\mathrm{x}}^{\mathrm{LDA}}}{\delta n(\mathbf{r})} = A_{\mathrm{x}} \frac{4}{3} n(\mathbf{r})^{1/3} . \tag{1.51}$$

Similarly,

$$\frac{\delta U[n]}{\delta n(\mathbf{r})} = u([n]; \mathbf{r}) , \tag{1.52}$$

where the right hand side is given by (1.3). Functional derivatives of various orders can be linked through the translational and rotational symmetries of empty space [27].

1.3.2 Density Variational Principle

We seek a density functional analog of (1.30). Instead of the original derivation of Hohenberg, Kohn and Sham [25,6], which was based upon "reductio ad absurdum", we follow the "constrained search" approach of Levy [28], which is in some respects simpler and more constructive.

Equation (1.30) tells us that the ground state energy can be found by minimizing $\langle \Psi | \hat{H} | \Psi \rangle$ over all normalized, antisymmetric N-particle wavefunctions:

$$E = \min_{\Psi} \langle \Psi | \hat{H} | \Psi \rangle . \tag{1.53}$$

We now separate the minimization of (1.53) into two steps. First we consider all wavefunctions Ψ which yield a given density $n(\mathbf{r})$, and minimize over those wavefunctions:

$$\min_{\Psi \to n} \langle \Psi | \hat{H} | \Psi \rangle = \min_{\Psi \to n} \langle \Psi | \hat{T} + \hat{V}_{\mathrm{ee}} | \Psi \rangle + \int \mathrm{d}^3 r \, v(\mathbf{r}) n(\mathbf{r}) , \tag{1.54}$$

where we have exploited the fact that all wavefunctions that yield the same $n(\mathbf{r})$ also yield the same $\langle \Psi | \hat{V}_{\mathrm{ext}} | \Psi \rangle$. Then we define the universal functional

$$F[n] = \min_{\Psi \to n} \langle \Psi | \hat{T} + \hat{V}_{\mathrm{ee}} | \Psi \rangle = \langle \Psi_n^{\mathrm{min}} | \hat{T} + \hat{V}_{\mathrm{ee}} | \Psi_n^{\mathrm{min}} \rangle , \tag{1.55}$$

where Ψ_n^{min} is that wavefunction which delivers the minimum for a given n. Finally we minimize over all N-electron densities $n(\mathbf{r})$:

$$E = \min_n E_v[n]$$

$$= \min_n \left\{ F[n] + \int \mathrm{d}^3 r \, v(\mathbf{r}) n(\mathbf{r}) \right\} , \tag{1.56}$$

where of course $v(\mathbf{r})$ is held fixed during the minimization. The minimizing density is then the ground-state density.

The constraint of fixed N can be handled formally through introduction of a Lagrange multiplier μ:

$$\delta \left\{ F[n] + \int \mathrm{d}^3 r \, v(\mathbf{r}) n(\mathbf{r}) - \mu \int \mathrm{d}^3 r \, n(\mathbf{r}) \right\} = 0 , \tag{1.57}$$

which is equivalent to the Euler equation

$$\frac{\delta F}{\delta n(\mathbf{r})} + v(\mathbf{r}) = \mu \, . \tag{1.58}$$

μ is to be adjusted until (1.5) is satisfied. Equation (1.58) shows that the external potential $v(\mathbf{r})$ is uniquely determined by the ground state density (or by any one of them, if the ground state is degenerate).

The functional $F[n]$ is defined via (1.55) for all densities $n(\mathbf{r})$ which are "N-representable", i.e., come from an antisymmetric N-electron wavefunction. We shall discuss the extension from wavefunctions to ensembles in Sect. 1.4.5. The functional derivative $\delta F/\delta n(\mathbf{r})$ is defined via (1.58) for all densities which are "v-representable", i.e., come from antisymmetric N-electron ground-state wavefunctions for some choice of external potential $v(\mathbf{r})$.

This formal development requires only the total density of (1.4), and not the separate spin densities $n_\uparrow(\mathbf{r})$ and $n_\downarrow(\mathbf{r})$. However, it is clear how to get to a spin-density functional theory: just replace the constraint of fixed n in (1.54) and subsequent equations by that of fixed $n_\uparrow$ and $n_\downarrow$. There are two practical reasons to do so: (1) This extension is required when the external potential is spin-dependent, i.e., $v(\mathbf{r}) \to v_\sigma(\mathbf{r})$, as when an external magnetic field couples to the z-component of electron spin. (If this field also couples to the current density $\mathbf{j}(\mathbf{r})$, then we must resort to a current-density functional theory.) (2) Even when $v(\mathbf{r})$ is spin-independent, we may be interested in the physical spin magnetization (e.g., in magnetic materials). (3) Even when neither (1) nor (2) applies, our local and semi-local approximations (see (1.11) and (1.12)) typically work better when we use $n_\uparrow$ and $n_\downarrow$ instead of n.

1.3.3 Kohn–Sham Non-interacting System

For a system of non-interacting electrons, $\hat{V}_{ee}$ of (1.16) vanishes so $F[n]$ of (1.55) reduces to

$$T_s[n] = \min_{\Psi \to n} \langle \Psi | \hat{T} | \Psi \rangle = \langle \Phi_n^{\min} | \hat{T} | \Phi_n^{\min} \rangle \, . \tag{1.59}$$

Although we can search over all antisymmetric N-electron wavefunctions in (1.59), the minimizing wavefunction $\Phi_n^{\min}$ for a given density will be a non-interacting wavefunction (a single Slater determinant or a linear combination of a few) for some external potential $\hat{V}_s$ such that

$$\frac{\delta T_s}{\delta n(\mathbf{r})} + v_s(\mathbf{r}) = \mu \, , \tag{1.60}$$

as in (1.58). In (1.60), the Kohn-Sham potential $v_s(\mathbf{r})$ is a functional of $n(\mathbf{r})$. If there were any difference between μ and μ_s, the chemical potentials for interacting and non-interacting systems of the same density, it could be absorbed into $v_s(\mathbf{r})$. We have assumed that $n(\mathbf{r})$ is both interacting and non-interacting v-representable.

Now we *define* the exchange-correlation energy $E_{xc}[n]$ by

$$F[n] = T_s[n] + U[n] + E_{xc}[n] , \qquad (1.61)$$

where $U[n]$ is given by (1.9). The Euler equations (1.58) and (1.60) are consistent with one another if and only if

$$v_s(\mathbf{r}) = v(\mathbf{r}) + \frac{\delta U[n]}{\delta n(\mathbf{r})} + \frac{\delta E_{xc}[n]}{\delta n(\mathbf{r})} . \qquad (1.62)$$

Thus we have derived the Kohn-Sham method [6] of Sect. 1.1.2.

The Kohn-Sham method treats $T_s[n]$ exactly, leaving only $E_{xc}[n]$ to be approximated. This makes good sense, for several reasons: (1) $T_s[n]$ is typically a very large part of the energy, while $E_{xc}[n]$ is a smaller part. (2) $T_s[n]$ is largely responsible for density oscillations of the shell structure and Friedel types, which are accurately described by the Kohn-Sham method. (3) $E_{xc}[n]$ is somewhat better suited to the local and semi-local approximations than is $T_s[n]$, for reasons to be discussed later. The price to be paid for these benefits is the appearance of orbitals. If we had a very accurate approximation for T_s directly in terms of n, we could dispense with the orbitals and solve the Euler equation (1.60) directly for $n(\mathbf{r})$.

The total energy of (1.6) may also be written as

$$E = \sum_{\alpha\sigma} \theta(\mu - \varepsilon_{\alpha\sigma})\varepsilon_{\alpha\sigma} - U[n] - \int d^3r \, n(\mathbf{r})v_{xc}([n];\mathbf{r}) + E_{xc}[n] , \qquad (1.63)$$

where the second and third terms on the right hand side simply remove contributions to the first term which do not belong in the total energy. The first term on the right of (1.63), the non-interacting energy E_{non}, is the only term that appears in the semi-empirical Hückel theory [26]. This first term includes most of the electronic shell structure effects which arise when $T_s[n]$ is treated exactly (but not when $T_s[n]$ is treated in a continuum model like the Thomas-Fermi approximation or the gradient expansion).

1.3.4   Exchange Energy and Correlation Energy

$E_{xc}[n]$ is the sum of distinct exchange and correlation terms:

$$E_{xc}[n] = E_x[n] + E_c[n] , \qquad (1.64)$$

where [29]

$$E_x[n] = \langle \Phi_n^{min} | \hat{V}_{ee} | \Phi_n^{min} \rangle - U[n] . \qquad (1.65)$$

When Φ_n^{min} is a single Slater determinant, (1.65) is just the usual Fock integral applied to the Kohn-Sham orbitals, i.e., it differs from the Hartree-Fock exchange energy only to the extent that the Kohn-Sham orbitals differ from the Hartree-Fock orbitals for a given system or density (in the same way that $T_s[n]$ differs from the Hartree-Fock kinetic energy). We note that

$$\langle \Phi_n^{\mathrm{min}} | \hat{T} + \hat{V}_{\mathrm{ee}} | \Phi_n^{\mathrm{min}} \rangle = T_s[n] + U[n] + E_{\mathrm{x}}[n] \,, \tag{1.66}$$

and that, in the one-electron ($\hat{V}_{\mathrm{ee}} = 0$) limit [9],

$$E_{\mathrm{x}}[n] = -U[n] \qquad (N = 1) \,. \tag{1.67}$$

The correlation energy is

$$\begin{aligned} E_{\mathrm{c}}[n] &= F[n] - \{T_s[n] + U[n] + E_{\mathrm{x}}[n]\} \\ &= \langle \Psi_n^{\mathrm{min}} | \hat{T} + \hat{V}_{\mathrm{ee}} | \Psi_n^{\mathrm{min}} \rangle - \langle \Phi_n^{\mathrm{min}} | \hat{T} + \hat{V}_{\mathrm{ee}} | \Phi_n^{\mathrm{min}} \rangle \,. \end{aligned} \tag{1.68}$$

Since Ψ_n^{min} is that wavefunction which yields density n and minimizes $\langle \hat{T} + \hat{V}_{\mathrm{ee}} \rangle$, (1.68) shows that

$$E_{\mathrm{c}}[n] \leq 0 \,. \tag{1.69}$$

Since Φ_n^{min} is that wavefunction which yields density n and minimizes $\langle \hat{T} \rangle$, (1.68) shows that $E_{\mathrm{c}}[n]$ is the sum of a positive kinetic energy piece and a negative potential energy piece. These pieces of E_c contribute respectively to the first and second terms of the virial theorem, (1.45). Clearly for any one-electron system [9]

$$E_{\mathrm{c}}[n] = 0 \qquad (N = 1) \,. \tag{1.70}$$

Equations (1.67) and (1.70) show that the exchange-correlation energy of a one-electron system simply cancels the spurious self-interaction $U[n]$. In the same way, the exchange-correlation potential cancels the spurious self-interaction in the Kohn-Sham potential [9]

$$\frac{\delta E_{\mathrm{x}}[n]}{\delta n(\mathbf{r})} = -u([n]; \mathbf{r}) \qquad (N = 1) \,, \tag{1.71}$$

$$\frac{\delta E_{\mathrm{c}}[n]}{\delta n(\mathbf{r})} = 0 \qquad (N = 1) \,. \tag{1.72}$$

Thus

$$\lim_{r \to \infty} \frac{\delta E_{\mathrm{xc}}[n]}{\delta n(\mathbf{r})} = -\frac{1}{r} \qquad (N = 1) \,. \tag{1.73}$$

The extension of these one-electron results to spin-density functional theory is straightforward, since a one-electron system is fully spin-polarized.

1.3.5 Coupling-Constant Integration

The definitions (1.65) and (1.68) are formal ones, and do not provide much intuitive or physical insight into the exchange and correlation energies, or much guidance for the approximation of their density functionals. These insights are provided by the coupling-constant integration [30–33] to be derived below.

Let us define $\Psi_n^{\min,\lambda}$ as that normalized, antisymmetric wavefunction which yields density $n(\mathbf{r})$ and minimizes the expectation value of $\hat{T} + \lambda \hat{V}_{ee}$, where we have introduced a non-negative coupling constant λ. When $\lambda = 1$, $\Psi_n^{\min,\lambda}$ is $\Psi_n^{\min}$, the interacting ground-state wavefunction for density n. When $\lambda = 0$, $\Psi_n^{\min,\lambda}$ is $\Phi_n^{\min}$, the non-interacting or Kohn-Sham wavefunction for density n. Varying λ at fixed $n(\mathbf{r})$ amounts to varying the external potential $v_\lambda(\mathbf{r})$: At $\lambda = 1$, $v_\lambda(\mathbf{r})$ is the true external potential, while at $\lambda = 0$ it is the Kohn-Sham effective potential $v_s(\mathbf{r})$. We normally assume a smooth, "adiabatic connection" between the interacting and non-interacting ground states as λ is reduced from 1 to 0.

Now we write (1.64), (1.65) and (1.68) as

$$
\begin{aligned}
E_{xc}&[n] \\
&= \langle \Psi_n^{\min,\lambda} | \hat{T} + \lambda \hat{V}_{ee} | \Psi_n^{\min,\lambda} \rangle \Big|_{\lambda=1} - \langle \Psi_n^{\min,\lambda} | \hat{T} + \lambda \hat{V}_{ee} | \Psi_n^{\min,\lambda} \rangle \Big|_{\lambda=0} - U[n] \\
&= \int_0^1 d\lambda \frac{d}{d\lambda} \langle \Psi_n^{\min,\lambda} | \hat{T} + \lambda \hat{V}_{ee} | \Psi_n^{\min,\lambda} \rangle - U[n] .
\end{aligned}
\tag{1.74}
$$

The Hellmann-Feynman theorem of Sect. 1.2.4 allows us to simplify (1.74) to

$$
E_{xc}[n] = \int_0^1 d\lambda \langle \Psi_n^{\min,\lambda} | \hat{V}_{ee} | \Psi_n^{\min,\lambda} \rangle - U[n] .
\tag{1.75}
$$

Equation (1.75) "looks like" a potential energy; the kinetic energy contribution to E_{xc} has been subsumed by the coupling-constant integration. We should remember, of course, that only $\lambda = 1$ is real or physical. The Kohn-Sham system at $\lambda = 0$, and all the intermediate values of λ, are convenient mathematical fictions.

To make further progress, we need to know how to evaluate the N-electron expectation value of a sum of one-body operators like $\hat{T}$, or a sum of two-body operators like $\hat{V}_{ee}$. For this purpose, we introduce one-electron (ρ_1) and two-electron (ρ_2) reduced density matrices [34] :

$$
\rho_1(\mathbf{r}'\sigma, \mathbf{r}\sigma) \equiv N \sum_{\sigma_2 \ldots \sigma_N} \int d^3 r_2 \ldots \int d^3 r_N
$$

$$
\Psi^*(\mathbf{r}'\sigma, \mathbf{r}_2\sigma_2, \ldots, \mathbf{r}_N\sigma_N) \Psi(\mathbf{r}\sigma, \mathbf{r}_2\sigma_2, \ldots, \mathbf{r}_N\sigma_N) , \tag{1.76}
$$

$$
\rho_2(\mathbf{r}', \mathbf{r}) \equiv N(N-1) \sum_{\sigma_1 \ldots \sigma_N} \int d^3 r_3 \ldots \int d^3 r_N
$$

$$
|\Psi(\mathbf{r}'\sigma_1, \mathbf{r}\sigma_2, \ldots, \mathbf{r}_N\sigma_N)|^2 . \tag{1.77}
$$

From (1.20),

$$n_\sigma(\mathbf{r}) = \rho_1(\mathbf{r}\sigma, \mathbf{r}\sigma) .$$ (1.78)

Clearly also

$$\langle \hat{T} \rangle = -\frac{1}{2} \sum_\sigma \int d^3r \, \frac{\partial}{\partial \mathbf{r}} \cdot \frac{\partial}{\partial \mathbf{r}} \rho_1(\mathbf{r}'\sigma, \mathbf{r}\sigma)\bigg|_{\mathbf{r}'=\mathbf{r}} ,$$ (1.79)

$$\langle \hat{V}_{ee} \rangle = \frac{1}{2} \int d^3r \int d^3r' \, \frac{\rho_2(\mathbf{r}', \mathbf{r})}{|\mathbf{r} - \mathbf{r}'|} .$$ (1.80)

We interpret the positive number $\rho_2(\mathbf{r}', \mathbf{r})d^3r'd^3r$ as the joint probability of finding an electron in volume element d^3r' at $\mathbf{r}'$, *and* an electron in d^3r at $\mathbf{r}$. By standard probability theory, this is the product of the probability of finding an electron in d^3r ($n(\mathbf{r})d^3r$) and the conditional probability of finding an electron in d^3r', given that there is one at $\mathbf{r}$ ($n_2(\mathbf{r}, \mathbf{r}')d^3r'$):

$$\rho_2(\mathbf{r}', \mathbf{r}) = n(\mathbf{r})n_2(\mathbf{r}, \mathbf{r}') .$$ (1.81)

By arguments similar to those used in Sect. 1.2.1, we interpret $n_2(\mathbf{r}, \mathbf{r}')$ as the average density of electrons at $\mathbf{r}'$, given that there is an electron at $\mathbf{r}$. Clearly then

$$\int d^3r' \, n_2(\mathbf{r}, \mathbf{r}') = N - 1 .$$ (1.82)

For the wavefunction $\Psi_n^{\min,\lambda}$, we write

$$n_2(\mathbf{r}, \mathbf{r}') = n(\mathbf{r}') + n_{xc}^\lambda(\mathbf{r}, \mathbf{r}') ,$$ (1.83)

an equation which defines $n_{xc}^\lambda(\mathbf{r}, \mathbf{r}')$, the density at $\mathbf{r}'$ of the exchange-correlation hole [33] about an electron at $\mathbf{r}$. Equations (1.5) and (1.83) imply that

$$\int d^3r' \, n_{xc}^\lambda(\mathbf{r}, \mathbf{r}') = -1 ,$$ (1.84)

which says that, if an electron is definitely at $\mathbf{r}$, it is missing from the rest of the system.

Because the Coulomb interaction $1/u$ is singular as $u = |\mathbf{r} - \mathbf{r}'| \to 0$, the exchange-correlation hole density has a cusp [35,34] around $u = 0$:

$$\frac{\partial}{\partial u} \int \frac{d\Omega_\mathbf{u}}{4\pi} \, n_{xc}^\lambda(\mathbf{r}, \mathbf{r} + \mathbf{u})\bigg|_{u=0} = \lambda \left[n(\mathbf{r}) + n_{xc}^\lambda(\mathbf{r}, \mathbf{r}) \right] ,$$ (1.85)

where $\int d\Omega_\mathbf{u}/(4\pi)$ is an angular average. This cusp vanishes when $\lambda = 0$, and also in the fully-spin-polarized and low-density limits, in which all other electrons are excluded from the position of a given electron: $n_{xc}^\lambda(\mathbf{r}, \mathbf{r}) = -n(\mathbf{r})$.

We can now rewrite (1.75) as [33]

$$E_{xc}[n] = \frac{1}{2} \int d^3r \int d^3r' \, \frac{n(\mathbf{r})\bar{n}_{xc}(\mathbf{r}, \mathbf{r}')}{|\mathbf{r} - \mathbf{r}'|} ,$$ (1.86)

where

$$\bar{n}_{xc}(\mathbf{r}, \mathbf{r}') = \int_0^1 d\lambda\, n_{xc}^\lambda(\mathbf{r}, \mathbf{r}')$$

(1.87)

is the coupling-constant averaged hole density. The exchange-correlation energy is just the electrostatic interaction between each electron and the coupling-constant-averaged exchange-correlation hole which surrounds it. The hole is created by three effects: (1) self-interaction correction, a classical effect which guarantees that an electron cannot interact with itself, (2) the Pauli exclusion principle, which tends to keep two electrons with parallel spins apart in space, and (3) the Coulomb repulsion, which tends to keep any two electrons apart in space. Effects (1) and (2) are responsible for the exchange energy, which is present even at $\lambda = 0$, while effect (3) is responsible for the correlation energy, and arises only for $\lambda \neq 0$.

If $\Psi_n^{\min,\lambda=0}$ is a single Slater determinant, as it typically is, then the one- and two-electron density matrices at $\lambda = 0$ can be constructed explicitly from the Kohn-Sham spin orbitals $\psi_{\alpha\sigma}(\mathbf{r})$:

$$\rho_1^{\lambda=0}(\mathbf{r}'\sigma, \mathbf{r}\sigma) = \sum_\alpha \theta(\mu - \varepsilon_{\alpha\sigma})\psi_{\alpha\sigma}^*(\mathbf{r}')\psi_{\alpha\sigma}(\mathbf{r}) ,$$

(1.88)

$$\rho_2^{\lambda=0}(\mathbf{r}', \mathbf{r}) = n(\mathbf{r})n(\mathbf{r}') + n(\mathbf{r})n_x(\mathbf{r}, \mathbf{r}') ,$$

(1.89)

where

$$n_x(\mathbf{r}, \mathbf{r}') = n_{xc}^{\lambda=0}(\mathbf{r}, \mathbf{r}') = -\sum_\sigma \frac{|\rho_1^{\lambda=0}(\mathbf{r}'\sigma, \mathbf{r}\sigma)|^2}{n(\mathbf{r})}$$

(1.90)

is the exact exchange-hole density. Equation (1.90) shows that

$$n_x(\mathbf{r}, \mathbf{r}') \leq 0 ,$$

(1.91)

so the exact exchange energy

$$E_x[n] = \frac{1}{2} \int d^3r \int d^3r' \frac{n(\mathbf{r})n_x(\mathbf{r}, \mathbf{r}')}{|\mathbf{r} - \mathbf{r}'|}$$

(1.92)

is also negative, and can be written as the sum of up-spin and down-spin contributions:

$$E_x = E_x^\uparrow + E_x^\downarrow < 0 .$$

(1.93)

Equation (1.84) provides a sum rule for the exchange hole:

$$\int d^3r'\, n_x(\mathbf{r}, \mathbf{r}') = -1 .$$

(1.94)

Equations (1.90) and (1.78) show that the "on-top" exchange hole density is [36]

$$n_x(\mathbf{r}, \mathbf{r}) = -\frac{n_\uparrow^2(\mathbf{r}) + n_\downarrow^2(\mathbf{r})}{n(\mathbf{r})} ,$$

(1.95)

which is determined by just the local spin densities at position $\mathbf{r}$ – suggesting a reason why local spin density approximations work better than local density approximations.

The correlation hole density is defined by

$$\bar{n}_{\mathrm{xc}}(\mathbf{r}, \mathbf{r}') = n_{\mathrm{x}}(\mathbf{r}, \mathbf{r}') + \bar{n}_{\mathrm{c}}(\mathbf{r}, \mathbf{r}') , \qquad (1.96)$$

and satisfies the sum rule

$$\int \mathrm{d}^3 r' \, \bar{n}_{\mathrm{c}}(\mathbf{r}, \mathbf{r}') = 0 , \qquad (1.97)$$

which says that Coulomb repulsion changes the shape of the hole but not its integral. In fact, this repulsion typically makes the hole deeper but more short-ranged, with a negative on-top correlation hole density:

$$\bar{n}_{\mathrm{c}}(\mathbf{r}, \mathbf{r}) \leq 0 . \qquad (1.98)$$

The positivity of (1.77) is equivalent via (1.81) and (1.83) to the inequality

$$\bar{n}_{\mathrm{xc}}(\mathbf{r}, \mathbf{r}') \geq -n(\mathbf{r}') , \qquad (1.99)$$

which asserts that the hole cannot take away electrons that were not there initially. By the sum rule (1.97), the correlation hole density $\bar{n}_{\mathrm{c}}(\mathbf{r}, \mathbf{r}')$ must have positive as well as negative contributions. Moreover, unlike the exchange hole density $n_x(\mathbf{r}, \mathbf{r}')$, the exchange-correlation hole density $\bar{n}_{xc}(\mathbf{r}, \mathbf{r}')$ can be positive.

To better understand E_{xc}, we can simplify (1.86) to the "real-space analysis" [37]

$$E_{\mathrm{xc}}[n] = \frac{N}{2} \int_0^\infty \mathrm{d}u \, 4\pi u^2 \frac{\langle \bar{n}_{\mathrm{xc}}(u) \rangle}{u} , \qquad (1.100)$$

where

$$\langle \bar{n}_{\mathrm{xc}}(u) \rangle = \frac{1}{N} \int \mathrm{d}^3 r \, n(\mathbf{r}) \int \frac{\mathrm{d}\Omega_{\mathbf{u}}}{4\pi} \, \bar{n}_{\mathrm{xc}}(\mathbf{r}, \mathbf{r} + \mathbf{u}) \qquad (1.101)$$

is the system- and spherical-average of the coupling-constant-averaged hole density. The sum rule of (1.84) becomes

$$\int_0^\infty \mathrm{d}u \, 4\pi u^2 \langle \bar{n}_{\mathrm{xc}}(u) \rangle = -1 . \qquad (1.102)$$

As u increases from 0, $\langle n_{\mathrm{x}}(u) \rangle$ rises analytically like $\langle n_{\mathrm{x}}(0) \rangle + \mathcal{O}(u^2)$, while $\langle \bar{n}_{\mathrm{c}}(u) \rangle$ rises like $\langle \bar{n}_{\mathrm{c}}(0) \rangle + \mathcal{O}(|u|)$ as a consequence of the cusp of (1.85). Because of the constraint of (1.102) and because of the factor $1/u$ in (1.100), E_{xc} typically becomes more negative as the on-top hole density $\langle \bar{n}_{\mathrm{xc}}(u) \rangle$ gets more negative.

1.4   Formal Properties of Functionals

1.4.1   Uniform Coordinate Scaling

The more we know of the exact properties of the density functionals $E_{xc}[n]$ and $T_s[n]$, the better we shall understand and be able to approximate these functionals. We start with the behavior of the functionals under a uniform coordinate scaling of the density, (1.40).

The Hartree electrostatic self-repulsion of the electrons is known exactly (see (1.9)), and has a simple coordinate scaling:

$$U[n_\gamma] = \frac{1}{2} \int d^3(\gamma r) \int d^3(\gamma r') \frac{n(\gamma \mathbf{r}) n(\gamma \mathbf{r}')}{|\mathbf{r} - \mathbf{r}'|}$$
$$= \gamma \frac{1}{2} \int d^3 r_1 \int d^3 r_1' \frac{n(\mathbf{r}_1) n(\mathbf{r}_1')}{|\mathbf{r}_1 - \mathbf{r}_1'|} = \gamma U[n] , \qquad (1.103)$$

where $\mathbf{r}_1 = \gamma \mathbf{r}$ and $\mathbf{r}_1' = \gamma \mathbf{r}'$.

Next consider the non-interacting kinetic energy of (1.59). Scaling all the wavefunctions Ψ in the constrained search as in (1.38) will scale the density as in (1.40) and scale each kinetic energy expectation value as in (1.43). Thus the constrained search for the unscaled density maps into the constrained search for the scaled density, and [38]

$$T_s[n_\gamma] = \gamma^2 \, T_s[n] . \qquad (1.104)$$

We turn now to the exchange energy of (1.65). By the argument of the last paragraph, $\Phi_{n_\gamma}^{\min}$ is the scaled version of $\Phi_n^{\min}$. Since also

$$\hat{V}_{ee}(\gamma \mathbf{r}_1, \dots, \gamma \mathbf{r}_N) = \gamma^{-1} \, \hat{V}_{ee}(\mathbf{r}_1, \dots, \mathbf{r}_N) , \qquad (1.105)$$

and with the help of (1.103), we find [38]

$$E_x[n_\gamma] = \gamma \, E_x[n] . \qquad (1.106)$$

In the high-density ($\gamma \to \infty$) limit, $T_s[n_\gamma]$ dominates $U[n_\gamma]$ and $E_x[n_\gamma]$. An example would be an ion with a fixed number of electrons N and a nuclear charge Z which tends to infinity; in this limit, the density and energy become essentially hydrogenic, and the effects of U and E_x become relatively negligible. In the low-density ($\gamma \to 0$) limit, $U[n_\gamma]$ and $E_x[n_\gamma]$ dominate $T_s[n_\gamma]$.

We can use coordinate scaling relations to fix the form of a local density approximation

$$F[n] = \int d^3 r \, f(n(\mathbf{r})) . \qquad (1.107)$$

If $F[n_\lambda] = \lambda^p F[n]$, then

$$\lambda^{-3} \int d^3(\lambda \mathbf{r}) \, f\left(\lambda^3 n(\lambda \mathbf{r})\right) = \lambda^p \int d^3 r \, f(n(\mathbf{r})) , \qquad (1.108)$$

or $f(\lambda^3 n) = \lambda^{p+3} f(n)$, whence

$$f(n) = n^{1+p/3} . \tag{1.109}$$

For the exchange energy of (1.106), $p = 1$ so (1.107) and (1.109) imply (1.49). For the non-interacting kinetic energy of (1.104), $p = 2$ so (1.107) and (1.109) imply the Thomas-Fermi approximation

$$T_0[n] = A_s \int d^3 r\, n^{5/3}(\mathbf{r}) . \tag{1.110}$$

$U[n]$ of (1.9) is too strongly nonlocal for any local approximation.

While $T_s[n]$, $U[n]$ and $E_x[n]$ have simple scalings, $E_c[n]$ of (1.68) does not. This is because $\Psi_{n_\gamma}^{\min}$, the wavefunction which via (1.55) yields the scaled density $n_\gamma(\mathbf{r})$ and minimizes the expectation value of $\hat{T} + \hat{V}_{ee}$, is *not* the scaled wavefunction $\gamma^{3N/2} \Psi_n^{\min}(\gamma\mathbf{r}_1, \ldots, \gamma\mathbf{r}_N)$. The scaled wavefunction yields $n_\gamma(\mathbf{r})$ but minimizes the expectation value of $\hat{T} + \gamma\hat{V}_{ee}$, and it is this latter expectation value which scales like γ^2 under wavefunction scaling. Thus [39]

$$E_c[n_\gamma] = \gamma^2 E_c^{1/\gamma}[n] , \tag{1.111}$$

where $E_c^{1/\gamma}[n]$ is the density functional for the correlation energy in a system for which the electron-electron interaction is not $\hat{V}_{ee}$ but $\gamma^{-1}\hat{V}_{ee}$.

To understand these results, let us assume that the Kohn-Sham non-interacting Hamiltonian has a non-degenerate ground state. In the high-density limit ($\gamma \to \infty$), $\Psi_{n_\gamma}^{\min}$ minimizes just $\langle\hat{T}\rangle$ and reduces to $\Phi_{n_\gamma}^{\min}$. Now we treat

$$\Delta \equiv \hat{V}_{ee} - \sum_{i=1}^{N} \left[\frac{\delta U[n]}{\delta n(\mathbf{r}_i)} + \frac{\delta E_x[n]}{\delta n(\mathbf{r}_i)} \right] \tag{1.112}$$

as a weak perturbation [40,41] on the Kohn-Sham non-interacting Hamiltonian, and find

$$E_c[n] = \sum_{k \neq 0} \frac{|\langle k|\Delta|0\rangle|^2}{E_0 - E_k} , \tag{1.113}$$

where the $|k\rangle$ are the eigenfunctions of the Kohn-Sham non-interacting Hamiltonian, and $|0\rangle$ is its ground state. Both the numerator and the denominator of (1.113) scale like γ^2, so [42]

$$\lim_{\gamma \to \infty} E_c[n_\gamma] = \text{constant} . \tag{1.114}$$

In the low-density limit, $\Psi_{n_\gamma}^{\min}$ minimizes just $\langle\hat{V}_{ee}\rangle$, and (1.68) then shows that [43]

$$E_c[n_\gamma] \approx \gamma D[n] \qquad (\gamma \to 0) , \tag{1.115}$$

with an appropriately chosen density functional $D[n]$.

Generally, we have a scaling inequality [38]

$$E_c[n_\gamma] > \gamma E_c[n] \qquad (\gamma > 1) , \qquad (1.116)$$

$$E_c[n_\gamma] < \gamma E_c[n] \qquad (\gamma < 1) . \qquad (1.117)$$

If we choose a density n, we can plot $E_c[n_\gamma]$ versus γ, and compare the result to the straight line $\gamma E_c[n]$. These two curves will drop away from zero as γ increases from zero (with different initial slopes), then cross at $\gamma = 1$. The convex $E_c[n_\gamma]$ will then approach a negative constant as $\gamma \to \infty$.

1.4.2 Local Lower Bounds

Because of the importance of local and semilocal approximations like (1.11) and (1.12), bounds on the exact functionals are especially useful when the bounds are themselves local functionals.

Lieb and Thirring [44] have conjectured that $T_s[n]$ is bounded from below by the Thomas-Fermi functional

$$T_s[n] \geq T_0[n] , \qquad (1.118)$$

where $T_0[n]$ is given by (1.110) with

$$A_s = \frac{3}{10} (3\pi^2)^{2/3} . \qquad (1.119)$$

We have already established that

$$E_x[n] \geq E_{xc}[n] \geq E_{xc}^{\lambda=1}[n] , \qquad (1.120)$$

where the final term of (1.120) is the integrand $E_{xc}^{\lambda}[n]$ of the coupling-constant integration of (1.75),

$$E_{xc}^{\lambda}[n] = \langle \Psi_n^{\min,\lambda} | \hat{V}_{ee} | \Psi_n^{\min,\lambda} \rangle - U[n] , \qquad (1.121)$$

evaluated at the upper limit $\lambda = 1$. Lieb and Oxford [45] have proved that

$$E_{xc}^{\lambda=1}[n] \geq 2.273 \, E_x^{\mathrm{LDA}}[n] , \qquad (1.122)$$

where $E_x^{\mathrm{LDA}}[n]$ is the local density approximation for the exchange energy, (1.49), with

$$A_x = -\frac{3}{4\pi} (3\pi^2)^{1/3} . \qquad (1.123)$$

1.4.3 Spin Scaling Relations

Spin scaling relations can be used to convert density functionals into spin-density functionals.

For example, the non-interacting kinetic energy is the sum of the separate kinetic energies of the spin-up and spin-down electrons:

$$T_s[n_\uparrow, n_\downarrow] = T_s[n_\uparrow, 0] + T_s[0, n_\downarrow] \ . \tag{1.124}$$

The corresponding density functional, appropriate to a spin-unpolarized system, is [46]

$$T_s[n] = T_s[n/2, n/2] = 2T_s[n/2, 0] \ , \tag{1.125}$$

whence $T_s[n/2, 0] = \frac{1}{2}T_s[n]$ and (1.124) becomes

$$T_s[n_\uparrow, n_\downarrow] = \frac{1}{2}T_s[2n_\uparrow] + \frac{1}{2}T_s[2n_\downarrow] \ . \tag{1.126}$$

Similarly, (1.93) implies [46]

$$E_x[n_\uparrow, n_\downarrow] = \frac{1}{2}E_x[2n_\uparrow] + \frac{1}{2}E_x[2n_\downarrow] \ . \tag{1.127}$$

For example, we can start with the local density approximations (1.110) and (1.49), then apply (1.126) and (1.127) to generate the corresponding local spin density approximations.

Because two electrons of anti-parallel spin repel one another coulombically, making an important contribution to the correlation energy, there is no simple spin scaling relation for E_c.

1.4.4 Size Consistency

Common sense tells us that the total energy E and density $n(\mathbf{r})$ for a system, comprised of two well-separated subsystems with energies E_1 and E_2 and densities $n_1(\mathbf{r})$ and $n_2(\mathbf{r})$, must be $E = E_1 + E_2$ and $n(\mathbf{r}) = n_1(\mathbf{r}) + n_2(\mathbf{r})$. Approximations which satisfy this expectation, such as the LSD of (1.11) or the GGA of (1.12), are properly size consistent [47]. Size consistency is not only a principle of physics, it is almost a principle of epistemology: How could we analyze or understand complex systems, if they could not be separated into simpler components?

Density functionals which are *not* size consistent are to be avoided. An example is the Fermi-Amaldi [48] approximation for the exchange energy,

$$E_x^{\mathrm{FA}}[n] = -U[n/N] \ , \tag{1.128}$$

where N is given by (1.5), which was constructed to satisfy (1.67).

1.4.5 Derivative Discontinuity

In Sect. 1.3, our density functionals were defined as constrained searches over wavefunctions. Because all wavefunctions searched have the same electron number, there is no way to make a number-nonconserving density variation $\delta n(\mathbf{r})$. The functional derivatives are defined only up to an arbitrary constant, which has no effect on (1.50) when $\int d^3r\, \delta n(\mathbf{r}) = 0$.

To complete the definition of the functional derivatives and of the chemical potential μ, we extend the constrained search from wavefunctions to ensembles [49,50]. An ensemble or mixed state is a set of wavefunctions or pure states and their respective probabilities. By including wavefunctions with different electron numbers in the same ensemble, we can develop a density functional theory for non-integer particle number. Fractional particle numbers can arise in an open system that shares electrons with its environment, and in which the electron number fluctuates between integers.

The upshot is that the ground-state energy $E(N)$ varies linearly between two adjacent integers, and has a derivative discontinuity at each integer. This discontinuity arises in part from the exchange-correlation energy (and entirely so in cases for which the integer does not fall on the boundary of an electronic shell or subshell, e.g., for $N = 6$ in the carbon atom but not for $N = 10$ in the neon atom).

By Janak's theorem [51], the highest partly-occupied Kohn-Sham eigenvalue $\varepsilon_{\mathrm{HO}}$ equals $\partial E/\partial N = \mu$, and so changes discontinuously [49,50] at an integer Z:

$$\varepsilon_{\mathrm{HO}} = \begin{cases} -I_Z & (Z - 1 < N < Z) \\ -A_Z & (Z < N < Z + 1) \end{cases}, \tag{1.129}$$

where I_Z is the first ionization energy of the Z-electron system (i.e., the least energy needed to remove an electron from this system), and A_Z is the electron affinity of the Z-electron system (i.e., $A_Z = I_{Z+1}$). If Z does not fall on the boundary of an electronic shell or subshell, all of the difference between $-I_Z$ and $-A_Z$ must arise from a discontinuous jump in the exchange-correlation potential $\delta E_{\mathrm{xc}}/\delta n(\mathbf{r})$ as the electron number N crosses the integer Z.

Since the asymptotic decay of the density of a finite system with Z electrons is controlled by I_Z, we can show that the exchange-correlation potential tends to zero as $|\mathbf{r}| \to \infty$ [52]:

$$\lim_{|\mathbf{r}|\to\infty} \frac{\delta E_{\mathrm{xc}}}{\delta n(\mathbf{r})} = 0 \quad (Z - 1 < N < Z), \tag{1.130}$$

or more precisely

$$\lim_{|\mathbf{r}|\to\infty} \frac{\delta E_{\mathrm{xc}}}{\delta n(\mathbf{r})} = -\frac{1}{r} \quad (Z - 1 < N < Z). \tag{1.131}$$

As N increases through the integer Z, $\delta E_{\mathrm{xc}}/\delta n(\mathbf{r})$ jumps up by a positive additive constant. With further increases in N above Z, this "constant" vanishes, first at very large $|\mathbf{r}|$ and then at smaller and smaller $|\mathbf{r}|$, until it is all gone in the limit where N approaches the integer $Z + 1$ from below.

Simple continuum approximations to $E_{xc}[n_\uparrow, n_\downarrow]$, such as the LSD of (1.11) or the GGA of (1.12), miss much or all the derivative discontinuity, and can at best average over it. For example, the highest occupied orbital energy for a neutral atom becomes approximately $-\frac{1}{2}(I_Z + A_Z)$, the average of (1.129) from the electron-deficient and electron-rich sides of neutrality. We must never forget, when we make these approximations, that we are fitting a round peg into a square hole. The areas (integrated properties) of a circle and a square can be matched, but their perimeters (differential properties) will remain stubbornly different.

1.5 Uniform Electron Gas

1.5.1 Kinetic Energy

Simple systems play an important paradigmatic role in science. For example, the hydrogen atom is a paradigm for all of atomic physics. In the same way, the uniform electron gas [24] is a paradigm for solid-state physics, and also for density functional theory. In this system, the electron density $n(\mathbf{r})$ is uniform or constant over space, and thus the electron number is infinite. The negative charge of the electrons is neutralized by a rigid uniform positive background. We could imagine creating such a system by starting with a simple metal, regarded as a perfect crystal of valence electrons and ions, and then smearing out the ions to make the uniform background of positive charge. In fact, the simple metal sodium is physically very much like a uniform electron gas.

We begin by evaluating the non-interacting kinetic energy (this section) and exchange energy (next section) per electron for a spin-unpolarized electron gas of uniform density n. The corresponding energies for the spin-polarized case can then be found from (1.126) and (1.127).

By symmetry, the Kohn-Sham potential $v_s(\mathbf{r})$ must be uniform or constant, and we take it to be zero. We impose boundary conditions within a cube of volume $\mathcal{V} \to \infty$, i.e., we require that the orbitals repeat from one face of the cube to its opposite face. (Presumably any choice of boundary conditions would give the same answer as $\mathcal{V} \to \infty$.) The Kohn-Sham orbitals are then plane waves $\exp(i\mathbf{k} \cdot \mathbf{r})/\sqrt{\mathcal{V}}$, with momenta or wavevectors $\mathbf{k}$ and energies $k^2/2$. The number of orbitals of both spins in a volume d^3k of wavevector space is $2[\mathcal{V}/(2\pi)^3]d^3k$, by an elementary geometrical argument [53].

Let $N = n\mathcal{V}$ be the number of electrons in volume $\mathcal{V}$. These electrons occupy the N lowest Kohn-Sham spin orbitals, i.e., those with $k < k_F$:

$$N = 2\sum_{\mathbf{k}} \theta(k_F - k) = 2\frac{\mathcal{V}}{(2\pi)^3} \int_0^{k_F} dk\, 4\pi k^2 = \mathcal{V}\frac{k_F^3}{3\pi^2} , \qquad (1.132)$$

where k_F is called the Fermi wavevector. The Fermi wavelength $2\pi/k_F$ is the shortest de Broglie wavelength for the non-interacting electrons. Clearly

$$n = \frac{k_F^3}{3\pi^2} = \frac{3}{4\pi r_s^3} , \qquad (1.133)$$

where we have introduced the Seitz radius r_s – the radius of a sphere which on average contains one electron.

The kinetic energy of an orbital is $k^2/2$, and the average kinetic energy per electron is

$$t_s(n) = \frac{2}{N} \sum_k \theta(k_F - k)\frac{k^2}{2} = \frac{2V}{N(2\pi)^3} \int_0^{k_F} dk\, 4\pi k^2 \frac{k^2}{2} = \frac{3}{5}\frac{k_F^2}{2} , \qquad (1.134)$$

or 3/5 of the Fermi energy. In other notation,

$$t_s(n) = \frac{3}{10}(3\pi^2 n)^{2/3} = \frac{3}{10}\frac{(9\pi/4)^{2/3}}{r_s^2} . \qquad (1.135)$$

All of this kinetic energy follows from the Pauli exclusion principle, i.e., from the fermion character of the electron.

1.5.2   Exchange Energy

To evaluate the exchange energy, we need the Kohn-Sham one-matrix for electrons of spin σ, as defined in (1.88):

$$
\begin{aligned}
\rho_1^{\lambda=0}(\mathbf{r} + \mathbf{u}\sigma, \mathbf{r}\sigma) &= \sum_k \theta(k_F - k)\frac{\exp(-i\mathbf{k}\cdot(\mathbf{r}+\mathbf{u}))}{\sqrt{V}}\frac{\exp(i\mathbf{k}\cdot\mathbf{r})}{\sqrt{V}} \\
&= \frac{1}{(2\pi)^3}\int_0^{k_F} dk\, 4\pi k^2 \int\frac{d\Omega_k}{4\pi}\exp(-i\mathbf{k}\cdot\mathbf{u}) \\
&= \frac{1}{2\pi^2}\int_0^{k_F} dk\, k^2\frac{\sin(ku)}{ku} \\
&= \frac{k_F^3}{2\pi^2}\frac{\sin(k_F u) - k_F u\cos(k_F u)}{(k_F u)^3} .
\end{aligned}
\qquad (1.136)
$$

The exchange hole density at distance u from an electron is, by (1.90),

$$n_x(u) = -2\frac{|\rho_1^{\lambda=0}(\mathbf{r}+\mathbf{u}\sigma,\mathbf{r}\sigma)|^2}{n} , \qquad (1.137)$$

which ranges from $-n/2$ at $u = 0$ (where all other electrons of the same spin are excluded by the Pauli principle) to 0 (like $1/u^4$) as $u \to \infty$. The exchange energy per electron is

$$e_x(n) = \int_0^\infty du\, 2\pi u n_x(u) = -\frac{3}{4\pi}k_F . \qquad (1.138)$$

In other notation,

$$e_{\mathrm{x}}(n) = -\frac{3}{4\pi}(3\pi^2 n)^{1/3} = -\frac{3}{4\pi}\frac{(9\pi/4)^{1/3}}{r_{\mathrm{s}}}. \tag{1.139}$$

Since the self-interaction correction vanishes for the diffuse orbitals of the uniform gas, all of this exchange energy is due to the Pauli exclusion principle.

1.5.3 Correlation Energy

Exact analytic expressions for $e_{\mathrm{c}}(n)$, the correlation energy per electron of the uniform gas, are known only in extreme limits. The high-density ($r_{\mathrm{s}} \to 0$) limit is also the weak-coupling limit, in which

$$e_{\mathrm{c}}(n) = c_0 \ln r_s - c_1 + c_2 r_s \ln r_s - c_3 r_s + \dots \quad (r_s \to 0) \tag{1.140}$$

from many-body perturbation theory [54]. The two positive constants $c_0 = 0.031091$ [54] and $c_1 = 0.046644$ [55] are known. Equation (1.140) does not quite tend to a constant when $r_{\mathrm{s}} \to 0$, as (1.114) would suggest, because the excited states of the non-interacting system lie arbitrarily close in energy to the ground state.

The low-density ($r_{\mathrm{s}} \to \infty$) limit is also the strong coupling limit in which the uniform fluid phase is unstable against the formation of a close-packed Wigner lattice of localized electrons. Because the energies of these two phases remain nearly degenerate as $r_{\mathrm{s}} \to \infty$, they have the same kind of dependence upon r_{s} [56]:

$$e_{\mathrm{c}}(n) \to -\frac{d_0}{r_{\mathrm{s}}} + \frac{d_1}{r_{\mathrm{s}}^{3/2}} + \dots \quad (r_{\mathrm{s}} \to \infty). \tag{1.141}$$

The constants d_0 and d_1 in (1.141) can be estimated from the Madelung electrostatic and zero-point vibrational energies of the Wigner crystal, respectively. The estimate

$$d_0 \approx -\frac{9}{10} \tag{1.142}$$

can be found from the electrostatic energy of a neutral spherical cell: Just add the electrostatic self-repulsion $3/5r_{\mathrm{s}}$ of a sphere of uniform positive background (with radius r_{s}) to the interaction $-3/2r_{\mathrm{s}}$ between this background and the electron at its center. The origin of the $r_{\mathrm{s}}^{-3/2}$ term in (1.141) is also simple: Think of the potential energy of the electron at small distance u from the center of the sphere as $-3/2r_{\mathrm{s}} + \frac{1}{2}ku^2$, where k is a spring constant. Since this potential energy must vanish for $u \approx r_{\mathrm{s}}$, we find that $k \sim r_{\mathrm{s}}^{-3}$ and thus the zero-point vibrational energy is $3\omega/2 = 1.5\sqrt{k/m} \sim r_{\mathrm{s}}^{-3/2}$.

An expression which encompasses both limits (1.140)and (1.141) is [8]

$$e_{\mathrm{c}}(n) = -2c_0(1 + \alpha_1 r_{\mathrm{s}}) \ln\left[1 + \frac{1}{2c_0(\beta_1 r_{\mathrm{s}}^{1/2} + \beta_2 r_{\mathrm{s}} + \beta_3 r_{\mathrm{s}}^{3/2} + \beta_4 r_{\mathrm{s}}^2)}\right], \tag{1.143}$$

where

$$\beta_1 = \frac{1}{2c_0} \exp\left(-\frac{c_1}{2c_0}\right), \tag{1.144}$$

$$\beta_2 = 2c_0\beta_1^2 . \tag{1.145}$$

The coefficients $\alpha_1 = 0.21370$, $\beta_3 = 1.6382$, and $\beta_4 = 0.49294$ are found by fitting to accurate Quantum Monte Carlo correlation energies [57] for $r_s = 2$, 5, 10, 20, 50, and 100.

The uniform electron gas is in equilibrium when the density n minimizes the total energy per electron, i.e., when

$$\frac{\partial}{\partial n}[t_s(n) + e_x(n) + e_c(n)] = 0 . \tag{1.146}$$

This condition is met at $r_s = 4.1$, close to the observed valence electron density of sodium. At any r_s, we have

$$\frac{\delta T_s}{\delta n(\mathbf{r})} = \frac{\partial}{\partial n}[n t_s(n)] = \frac{1}{2}k_F^2 , \tag{1.147}$$

$$\frac{\delta E_x}{\delta n(\mathbf{r})} = \frac{\partial}{\partial n}[n e_x(n)] = -\frac{1}{\pi}k_F . \tag{1.148}$$

Equation (1.143) with the parameters listed above provides a representation of $e_c(n_\uparrow, n_\downarrow)$ for $n_\uparrow = n_\downarrow = n/2$; other accurate representations are also available [9,10]. Equation (1.143) with different parameters ($c_0 = 0.015545$, $c_1 = 0.025599$, $\alpha_1 = 0.20548$, $\beta_3 = 3.3662$, $\beta_4 = 0.62517$) can represent $e_c(n_\uparrow, n_\downarrow)$ for $n_\uparrow = n$ and $n_\downarrow = 0$, the correlation energy per electron for a fully spin-polarized uniform gas. But we shall need $e_c(n_\uparrow, n_\downarrow)$ for arbitrary relative spin polarization

$$\zeta = \frac{(n_\uparrow - n_\downarrow)}{(n_\uparrow + n_\downarrow)} , \tag{1.149}$$

which ranges from 0 for an unpolarized system to ± 1 for a fully-spin-polarized system. A useful interpolation formula, based upon a study of the random phase approximation, is [10]

$$e_c(n_\uparrow, n_\downarrow) = e_c(n) + \alpha_c(n)\frac{f(\zeta)}{f''(0)}(1 - \zeta^4) + [e_c(n,0) - e_c(n)]f(\zeta)\zeta^4$$

$$= e_c(n) + \alpha_c(n)\zeta^2 + \mathcal{O}(\zeta^4) , \tag{1.150}$$

where

$$f(\zeta) = \frac{[(1+\zeta)^{4/3} + (1-\zeta)^{4/3} - 2]}{(2^{4/3} - 2)} . \tag{1.151}$$

In (1.150), $\alpha_c(n)$ is the correlation contribution to the spin stiffness. Roughly $\alpha_c(n) \approx e_c(n,0) - e_c(n)$, but more precisely $-\alpha_c(n)$ can be parametrized in the form of (1.143) (with $c_0 = 0.016887$, $c_1 = 0.035475$, $\alpha_1 = 0.11125$, $\beta_3 = 0.88026$, $\beta_4 = 0.49671$).

For completeness, we note that the spin-scaling relations (1.126) and (1.127) imply that

$$e_x(n_\uparrow, n_\downarrow) = e_x(n)\frac{\left[(1+\zeta)^{4/3} + (1-\zeta)^{4/3}\right]}{2} , \tag{1.152}$$

$$t_s(n_\uparrow, n_\downarrow) = t_s(n)\frac{\left[(1+\zeta)^{5/3} + (1-\zeta)^{5/3}\right]}{2} . \tag{1.153}$$

The exchange-hole density of (1.137) can also be spin scaled. Expressions for the exchange and correlation holes for arbitrary r_s and ζ are given in [58].

1.5.4 Linear Response

We now discuss the linear response of the spin-unpolarized uniform electron gas to a weak, static, external potential $\delta v(\mathbf{r})$. This is a well-studied problem [59], and a practical one for the local-pseudopotential description of a simple metal [60].

Because the unperturbed system is homogeneous, we find that, to first order in $\delta v(\mathbf{r})$, the electron density response is

$$\delta n(\mathbf{r}) = \int d^3 r'\, \chi(|\mathbf{r} - \mathbf{r}'|)\delta v(\mathbf{r}') \tag{1.154}$$

where χ is a linear response function. If

$$\delta v(\mathbf{r}) = \delta v(\mathbf{q})\exp(i\mathbf{q} \cdot \mathbf{r}) \tag{1.155}$$

is a wave of wavevector $\mathbf{q}$ and small amplitude $\delta v(\mathbf{q})$, then (1.154) becomes $\delta n(\mathbf{r}) = \delta n(\mathbf{q})\exp(i\mathbf{q} \cdot \mathbf{r})$, where

$$\delta n(\mathbf{q}) = \chi(\mathbf{q})\delta v(\mathbf{q}) , \tag{1.156}$$

and

$$\chi(\mathbf{q}) = \int d^3 x\, \exp(-i\mathbf{q} \cdot \mathbf{x})\chi(|\mathbf{x}|) \tag{1.157}$$

is the Fourier transform of $\chi(|\mathbf{r} - \mathbf{r}'|)$ with respect to $\mathbf{x} = \mathbf{r} - \mathbf{r}'$. (In (1.155), the real part of the complex exponential $\exp(i\alpha) = \cos(\alpha) + i\sin(\alpha)$ is understood.)

By the Kohn-Sham theorem, we also have

$$\delta n(\mathbf{q}) = \chi_s(q)\delta v_s(\mathbf{q}) , \tag{1.158}$$

where $\delta v_s(\mathbf{q})$ is the change in the Kohn-Sham effective one-electron potential of (1.62), and

$$\chi_s(q) = -\frac{k_F}{\pi^2}F(q/2k_F) \tag{1.159}$$

is the density response function for the non-interacting uniform electron gas. The Lindhard function

$$F(x) = \frac{1}{2} + \frac{1-x^2}{4x} \ln \left| \frac{1+x}{1-x} \right| \tag{1.160}$$

equals $1 - x^2/3 - x^4/15$ as $x \to 0$, $1/2$ at $x = 1$, and $1/(3x^2) + 1/(15x^4)$ as $x \to \infty$. dF/dx diverges logarithmically as $x \to 1$.

Besides $\delta v(\mathbf{r})$, the other contributions to $\delta v_s(\mathbf{r})$ of (1.62) are

$$\delta \left(\frac{\delta U}{\delta n(\mathbf{r})} \right) = \int d^3 r' \frac{\delta n(\mathbf{r}')}{|\mathbf{r} - \mathbf{r}'|} , \tag{1.161}$$

$$\delta \left(\frac{\delta E_{xc}}{\delta n(\mathbf{r})} \right) = \int d^3 r' \frac{\delta^2 E_{xc}}{\delta n(\mathbf{r}) \delta n(\mathbf{r}')} \delta n(\mathbf{r}') . \tag{1.162}$$

In other words,

$$\delta v_s(\mathbf{q}) = \delta v(\mathbf{q}) + \frac{4\pi}{q^2} \delta n(\mathbf{q}) - \frac{\pi}{k_F^2} \gamma_{xc}(q) \delta n(\mathbf{q}) , \tag{1.163}$$

where the coefficient of the first $\delta n(\mathbf{q})$ is the Fourier transform of the Coulomb interaction $1/|\mathbf{r} - \mathbf{r}'|$, and the coefficient of the second $\delta n(\mathbf{q})$ is the Fourier transform of $\delta^2 E_{xc}/\delta n(\mathbf{r})\delta n(\mathbf{r}')$.

We re-write (1.163) as

$$\delta v_s(\mathbf{q}) = \delta v(\mathbf{q}) + \frac{4\pi}{q^2} [1 - G_{xc}(q)] \delta n(\mathbf{q}) , \tag{1.164}$$

where

$$G_{xc}(q) = \gamma_{xc}(q) \left(\frac{q}{2k_F} \right)^2 \tag{1.165}$$

is the so-called local-field factor. Then we insert (1.158) into (1.164) and find

$$\delta v_s(\mathbf{q}) = \frac{\delta v(\mathbf{q})}{\epsilon_s(q)} \tag{1.166}$$

where

$$\epsilon_s(q) = 1 - \frac{4\pi}{q^2} [1 - G_{xc}(q)] \chi_s(q) . \tag{1.167}$$

In other words, the density response function of the interacting uniform electron gas is

$$\chi(q) = \frac{\chi_s(q)}{\epsilon_s(q)} . \tag{1.168}$$

These results are particularly simple in the long-wavelength ($q \to 0$) limit, in which $\gamma_{xc}(q)$ tends to a constant and

$$\epsilon_s(q) \to 1 - \frac{\gamma_{xc}(q=0)}{\pi k_F} + \frac{k_s^2}{q^2} \quad (q \to 0) , \tag{1.169}$$

where

$$k_s = \left(\frac{4k_F}{\pi}\right)^{1/2} = \left(\frac{4}{\pi}\right)^{1/2}\left(\frac{9\pi}{4}\right)^{1/6}\frac{1}{r_s^{1/2}} \tag{1.170}$$

is the inverse of the Thomas-Fermi screening length – the characteristic distance over which an external perturbation is screened out. Equations (1.166) and (1.167) show that a slowly-varying external perturbation $\delta v(\mathbf{q})$ is strongly "screened out" by the uniform electron gas, leaving only a very weak Kohn-Sham potential $\delta v_s(\mathbf{q})$. Equation (1.168) shows that the response function $\chi(q)$ is weaker than $\chi_s(q)$ by a factor $(q/k_s)^2$ in the limit $q \to 0$.

In (1.166), $\epsilon_s(q)$ is a kind of dielectric function, but it is not the standard dielectric function $\epsilon(q)$ which predicts the response of the electrostatic potential alone:

$$\delta v(\mathbf{q}) + \frac{4\pi}{q^2}\delta n(\mathbf{q}) = \frac{\delta v(\mathbf{q})}{\epsilon(q)} . \tag{1.171}$$

By inserting (1.156) into (1.171), we find

$$\frac{1}{\epsilon(q)} = 1 + \frac{4\pi}{q^2}\chi(q) . \tag{1.172}$$

It is only when we neglect exchange and correlation that we find the simple Lindhard result

$$\epsilon(q) \to \epsilon_s(q) \to \epsilon_L(q) = 1 - \frac{4\pi}{q^2}\chi_s(q) \quad (\gamma_{xc} \to 0) . \tag{1.173}$$

Neglecting correlation, γ_x is a numerically-tabulated function of $(q/2k_F)$ with the small-q expansion [61]

$$\gamma_x(q) = 1 + \frac{5}{9}\left(\frac{q}{2k_F}\right)^2 + \frac{73}{225}\left(\frac{q}{2k_F}\right)^4 \quad (q \to 0) . \tag{1.174}$$

When correlation is included, $\gamma_{xc}(q)$ depends upon r_s as well as $(q/2k_F)$, in a way that is known from Quantum Monte Carlo studies [62] of the weakly-perturbed uniform gas.

The second-order change δE in the total energy may be found from the Hellmann-Feynman theorem of Sect. 1.2.4. Replace $\delta v(\mathbf{r})$ by $v_\lambda(\mathbf{r}) = \lambda\delta v(\mathbf{r})$ and $\delta n(\mathbf{r})$ by $\lambda\delta n(\mathbf{r})$, to find

$$\begin{aligned}
\delta E &= \int_0^1 d\lambda \int d^3r\, n_\lambda(\mathbf{r})\frac{d}{d\lambda}v_\lambda(\mathbf{r}) \\
&= \int_0^1 d\lambda \int d^3r\, [n + \lambda\delta n(\mathbf{r})]\delta v(\mathbf{r}) \\
&= \frac{1}{2}\int d^3r\, \delta n(\mathbf{r})\delta v(\mathbf{r}) \\
&= \frac{1}{2}\delta n(-\mathbf{q})\delta v(\mathbf{q}) .
\end{aligned} \tag{1.175}$$

1.5.5 Clumping and Adiabatic Connection

The uniform electron gas for $r_s \leq 30$ provides a nice example of the adiabatic connection discussed in Sect. 1.3.5. As the coupling constant λ turns on from 0 to 1, the ground state wavefunction evolves continuously from the Kohn-Sham determinant of plane waves to the ground state of interacting electrons in the presence of the external potential, while the density remains fixed. (One should of course regard the infinite system as the infinite-volume limit of a finite chunk of uniform background neutralized by electrons.)

The adiabatic connection between non-interacting and interacting uniform-density ground states could be destroyed by any tendency of the density to clump. A fictitious attractive interaction between electrons would yield such a tendency. Even in the absence of attractive interactions, clumping appears in the very-low-density electron gas as a charge density wave or Wigner crystallization [56,59]. Then there is probably no external potential which will hold the interacting system in a uniform-density ground state, but one can still find the energy of the uniform state by imposing density uniformity as a constraint on a trial interacting wavefunction.

The uniform phase becomes unstable against a charge density wave of wavevector $\mathbf{q}$ and infinitesimal amplitude when $\epsilon_s(q)$ of (1.167) vanishes [59]. This instability for $q \approx 2k_F$ arises at low density as a consequence of exchange and correlation.

1.6 Local, Semi-local and Non-local Approximations

1.6.1 Local Spin Density Approximation

The local spin density approximation (LSD) for the exchange-correlation energy, (1.11), was proposed in the original work of Kohn and Sham [6], and has proved to be remarkably accurate, useful, and hard to improve upon. The generalized gradient approximation (GGA) of (1.12), a kind of simple extension of LSD, is now more widely used in quantum chemistry, but LSD remains the most popular way to do electronic-structure calculations in solid state physics. Tables 1.1 and 1.2 provide a summary of typical errors for LSD and GGA, while Tables 1.3 and 1.4 make this comparison for a few specific atoms and molecules. The LSD is parametrized as in Sect. 1.5, while the GGA is the non-empirical one of Perdew, Burke, and Ernzerhof [20], to be presented later.

The LSD approximation to any energy component G is

$$G^{\mathrm{LSD}}[n_\uparrow, n_\downarrow] = \int \mathrm{d}^3 r \, n(\mathbf{r}) g(n_\uparrow(\mathbf{r}), n_\downarrow(\mathbf{r})) \,, \tag{1.176}$$

where $g(n_\uparrow, n_\downarrow)$ is that energy component per particle in an electron gas with uniform spin densities $n_\uparrow$ and $n_\downarrow$, and $n(\mathbf{r})\mathrm{d}^3 r$ is the average number of

Table 1.3. Exchange-correlation energies of atoms, in hartree

Atom	LSD	GGA	Exact
H	-0.29	-0.31	-0.31
He	-1.00	-1.06	-1.09
Li	-1.69	-1.81	-1.83
Be	-2.54	-2.72	-2.76
N	-6.32	-6.73	-6.78
Ne	-11.78	-12.42	-12.50

Table 1.4. Atomization energies of molecules, in eV. (1 hartree $= 27.21\,\text{eV}$). From [20]

Molecule	LSD	GGA	Exact
H_2	4.9	4.6	4.7
CH_4	20.0	18.2	18.2
NH_3	14.6	13.1	12.9
H_2O	11.6	10.1	10.1
CO	13.0	11.7	11.2
O_2	7.6	6.2	5.2

electrons in volume element d^3r. Sections 1.5.1–1.5.3 provide the ingredients for $T_s^{\text{LSD}} = T_0$, E_x^{LSD}, and E_c^{LSD}. The functional derivative of (1.176) is

$$\frac{\delta G^{\text{LSD}}}{\delta n_\sigma(\mathbf{r})} = \frac{\partial}{\partial n_\sigma} \left[(n_\uparrow + n_\downarrow) g(n_\uparrow, n_\downarrow) \right] . \tag{1.177}$$

By construction, LSD is exact for a uniform density, or more generally for a density that varies slowly over space [6]. More precisely, LSD should be valid when the length scale of the density variation is large in comparison with length scales set by the local density, such as the Fermi wavelength $2\pi/k_F$ or the screening length $1/k_s$. This condition is rarely satisfied in real electronic systems, so we must look elsewhere to understand why LSD works.

We need to understand why LSD works, for three reasons: to justify LSD calculations, to understand the physics, and to develop improved density functional approximations. Thus we will start with the good news about LSD, proceed to the mixed good/bad news, and close with the bad news.

LSD has *many* correct formal features. It is exact for uniform densities and nearly-exact for slowly-varying ones, a feature that makes LSD well suited at least to the description of the crystalline simple metals. It satisfies the inequalities $E_x < 0$ (see (1.93)) and $E_c < 0$ (see (1.69)), the correct uniform coordinate scaling of E_x (see (1.106)), the correct spin scaling of E_x (see (1.127)), the correct coordinate scaling for E_c (see (1.111), (1.116), (1.117)), the correct low-density behavior of E_c (see (1.115)), and the correct Lieb-Oxford bound on E_{xc} (see (1.120) and (1.122)). LSD is properly size-consistent (Sect. 1.4.4).

LSD provides a surprisingly good account of the linear response of the spin-unpolarized uniform electron gas (Sect. 1.5.4). Since

$$\frac{\delta^2 E_{xc}^{LSD}}{\delta n(\mathbf{r})\delta n(\mathbf{r}')} = \delta(\mathbf{r} - \mathbf{r}')\frac{\partial^2 [n e_{xc}(n)]}{\partial n^2} , \qquad (1.178)$$

where $\delta(\mathbf{r} - \mathbf{r}')$ is the Dirac delta function, we find

$$\gamma_{xc}^{LSD}(q) = 1 - \frac{k_F^2}{\pi}\frac{\partial^2}{\partial n^2}[n e_c(n)] , \qquad (1.179)$$

a constant independent of q, which must be the exact $q \to 0$ or slowly-varying limit of $\gamma_{xc}(q)$. Figure 1 of [20] shows that the "exact" $\gamma_{xc}(q)$ from a Quantum Monte Carlo calculation [62] for $r_s = 4$ is remarkably close to the LSD prediction for $q \leq 2k_F$. The same is true over the whole valence-electron density range $2 \leq r_s \leq 5$, and results from a strong cancellation between the nonlocalities of exchange and correlation. Indeed the exact result for exchange (neglecting correlation), equation (1.174), is strongly q-dependent or nonlocal. The displayed terms of (1.174) suffice for $q \leq 2k_F$.

Powerful reasons for the success of LSD are provided by the coupling constant integration of Sect. 1.3.5. Comparison of (1.86) and (1.11) reveals that the LSD approximations for the exchange and correlation holes of an inhomogeneous system are

$$n_x^{LSD}(\mathbf{r}, \mathbf{r}') = n_x^{unif}(n_\uparrow(\mathbf{r}), n_\downarrow(\mathbf{r}); |\mathbf{r} - \mathbf{r}'|) , \qquad (1.180)$$

$$n_c^{LSD}(\mathbf{r}, \mathbf{r}') = n_c^{unif}(n_\uparrow(\mathbf{r}), n_\downarrow(\mathbf{r}); |\mathbf{r} - \mathbf{r}'|) , \qquad (1.181)$$

where $n_{xc}^{unif}(n_\uparrow, n_\downarrow; u)$ is the hole in an electron gas with uniform spin densities $n_\uparrow$ and $n_\downarrow$. Since the uniform gas is a possible physical system, (1.180) and (1.181) obey the exact constraints of (1.91) (negativity of n_x), (1.94) (sum rule on n_x), (1.95), (1.97) (sum rule on $\bar{n}_c$), (1.98), and (1.85) (cusp condition).

By (1.95), the LSD on-top exchange hole $n_x^{LSD}(\mathbf{r}, \mathbf{r})$ is exact, at least when the Kohn-Sham wavefunction is a single Slater determinant. The LSD on-top correlation hole $\bar{n}_c^{LSD}(\mathbf{r}, \mathbf{r})$ is *not* exact [63] (except in the high-density, low-density, fully spin-polarized, or slowly-varying limit), but it is often quite realistic [64]. By (1.85), its cusp is then also realistic.

Because it satisfies all these constraints, the LSD model for the system-, spherically-, and coupling-constant-averaged hole of (1.101),

$$\langle \bar{n}_{xc}^{LSD}(u)\rangle = \frac{1}{N}\int d^3r\, n(\mathbf{r})\bar{n}_{xc}^{unif}(n_\uparrow(\mathbf{r}), n_\downarrow(\mathbf{r}); u) , \qquad (1.182)$$

can be very physical. Moreover, the system average in (1.182) "unweights" regions of space where LSD is expected to be least reliable, such as near a nucleus or in the evanescent tail of the electron density [65,64].

Since correlation makes $\langle \bar{n}_{xc}(u = 0) \rangle$ deeper, and thus by (1.102) makes $\langle \bar{n}_{xc}(u) \rangle$ more short-ranged, E_{xc} can be "more local" than either E_x or E_c. In other words, LSD often benefits from a cancellation of errors between exchange and correlation.

Mixed good and bad news about LSD is the fact that selfconsistent LSD calculations can break exact spin symmetries. As an example, consider "stretched H_2", the hydrogen molecule ($N = 2$) with a very large separation between the two nuclei. The exact ground state is a spin singlet ($S = 0$), with $n_\uparrow(\mathbf{r}) = n_\downarrow(\mathbf{r}) = n(\mathbf{r})/2$. But the LSD ground state localizes all of the spin-up density on one of the nuclei, and all of the spin-down density on the other. Although (or rather *because*) the LSD spin densities are wrong, the LSD total energy is correctly the sum of the energies of two isolated hydrogen atoms, so this symmetry breaking is by no means entirely a bad thing [66,67]. The selfconsistent LSD on-top hole density $\langle \bar{n}_{xc}(0) \rangle = -\langle n \rangle$ is also right: Heitler-London correlation ensues that two electrons are never found near one another, or on the same nucleus at the same time.

Finally, we present the bad news about LSD: (1) LSD does not incorporate known inhomogeneity or gradient corrections to the exchange-correlation hole near the electron (Sect. 1.6.2) (2) It does not satisfy the high-density correlation scaling requirement of (1.114), but shows a $\ln \gamma$ divergence associated with the $\ln r_s$ term of (1.140). (3) LSD is not exact in the one-electron limit, i.e., does not satisfy (1.67), and (1.70)–(1.73). Although the "self-interaction error" is small for the exchange-correlation energy, it is more substantial for the exchange-correlation potential and orbital eigenvalues. (4) As a "continuum approximation", based as it is on the uniform electron gas and its continuous one-electron energy spectrum, LSD misses the derivative discontinuity of Sect. 1.4.5. Effectively, LSD averages over the discontinuity, so its highest occupied orbital energy for a Z-electron system is not (1.129) but $\varepsilon^{HO} \approx -(I_Z + A_Z)/2$. A second consequence is that LSD predicts an incorrect dissociation of a hetero-nuclear molecule or solid to fractionally charged fragments. (In LSD calculations of atomization energies, the dissociation products are constrained to be neutral atoms, and not these unphysical fragments.) (5) LSD does not guarantee satisfaction of (1.99), an inherently nonlocal constraint.

The GGA to be derived in Sect. 1.6.4 will preserve all the good or mixed features of LSD listed above, while eliminating bad features (1) and (2) but not (3)–(5). Elimination of (3)–(5) will probably require the construction of $E_{xc}[n_\uparrow, n_\downarrow]$ from the Kohn-Sham orbitals (which are themselves highly-nonlocal functionals of the density). For example, the self-interaction correction [9,68] to LSD eliminates most of the bad features (3) and (4), but not in an entirely satisfactory way.

1.6.2   Gradient Expansion

Gradient expansions [6,69], which offer systematic corrections to LSD for electron densities that vary slowly over space, might appear to be the natural next step beyond LSD. As we shall see, they are not; understanding why not will light the path to the generalized gradient approximations of Sect. 1.6.3.

As a first measure of inhomogeneity, we define the reduced density gradient

$$s = \frac{|\nabla n|}{2k_F n} = \frac{|\nabla n|}{2(3\pi^2)^{1/3} n^{4/3}} = \frac{3}{2}\left(\frac{4}{9\pi}\right)^{1/3}|\nabla r_s|, \qquad (1.183)$$

which measures how fast and how much the density varies on the scale of the local Fermi wavelength $2\pi/k_F$. For the energy of an atom, molecule, or solid, the range $0 \le s \le 1$ is very important. The range $1 \le s \le 3$ is somewhat important, more so in atoms than in solids, while $s > 3$ (as in the exponential tail of the density) is unimportant [70,71].

Other measures of density inhomogeneity, such as $p = \nabla^2 n/(2k_F)^2 n$, are also possible. Note that s and p are small not only for a slow density variation but also for a density variation of small amplitude (as in Sect. 1.5.4). The slowly-varying limit is one in which p/s is also small [6].

Under the uniform density scaling of (1.40), $s(\mathbf{r}) \to s_\gamma(\mathbf{r}) = s(\gamma\mathbf{r})$. The functionals $T_s[n]$ and $E_x[n]$ must scale as in (1.104) and (1.106), so their gradient expansions are

$$T_s[n] = A_s \int d^3r\, n^{5/3}[1 + \alpha s^2 + \ldots], \qquad (1.184)$$

$$E_x[n] = A_x \int d^3r\, n^{4/3}[1 + \mu s^2 + \ldots], \qquad (1.185)$$

Because there is no special direction in the uniform electron gas, there can be no term linear in ∇n. Moreover, terms linear in $\nabla^2 n$ can be recast as s^2 terms, since

$$\int d^3r\, f(n)\nabla^2 n = -\int d^3r\, \left(\frac{\partial f}{\partial n}\right)|\nabla n|^2 \qquad (1.186)$$

via integration by parts. Neglecting the dotted terms in (1.184) and (1.185), which are fourth or higher-order in ∇, amounts to the second-order gradient expansion, which we call the gradient expansion approximation (GEA).

Correlation introduces a second length scale, the screening length $1/k_s$, and thus another reduced density gradient

$$t = \frac{|\nabla n|}{2k_s n} = \left(\frac{\pi}{4}\right)^{1/2}\left(\frac{9\pi}{4}\right)^{1/6}\frac{s}{r_s^{1/2}}. \qquad (1.187)$$

In the high-density ($r_s \to 0$) limit, the screening length ($1/k_s \sim r_s^{1/2}$) is the only important length scale for the correlation hole. The gradient expansion of the correlation energy is

$$E_c[n] = \int d^3r\, n \left[e_c(n) + \beta(n)t^2 + \ldots \right] . \tag{1.188}$$

While $e_c(n)$ does not quite approach a constant as $n \to \infty$, $\beta(n)$ does [69].

While the form of the gradient expansion is easy to guess, the coefficients can only be calculated by hard work. Start with the uniform electron gas, in either its non-interacting (T_s, E_x) or interacting (E_c) ground state, and apply a weak external perturbation $\delta v_s(\mathbf{q}) \exp(i\mathbf{q}\cdot\mathbf{r})$ or $\delta v(\mathbf{q}) \exp(i\mathbf{q}\cdot\mathbf{r})$, respectively. Find the linear response $\delta n(\mathbf{q})$ of the density, and the second-order response δG of the energy component G of interest. Use the linear response of the density (as in (1.157) or (1.156)) to express δG entirely in terms of $\delta n(\mathbf{q})$. Finally, expand δG in powers of q^2, observing that $|\nabla n|^2 \sim q^2 |\delta n(\mathbf{q})|^2$, and extract the gradient coefficient.

In this way, Kirznits [72] found the gradient coefficient for T_s,

$$\alpha = \frac{5}{27} \tag{1.189}$$

(which respects the conjectured bound of (1.118)), Sham [73] found the coefficient of E_x,

$$\mu_{\text{Sham}} = \frac{7}{81} , \tag{1.190}$$

and Ma and Brueckner [69] found the high-density limit of $\beta(n)$:

$$\beta_{\text{MB}} = 0.066725 . \tag{1.191}$$

The weak density dependence of $\beta(n)$ is also known [74], as is its spin-dependence [75]. Neglecting small $\nabla\zeta$ contributions, the gradient coefficients (coefficients of $|\nabla n|^2/n^{4/3}$) for both exchange and correlation at arbitrary relative spin polarization ζ are found from those for $\zeta = 0$ through multiplication by [76]

$$\phi(\zeta) = \frac{1}{2}\left[(1+\zeta)^{2/3} + (1-\zeta)^{2/3} \right] . \tag{1.192}$$

For exchange, this is easily verified by applying the spin-scaling relation of (1.127) to (1.185) and (1.183).

There is another interesting similarity between the gradient coefficients for exchange and correlation. Generalize the definition of t (see (1.187)) to

$$t = \frac{|\nabla n|}{2\phi k_s n} = \left(\frac{\pi}{4}\right)^{1/2} \left(\frac{9\pi}{4}\right)^{1/6} \frac{s}{\phi r_s^{1/2}} . \tag{1.193}$$

Then

$$\beta_{\text{MB}}\phi^3 n t^2 = \mu C_x \phi n^{4/3} s^2 , \tag{1.194}$$

where

$$\mu = \beta_{\text{MB}} \frac{\pi^2}{3} = 0.21951 . \tag{1.195}$$

Sham's derivation [73] of (1.190) starts with a screened Coulomb interaction $(1/u)\exp(-\kappa u)$, and takes the limit $\kappa \to 0$ at the end of the calculation. Antoniewicz and Kleinman [77] showed that the correct gradient coefficient for the unscreened Coulomb interaction is not μ_{Sham} but

$$\mu_{\text{AK}} = \frac{10}{81} . \tag{1.196}$$

It is believed [78] that a similar order-of-limits problem exists for β, in such a way that the combination of Sham's exchange coefficient with the Ma-Brueckner [69] correlation coefficient yields the correct gradient expansion of E_{xc} in the slowly-varying high-density limit.

Numerical tests of these gradient expansions for atoms show that the second-order gradient term provides a useful correction to the Thomas-Fermi or local density approximation for T_{s}, and a modestly useful correction to the local density approximation for E_{x}, but seriously worsens the local spin density results for E_{c} and E_{xc}. In fact, the GEA correlation energies are positive! The latter fact was pointed out in the original work of Ma and Brueckner [69], who suggested the first generalized gradient approximation as a remedy.

The local spin density approximation to E_{xc}, which is the leading term of the gradient expansion, provides rather realistic results for atoms, molecules, and solids. But the second-order term, which is the next systematic correction for slowly-varying densities, makes E_{xc} worse.

There are two answers to the seeming paradox of the previous paragraph. The first is that realistic electron densities are *not* very close to the slowly-varying limit ($s \ll 1$, $p/s \ll 1$, $t \ll 1$, etc.). The second is this: The LSD approximation to the exchange-correlation hole is the hole of a possible physical system, the uniform electron gas, and so satisfies many exact constraints, as discussed in Sect. 1.6.1. The second-order gradient expansion or GEA approximation to the hole is not, and does not.

The second-order gradient expansion or GEA models are known for both the exchange hole [12,13] $n_{\text{x}}(\mathbf{r}, \mathbf{r}+\mathbf{u})$ and the correlation hole $\bar{n}_{\text{c}}(\mathbf{r}, \mathbf{r}+\mathbf{u})$ [79]. They appear to be more realistic than the corresponding LSD models at small u, but far less realistic at large u, where several spurious features appear: $n_{\text{x}}(\mathbf{r}, \mathbf{r} + \mathbf{u})_{\text{GEA}}$ has an undamped $\cos(2k_{\text{F}}u)$ oscillation which violates the negativity constraint of (1.91), and integrates to -1 (see (1.94)) only with the help of a convergence factor $\exp(-\kappa u)$ ($\kappa \to 0$). $\bar{n}_{\text{c}}(\mathbf{r}, \mathbf{r} + \mathbf{u})_{\text{GEA}}$ has a positive u^{-4} tail, and integrates not to zero (see (1.97)) but to a positive number $\sim s^2$. These spurious large-u behaviors are sampled by the long range of the Coulomb interaction $1/u$, leading to unsatisfactory energies for real systems.

The gradient expansion for the exchange hole density is known [80] to third order in ∇, and suggests the following interpretation of the gradient expansion: When the density does not vary too rapidly over space (e.g., in the weak-pseudopotential description of a simple metal), the addition of each successive order of the gradient expansion improves the description of the hole at small u while worsening it at large u. The bad large-u behavior thwarts our expectation that the hole will remain normalized to each order in ∇.

The non-interacting kinetic energy T_s does not sample the spurious large-u part of the gradient expansion, so its gradient expansion (see (1.184) and (1.189)) works reasonably well even for realistic electron densities. In fact, we can use (1.79) to show that

$$T_s[n] = -\frac{1}{2}\sum_\sigma \int d^3r\, \frac{\partial}{\partial \mathbf{r}} \cdot \frac{\partial}{\partial \mathbf{r}} \rho_1^{\lambda=0}(\mathbf{r}'\sigma, \mathbf{r}\sigma)\bigg|_{\mathbf{r}'=\mathbf{r}} \tag{1.197}$$

samples only the small-u part of the gradient expansion of the Kohn-Sham one-electron reduced density matrix, while $E_x[n]$ of (1.90) and (1.92) also samples large values of u. The GEA for $T_s[n]$ is, in a sense, its own GGA [81]. Moreover, the sixth-order gradient expansion of T_s is also known: it diverges for finite systems, but provides accurate monovacancy formation energies for jellium [82].

The GEA form of (1.184), (1.185), and (1.188) is a special case of the GGA form of (1.12). To find the functional derivative, note that

$$\delta F = \int d^3r\, \delta f(n_\uparrow, n_\downarrow, \nabla n_\uparrow, \nabla n_\downarrow)$$

$$= \sum_\sigma \int d^3r\, \left[\frac{\partial f}{\partial n_\sigma(\mathbf{r})} \delta n_\sigma(\mathbf{r}) + \frac{\partial f}{\partial \nabla n_\sigma(\mathbf{r})} \cdot \nabla \delta n_\sigma(\mathbf{r}) \right]$$

$$= \sum_\sigma \int d^3r\, \frac{\delta F}{\delta n_\sigma(\mathbf{r})} \delta n_\sigma(\mathbf{r}) . \tag{1.198}$$

Integration by parts gives

$$\frac{\delta F}{\delta n_\sigma(\mathbf{r})} = \frac{\partial f}{\partial n_\sigma(\mathbf{r})} - \nabla \cdot \frac{\partial f}{\partial \nabla n_\sigma(\mathbf{r})} . \tag{1.199}$$

For example, the functional derivative of the gradient term in the spin-unpolarized high-density limit is

$$\frac{\delta}{\delta n(\mathbf{r})} \int d^3r\, C_{xc} \frac{|\nabla n(\mathbf{r})|^2}{n^{4/3}} = C_{xc} \left[\frac{4}{3} \frac{|\nabla n(\mathbf{r})|^2}{n^{7/3}} - 2\frac{\nabla^2 n}{n^{4/3}} \right] , \tag{1.200}$$

which involves second as well as first derivatives of the density.

The GEA for the linear response function $\gamma_{xc}(q)$ of (1.163) is found by inserting $n(\mathbf{r}) = n + \delta n(\mathbf{q})\exp(i\mathbf{q}\cdot\mathbf{r})$ into (1.199) and linearizing in $\delta n(\mathbf{q})$:

$$\gamma_{xc}^{GEA}(q) = \gamma_{xc}^{LSD} - 24\pi(3\pi^2)^{1/3} C_{xc} \left(\frac{q}{2k_F} \right)^2 . \tag{1.201}$$

For example, the Antoniewicz-Kleinman gradient coefficient [77] for exchange of (1.196), inserted into (1.200) and (1.201), yields the q^2 term of (1.174).

1.6.3   History of Several Generalized Gradient Approximations

In 1968, Ma and Brueckner [69] derived the second-order gradient expansion for the correlation energy in the high-density limit, (1.188) and (1.191). In numerical tests, they found that it led to improperly positive correlation energies for atoms, because of the large size of the positive gradient term. As a remedy, they proposed the first GGA,

$$E_c^{MB}[n] = \int d^3r \, n e_c(n) \left[1 - \frac{\beta_{MB} t^2}{\nu n e_c(n)} \right]^{-\nu} , \qquad (1.202)$$

where $\nu \approx 0.32$ was fitted to known correlation energies. Equation (1.202) reduces to (1.188) and (1.191) in the slowly-varying ($t \to 0$) limit, but provides a strictly negative "energy density" which tends to zero as $t \to \infty$. In this respect, it is strikingly like the nonempirical GGA's that were developed in 1991 or later, differing from them mainly in the presence of an empirical parameter, the absence of a spin-density generalization, and a less satisfactory high-density limit.

Under the uniform scaling of (1.40), $n(\mathbf{r}) \to n_\gamma(\mathbf{r})$, we find $r_s(\mathbf{r}) \to \gamma^{-1} r_s(\gamma \mathbf{r})$, $\zeta(\mathbf{r}) \to \zeta(\gamma \mathbf{r})$, $s(\mathbf{r}) \to s(\gamma \mathbf{r})$, and $t(\mathbf{r}) \to \gamma^{1/2} t(\gamma \mathbf{r})$. Thus $E_c^{MB}[n_\gamma]$ tends to $E_c^{LSD}[n_\gamma]$ as $\gamma \to \infty$, and not to a negative constant as required by (1.114).

In 1980, Langreth and Perdew [83] explained the failure of the second-order gradient expansion (GEA) for E_c. They made a complete wavevector analysis of E_{xc}, i.e., they replaced the Coulomb interaction $1/u$ in (1.100) by its Fourier transform and found

$$E_{xc}[n] = \frac{N}{2} \int_0^\infty dk \, \frac{4\pi k^2}{(2\pi)^3} \langle \bar{n}_{xc}(k) \rangle \frac{4\pi}{k^2} , \qquad (1.203)$$

where

$$\langle \bar{n}_{xc}(k) \rangle = \int_0^\infty du \, 4\pi u^2 \langle \bar{n}_{xc}(u) \rangle \frac{\sin(ku)}{ku} \qquad (1.204)$$

is the Fourier transform of the system- and spherically-averaged exchange-correlation hole. In (1.203), E_{xc} is decomposed into contributions from dynamic density fluctuations of various wavevectors k.

The sum rule of (1.102) should emerge from (1.204) in the $k \to 0$ limit (since $\sin(x)/x \to 1$ as $x \to 0$), and does so for the exchange energy at the GEA level. But the $k \to 0$ limit of $\bar{n}_c^{GEA}(k)$ turns out to be a positive number proportional to t^2, and not zero. The reason seems to be that the GEA correlation hole is only a truncated expansion, and not the exact hole for any physical system, so it can and does violate the sum rule.

Langreth and Mehl [11] (1983) proposed a GGA based upon the wavevector analysis of (1.203). They introduced a sharp cutoff of the spurious small-k contributions to E_c^{GEA}: All contributions were set to zero for $k < k_c =$

$f|\nabla n/n|$, where $f \approx 0.15$ is only semi-empirical since $f \approx 1/6$ was estimated theoretically. Extension of the Langreth-Mehl E_c^{GGA} beyond the random phase approximation was made by Perdew [14] in 1986.

The errors of the GEA for the exchange energy are best revealed in real space (see (1.100)), not in wavevector space (see (1.203)). In 1985, Perdew [12] showed that the GEA for the exchange hole density $n_x(\mathbf{r}, \mathbf{r} + \mathbf{u})$ contains a spurious undamped $\cos(2k_F u)$ oscillation as $u \to \infty$, which violates the negativity constraint of (1.91) and respects the sum rule of (1.94) only with the help of a convergence factor (e.g., $\exp(-\kappa u)$ as $\kappa \to 0$). This suggested that the required cutoffs should be done in real space, not in wavevector space. The GEA hole density $n_x^{\mathrm{GEA}}(\mathbf{r}, \mathbf{r} + \mathbf{u})$ was replaced by zero for all $\mathbf{u}$ where n_x^{GEA} was positive, and for all $u > u_x(\mathbf{r})$ where the cutoff radius $u_x(\mathbf{r})$ was chosen to recover (1.94). Equation (1.92) then provided a numerically-defined GGA for E_x, which turned out to be more accurate than either LSD or GEA. In 1986, Perdew and Wang [13] simplified this GGA in two ways: (1) They replaced $n_x^{\mathrm{GEA}}(\mathbf{r}, \mathbf{r}+\mathbf{u})$, which depends upon both first and second derivatives of $n(\mathbf{r})$, by $\tilde{n}_x^{\mathrm{GEA}}(\mathbf{r}, \mathbf{r}+\mathbf{u})$, an equivalent expression found through integration by parts, which depends only upon $\nabla n(\mathbf{r})$. (2) The resulting numerical GGA has the form

$$E_x^{\mathrm{GGA}}[n] = A_x \int \mathrm{d}^3 r \, n^{4/3} F_x(s) , \qquad (1.205)$$

which scales properly as in (1.106). The function $F_x(s)$ was plotted and fitted by an analytic form. The spin-scaling relation (1.127) was used to generate a spin-density generalization. Perdew and Wang [13] also coined the term "generalized gradient approximation".

A parallel but more empirical line of GGA development arose in quantum chemistry around 1986. Becke [15,16] showed that a GGA for E_x could be constructed with the help of one or two parameters fitted to exchange energies of atoms, and demonstrated numerically that these functionals could greatly reduce the LSD overestimate of atomization energies of molecules. Lee, Yang, and Parr [17] transformed the Colle-Salvetti [84] expression for the correlation energy from a functional of the Kohn-Sham one-particle density matrix into a functional of the density. This functional contains one empirical parameter and works well in conjunction with Becke [16] exchange for many atoms and molecules, although it underestimates the correlation energy of the uniform electron gas by about a factor of two at valence-electron densities.

The real-space cutoff of the GEA hole provides a powerful nonempirical way to construct GGA's. Since exchange and correlation should be treated in a balanced way, there was a need to extend the 1986 real-space cutoff construction [13] from exchange to correlation with the help of a second cutoff radius $u_c(\mathbf{r})$ chosen to satisfy (1.97). Without accurate formulas for the correlation hole of the uniform electron gas, this extension had to wait until 1991, when it led to the Perdew-Wang 1991 (PW91) [18,79] GGA for E_{xc}. For most practical purposes, PW91 is equivalent to the Perdew-Burke-

Ernzerhof [20,21] "GGA made simple", which will be derived, presented, and discussed in the next two sections.

1.6.4 Construction of a "GGA Made Simple"

The PW91 GGA and its construction [18,79] are simple in principle, but complicated in practice by a mass of detail. In 1996, Perdew, Burke and Ernzerhof [20,21] (PBE) showed how to construct essentially the same GGA in a much simpler form and with a much simpler derivation.

Ideally, an approximate density functional $E_{xc}[n_\uparrow, n_\downarrow]$ should have all of the following features: (1) a non-empirical derivation, since the principles of quantum mechanics are well-known and sufficient; (2) universality, since in principle one functional should work for diverse systems (atoms, molecules, solids) with different bonding characters (covalent, ionic, metallic, hydrogen, and van der Waals); (3) simplicity, since this is our only hope for intuitive understanding and our best hope for practical calculation; and (4) accuracy enough to be useful in calculations for real systems.

The LSD of (1.11) and the non-empirical GGA of (1.12) nicely balance these desiderata. Both are exact only for the electron gas of uniform density, and represent controlled extrapolations away from the slowly-varying limit (unlike the GEA of Sect. 1.6.2, which is an uncontrolled extrapolation). LSD is a controlled extrapolation because, even when applied to a density that varies rapidly over space, it preserves many features of the exact E_{xc}, as discussed in Sect. 1.6.1. LSD has worked well in solid state applications for thirty years.

Our conservative philosophy of GGA construction is to try to retain all the correct features of LSD, while adding others. In particular, we retain the correct uniform-gas limit, for two reasons: (1) This is the only limit in which the restricted GGA form *can* be exact. (2) Nature's data set includes the crystalline simple metals like Na and Al. The success of the stabilized jellium model [85] reaffirms that the valence electrons in these systems are correlated very much as in a uniform gas. Among the welter of possible conditions which could be imposed to construct a GGA, the most natural and important are those respected by LSD or by the real-space cutoff construction of PW91, and these are the conditions chosen in the PBE derivation [20] below. The resulting GGA is one in which all parameters (other than those in LSD) are fundamental constants.

We start by writing the correlation energy in the form

$$E_c^{GGA}[n_\uparrow, n_\downarrow] = \int d^3r\, n[e_c(r_s, \zeta) + H(r_s, \zeta, t)]\,, \qquad (1.206)$$

where the local density parameters r_s and ζ are defined in (1.133) and (1.149), and the reduced density gradient t in (1.193). The small-t behavior of nH should be given by the left-hand side of (1.194), which emerges naturally from the real-space cutoff construction of PW91 [79]. In the opposite or $t \to \infty$ limit, we expect that $H \to -e_c(r_s, \zeta)$, the correlation energy per electron of the uniform gas, as it does in the PW91 construction or in the Ma-Brueckner GGA of (1.202). Finally, under the uniform scaling of (1.40) to the high-density ($\gamma \to \infty$) limit, (1.206) should tend to a negative constant, as in (1.114) or in the numerically-constructed PW91. This means that H must cancel the logarithmic singularity of e_c (see (1.140)) in this limit.

A simple function which meets these expectations is

$$H = c_0 \phi^3 \ln \left\{ 1 + \frac{\beta_{\mathrm{MB}}}{c_0} t^2 \left[\frac{1 + At^2}{1 + At^2 + A^2 t^4} \right] \right\} , \qquad (1.207)$$

where ϕ is given by (1.192) and

$$A = \frac{\beta_{\mathrm{MB}}}{c_0} \frac{1}{\exp\left[-e_c(r_s, \zeta)/c_0 \phi^3 \right] - 1} . \qquad (1.208)$$

We now check the required limits:

$$t \to 0: \ H \to c_0 \phi^3 \ln \left\{ 1 + \frac{\beta_{\mathrm{MB}}}{c_0} t^2 \right\}$$

$$\to \beta_{\mathrm{MB}} \phi^3 t^2 . \qquad (1.209)$$

$$t \to \infty: \ H \to c_0 \phi^3 \ln \left\{ 1 + \frac{\beta_{\mathrm{MB}}}{c_0 A} \right\}$$

$$\to c_0 \phi^3 \ln \left\{ \exp \left[-\frac{e_c(r_s, \zeta)}{c_0 \phi^3} \right] \right\}$$

$$\to -e_c(r_s, \zeta) . \qquad (1.210)$$

$$r_s \to 0 \text{ at fixed } s: \quad H \to c_0 \phi^3 \ln t^2 \to -c_0 \phi^3 \ln r_s . \qquad (1.211)$$

To a good approximation, (1.140) can be generalized to

$$e_c(r_s, \zeta) = \phi^3 [c_0 \ln r_s - c_1 + \ldots] , \qquad (1.212)$$

which cancels the log singularity of (1.211).

Under uniform density scaling to the high-density limit, we find

$$\gamma \to \infty: \ E_c^{\mathrm{GGA}}[n_\gamma] \to -c_0 \int d^3 r \, n \, \phi^3 \ln \left[1 + \frac{1}{\chi s^2/\phi^2 + (\chi s^2/\phi^2)^2} \right] \quad (1.213)$$

(where s is defined by (1.183)), a negative constant as required by (1.114), with

$$\chi = \left(\frac{3\pi^2}{16} \right)^{2/3} \frac{\beta_{\mathrm{MB}}}{c_0} \exp(-c_1/c_0) . \qquad (1.214)$$

For a two-electron ion of nuclear charge Z in the limit $Z \to \infty$, (1.213) is -0.0479 hartree and the exact value is -0.0467. Realistic results from (1.213) in the $Z \to \infty$ limit have also be found [86] for ions with 3, 9, 10, and 11 electrons.

Now we turn to the construction of a GGA for the exchange energy. Because of the spin-scaling relation (1.127), we only need to construct $E_x^{GGA}[n]$, which must be of the form of (1.205). To recover the good LSD description of the linear response of the uniform gas (Sect. 1.5.4), we choose the gradient coefficient for exchange to cancel that for correlation, i.e., we take advantage of (1.194) to write

$$s \to 0: \quad F_x(s) = 1 + \mu s^2 . \tag{1.215}$$

Then the gradient coefficients for exchange and correlation will cancel for *all* r_s and ζ, apart from small $\nabla \zeta$ contributions to E_x^{GGA}, as discussed in the next section.

The value of μ of (1.195) is 1.78 times bigger than μ_{AK} of (1.196), the proper gradient coefficient for exchange in the slowly-varying limit. But this choice can be justified in two other ways as well: (a) It provides a decent fit to the results of the real-space cutoff construction [79] of the PW91 exchange energy, which does *not* recover μ_{AK} in the slowly-varying limit. (b) It provides a reasonable emulation of the exact-exchange linear response function of (1.174) over the important range of $0 < q/2k_F \leq 1$ (but not of course in the limit $q \to 0$, where μ_{AK} is needed).

Finally, we want to satisfy the Lieb-Oxford bound of (1.120) and (1.122), which LSD respects. We can achieve this, and also recover the limit of (1.215), with the simple form

$$F_x(s) = 1 + \kappa - \frac{\kappa}{(1 + \mu s^2/\kappa)} , \tag{1.216}$$

where κ is a constant less than or equal to 0.804. Taking $\kappa = 0.804$ gives a GGA which is virtually identical to PW91 over the range of densities and reduced density gradients important in most real systems. We shall complete the discussion of this paragraph in the next section.

1.6.5   GGA Nonlocality: Its Character, Origins, and Effects

A useful way to visualize and think about gradient-corrected nonlocality, or to compare one GGA with another, is to write [19,87]

$$E_{xc}^{GGA}[n_\uparrow, n_\downarrow] \approx \int d^3 r \, n \left(-\frac{c}{r_s} \right) F_{xc}(r_s, \zeta, s) , \tag{1.217}$$

where $c = (3/4\pi)(9\pi/4)^{1/3}$ and $-c/r_s = e_x(r_s, \zeta = 0)$ is the exchange energy per electron of a spin-unpolarized uniform electron gas. The enhancement factor $F_{xc}(r_s, \zeta, s)$ shows the effects of correlation (through its r_s dependence), spin polarization (ζ), and inhomogeneity or nonlocality (s). F_{xc} is the analog of $3\alpha/2$ in Slater's $X\alpha$ method [88], so its variation is bounded and plottable. Figure 1.1 shows $F_{xc}(r_s, \zeta = 0, s)$, the enhancement factor for a spin-unpolarized system. Figure 1.2 shows $F_{xc}(r_s, \zeta = 1, s) - F_{xc}(r_s, \zeta = 0, s)$, the enhancement factor for the spin polarization energy. (Roughly, $F_{xc}(r_s, \zeta, s) \approx F_{xc}(r_s, \zeta = 0, s) + \zeta^2[F_{xc}(r_s, \zeta = 1, s) - F_{xc}(r_s, \zeta = 0, s)]$). The nonlocality is the s-dependence, and

$$F_{xc}^{LSD}(r_s, \zeta, s) = F_{xc}(r_s, \zeta, s = 0) \tag{1.218}$$

is visualized as a set of horizontal straight lines coinciding with the GGA curves in the limit $s \to 0$.

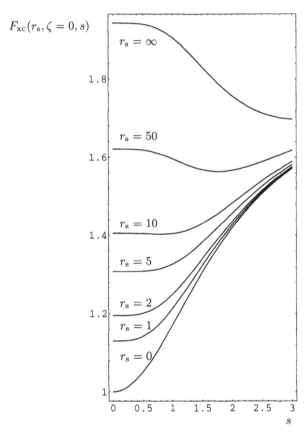

Fig. 1.1. The enhancement factor F_{xc} of (1.217) for the GGA of Perdew, Burke, and Ernzerhof [20], as a function of the reduced density gradient s of (1.183), for $\zeta = 0$. The local density parameter r_s and the relative spin polarization ζ are defined in (1.133) and (1.149), respectively

$$F_{xc}(r_s, \zeta = 1, s) - F_{xc}(r_s, \zeta = 0, s)$$

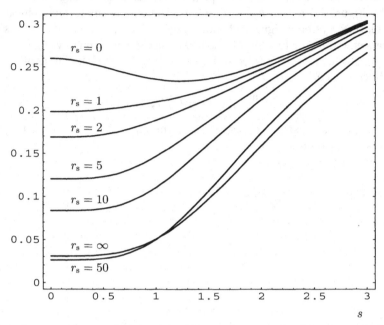

Fig. 1.2. Same as Fig. 1.1, but for the difference between the fully spin-polarized ($\zeta = 1$) and unpolarized ($\zeta = 0$) enhancement factors

Clearly, the correlation energy of (1.206) can be written in the form of (1.217). To get the exchange energy into this form, apply the spin-scaling relation (1.127) to (1.205), then drop small ∇s contributions to find

$$F_x(\zeta, s) = \frac{1}{2}(1+\zeta)^{4/3} F_x\left(s/(1+\zeta)^{1/3}\right) + \frac{1}{2}(1-\zeta)^{4/3} F_x\left(s/(1-\zeta)^{1/3}\right)$$

$$= \frac{1}{2}\left[(1+\zeta)^{4/3} + (1-\zeta)^{4/3}\right] + \mu\phi s^2 + \dots \tag{1.219}$$

Now

$$F_{xc}(r_s, \zeta, s) = F_x(\zeta, s) + F_c(r_s, \zeta, s) , \tag{1.220}$$

where

$$\lim_{r_s \to 0} F_c(r_s, \zeta, s) = 0 \tag{1.221}$$

by (1.106) and (1.114). Thus the $r_s = 0$ or high-density-limit curve in each figure is the exchange-only enhancement factor. Clearly $F_x > 0$, $F_c > 0$, and $F_x(\zeta = 0, s = 0) = 1$ by definition.

The Lieb-Oxford bound of (1.122) will be satisfied for all densities $n(\mathbf{r})$ if and only if

$$F_{xc}(r_s, \zeta, s) \geq 2.273 . \tag{1.222}$$

For the PBE GGA of (1.206) and (1.216), this requires that

$$2^{1/3} F_x(s/2^{1/3}) \leq 2.273 , \tag{1.223}$$

or

$$\kappa \leq 2.273/2^{1/3} - 1 = 0.804 , \tag{1.224}$$

as stated in Sect. 1.6.4.

There is much to be seen and explained [21] in (1.217) and Figs. 1.1 and 1.2. However, the main qualitative features are simply stated: When we make a density variation in which r_s decreases, ζ increases, or s increases everywhere, we find that $|E_x|$ increases and $|E_c/E_x|$ decreases.

To understand this pattern [21], we note that the second-order gradient expansion for the non-interacting kinetic energy $T_s[n_\uparrow, n_\downarrow]$, which is arguably its own GGA [81], can be written as

$$T_s^{GGA}[n_\uparrow, n_\downarrow] = \int d^3 r \, n \frac{3}{10} \frac{\left(\frac{9\pi}{4}\right)^{2/3}}{r_s^2} G(\zeta, s) , \tag{1.225}$$

$$G(\zeta, s) = \frac{1}{2} \left[(1 + \zeta)^{5/3} + (1 - \zeta)^{5/3} \right] + \frac{5}{27} s^2 , \tag{1.226}$$

using approximate spin scaling (see (1.126) plus neglect of $\nabla \zeta$ contributions). Equations (1.225) and (1.226) respect (1.104) and confirm our intuition based upon the Pauli exclusion and uncertainty principles: Under a density variation in which r_s decreases, ζ increases, or s increases everywhere, we find that $T_s[n_\uparrow, n_\downarrow]$ increases.

The first effect of such an increase in T_s is an increase in $|E_x|$. T_s and $|E_x|$ are "conjoint" [89], in the sense that both can be constructed from the occupied Kohn-Sham orbitals (see (1.7), (1.88), (1.90) and (1.92)). With more kinetic energy, these occupied orbitals will have shorter de Broglie wavelengths. By the uncertainty principle, they can then dig a more short-ranged and deeper exchange hole with a more negative exchange energy. Thus exchange turns on when we decrease r_s, increase ζ, or increase s.

The second effect of such an increase in T_s is to strengthen the Kohn-Sham Hamiltonian which holds non-interacting electrons at the spin densities $n_\uparrow(\mathbf{r})$ and $n_\downarrow(\mathbf{r})$. This makes the electron-electron repulsion of (1.112) a relatively weaker perturbation on the Kohn-Sham problem, and so reduces the ratio $|E_c/E_x|$. Thus correlation turns off relative to exchange when we decrease r_s, increase ζ, or increase s.

We note in particular that $F_x(r_s, \zeta, s)$ increases while $F_c(r_s, \zeta, s)$ decreases with increasing s. The nonlocalities of exchange and correlation are opposite, and tend to cancel for valence-electron densities ($1 \leq r_s \leq 10$) in the range $0 \leq s \leq 1$. The same remarkable cancellation occurs [62,21] in the linear response function for the uniform gas of (1.163), i.e., $\gamma_{xc}(q) \approx \gamma_{xc}^{LSD}(q) = \gamma_{xc}(q = 0)$ for $0 \leq q/2k_F \leq 1$.

The core electrons in any system, and the valence electrons in solids, sample primarily the range $0 \leq s \leq 1$. The high-density core electrons see a strong, exchange-like nonlocality of E_{xc} which provides an important correction to the LSD total energy. But the valence electrons in solids see an almost-complete cancellation between the nonlocalities of exchange and correlation. This helps to explain why LSD has been so successful in solid state physics, and why the small residue of GGA nonlocality in solids does not provide a universally-better description than LSD.

The valence electrons in atoms and molecules see $0 \leq s \leq \infty$, when s diverges in the exponential tail of the density, but the energetically-important range is $0 \leq s \leq 3$ [70,71]. Figures 1.1 and 1.2 show that GGA nonlocality *is* important in this range, so GGA is almost-always better than LSD for atoms and molecules.

For $r_s \leq 10$, the residual GGA nonlocality is exchange-like, i.e., exchange and correlation together turn on stronger with increasing inhomogeneity. It can then be seen from (1.217) that gradient corrections will favor greater density inhomogeneity and higher density [70]. Defining average density parameters $\langle r_s \rangle$, $\langle \zeta \rangle$, and $\langle s \rangle$ as in [70], we find that gradient corrections favor changes $d\langle s \rangle > 0$ and $d\langle r_s \rangle < 0$. Gradient corrections tend to drive a process forward when [70]

$$\frac{d\langle s \rangle}{\langle s \rangle} \geq \frac{d\langle r_s \rangle}{\langle r_s \rangle} . \tag{1.227}$$

In a typical process (bond stretching, transition to a more open structure, fragmentation, or atomization), one has $d\langle s \rangle > 0$ and $d\langle r_s \rangle > 0$. Thus, by (1.227), these effects compete – another reason why LSD has met with some success. In most such cases, the left-hand side of (1.227) is bigger than the right, so typically gradient corrections favor larger bond lengths or lattice constants (and thus softer vibration frequencies), more open structures, fragmentation of a highly-bonded transition state, or atomization of a molecule. In the case of bond stretching in H_2, however, the right hand side of (1.227) exceeds the left, so gradient corrections actually and correctly shrink the equilibrium bond length relative to LSD.

There have been many interesting tests and applications of GGA to a wide range of atoms, molecules, and solids. Some references will be found in [19,90,79,21].

We close by discussing those situations in which LSD or GGA can fail badly. They seem to be of two types: (1) When the Kohn-Sham non-interacting wavefunction is not a single Slater determinant, or when the non-interacting energies are nearly degenerate, the LSD and GGA exchange-correlation holes can be unrealistic even very close to or on top of the electron [36,91,66]. (2) In an extended system, the exact hole may display a diffuse long-range tail which is not properly captured by either LSD or GGA. To a limited extent, this effect could be mimicked by reducing the parameter κ in (1.216). An example of a diffuse hole arises in the calculation of the surface energy of a metal [19,32]: When an electron wanders out into the vacuum region, the exchange-correlation hole around it can extend significantly backward into the interior of the metal. A more extreme example is "stretched H_2^+", the ground state of one electron in the presence of two protons at very large separation: Half of the exact hole is localized on each proton, a situation which has no analog in the electron gas of uniform or slowly-varying density, and for which LSD and GGA make large self-interaction errors [9,92,68].

"Stretched H_2^+" and related systems are of course unusual. In most systems, the exact exchange-correlation hole is reasonably localized around its electron, as it is in LSD or GGA – and that fact is one of the reasons [93] why LSD and GGA work as well as they do.

1.6.6   Hybrid Functionals

At the lower limit ($\lambda = 0$) of the coupling constant integration of (1.87) is the exact exchange hole. This observation led Becke [94,95] to conclude that a fraction of exact exchange should be mixed with GGA exchange and correlation. The simplest such hybrid functional is

$$E_{xc}^{hyb} = aE_x^{exact} + (1-a)E_x^{GGA} + E_c^{GGA} , \qquad (1.228)$$

where the constant a can be fitted empirically or estimated theoretically [96–98] as $a \approx 1/4$ for molecules.

The mixing coefficient a is not equal to or close to 1, because full exact exchange is incompatible with GGA correlation. The exact exchange hole in a molecule can have a highly nonlocal, multi-center character which is largely cancelled by an almost equal-but-opposite nonlocal, multicenter character in the exact correlation hole. The GGA exchange and correlation holes are more local, and more localized around the reference electron.

Equation (1.228) can be re-written as

$$E_{xc}^{hyb} = E_x^{exact} + (1-a)(E_x^{GGA} - E_x^{exact}) + E_c^{GGA} . \qquad (1.229)$$

In this form, we can think of the correlation energy as the sum of two pieces: The dynamic correlation energy modelled by E_c^{GGA} results from the tendency of electrons to avoid one another by "swerving" upon close approach, while the static correlation energy modelled by $(1-a)(E_x^{GGA} - E_x^{exact})$ results from the tendency of electrons to avoid one another by sitting on different atomic sites [99]. This model for static correlation must fail in the high-density limit, since it does not satisfy (1.114).

Hybrid functionals are perhaps the most accurate density functionals in use for quantum chemical calculations. Although based upon a valid physical insight, they do not satisfy any exact constraints that their underlying GGA's do not satisfy.

1.6.7 Meta-generalized Gradient Approximations

While GGA's take the form

$$E_{xc}^{GGA} = \int d^3r \; n \; e_{xc}^{GGA}(n_\uparrow, n_\downarrow, \nabla n_\uparrow, \nabla n_\downarrow), \qquad (1.230)$$

meta-GGA's take the more general form

$$E_{xc}^{MGGA} = \int d^3r \; n \; e_{xc}^{MGGA}(n_\uparrow, n_\downarrow, \nabla n_\uparrow, \nabla n_\downarrow, \nabla^2 n_\uparrow, \nabla^2 n_\downarrow, \tau_\uparrow, \tau_\downarrow), \quad (1.231)$$

where $\tau_\sigma(\mathbf{r}) = \frac{1}{2}\sum_\alpha \theta(\mu - \epsilon_{\alpha\sigma})|\nabla\psi_{\alpha\sigma}(\mathbf{r})|^2$ is the Kohn-Sham orbital kinetic energy density for electrons of spin σ. The added ingredients are natural ones from several points of view:

Becke [100,101] noted that, while the on-top ($\mathbf{r} = \mathbf{r}'$) exchange hole $n_x(\mathbf{r}, \mathbf{r}')$ is determined by $n_\uparrow(\mathbf{r})$ and $n_\downarrow(\mathbf{r})$, the leading correction for small $|\mathbf{r} - \mathbf{r}'|$ depends upon all the ingredients in (1.231). He also observed that one-electron regions of space can be recognized by the condition $\tau_\sigma(\mathbf{r}) = \tau_\sigma^W(\mathbf{r})$ (where $\tau_\sigma^W(\mathbf{r}) = |\nabla n_\sigma(\mathbf{r})|^2/[8n_\sigma(\mathbf{r})]$), and $n_\sigma/n = 1$, and that the correlation energy density can be zeroed out in these regions [102], achieving satisfaction of the exact condition of (1.70).

Several meta-GGA's have been constructed by a combination of theoretical constraints and fitting to chemical data [103–107]. While some of these functionals use up to 20 fit parameters, there is only one empirical parameter in the meta-GGA of Perdew, Kurth, Zupan, and Blaha (PKZB) [107], who realized that the extra meta-GGA ingredients could be used to recover the fourth-order gradient expansion for the exchange energy, and that the self-interaction correction to GGA could be made without destroying the correct second-order gradient expansion for the correlation energy.

The PKZB meta-GGA achieves very accurate atomization energies of molecules, surface energies of metals, and lattice constants of solids [108]. These properties are greatly improved over GGA. On the other hand, meta-GGA's that are heavily fitted to molecular properties tend to give surface energies and lattice constants that are less accurate than those of non-empirical GGA's or even LSD [108].

The PKZB self-correlation correction to the PBE GGA has a remarkable feature: Under uniform scaling to the low-density or strongly-interacting limit (see (1.115)), it yields essentially correct correlation energies while LSD and GGA yield correlation energies that are much too negative [109].

There are two problems with the PKZB meta-GGA: (1) It depends upon one empirical parameter, which is one too many in the view of the authors. (2) It predicts bond lengths for molecules which are typically longer and less accurate than those of GGA [110]. These problems have been eliminated in a new, fully-nonempirical meta-GGA of Perdew and Tao [111]. While the PKZB correlation is merely refined in this work, the PKZB exchange is revised to reflect exact constraints on iso-orbital densities, i.e., those where the density and kinetic-energy density are dominated by a single orbital shape ($\tau = \tau^{\mathrm{W}} = \frac{1}{8}|\nabla n|^2/n$), such as one- and two-electron ground states.

The PKZB and Perdew-Tao meta-GGA's do not make use of the Laplacians $\nabla^2 n_\uparrow$ and $\nabla^2 n_\downarrow$. This has two advantages: (a) it avoids the singularities of these Laplacians at the nucleus, and (b) it reduces the number of ingredients, making the functionals easier to visualize [108].

1.6.8 Jacob's Ladder of Density Functional Approximations

The main line of development of density functionals for the exchange-correlation energy suggests a Jacob's Ladder stretching from the Hartree world up to the heaven of chemical accuracy [112]. This ladder has five rungs, corresponding to increasingly complex choices for the ingredients of the "energy density":

(1) The local spin density approximation, the "mother of all approximations", constitutes the lowest and most basic rung, using only $n_\uparrow(\mathbf{r})$ and $n_\downarrow(\mathbf{r})$ as its ingredients.

(2) The generalized gradient approximation adds the ingredients $\nabla n_\uparrow$ and $\nabla n_\downarrow$.

(3) The meta-GGA adds the further ingredients $\nabla^2 n_\uparrow$, $\nabla^2 n_\downarrow$, $\tau_\uparrow$, and $\tau_\downarrow$, or at least some of them. While $\tau_\uparrow$ and $\tau_\downarrow$ are fully nonlocal functionals of the density, they are semi-local functionals of the occupied orbitals which are available in any Kohn-Sham calculation.

(4) The hyper-GGA [112] adds another ingredient: the exact exchange energy density, a fully nonlocal functional of the occupied Kohn-Sham orbitals. The hybrid functionals of Sect. 1.6.6 are in a sense hyper-GGA's, but hyper-GGA's can also make use of full exact exchange and a fully nonlocal correlation functional which incorporates the exact exchange energy density [112], achieving an E_{xc} with full freedom from self-interaction error and the correct high-density limit under uniform scaling.

(5) Exact exchange can be combined with exact partial correlation, making use not only of the occupied Kohn-Sham orbitals but also of the unoccupied ones. Examples are the random phase approximation using Kohn-Sham orbitals [83,113–115], with or without a correction for short-range correlation [116,117], or the interaction strength interpolation [118].

All of these approximations are density functionals, because the Kohn-Sham orbitals are implicit functionals of the density. Finding the exchange-correlation potential for rungs (3)–(5) requires the construction of the optimized effective potential [119], which is now practical even for fully three-dimensional densities [120]. For many purposes a non-selfconsistent implementation of rungs (3)–(5) using GGA orbitals will suffice.

Acknowledgements

Work supported in part by the U.S. National Science Foundation under Grant No. DMR95-21353 and DMR01-35678. We thank Matthias Ernzerhof for help with the figures.

References

1. D. Park, *Introduction to the Quantum Theory* (McGraw-Hill, New York, 1974).
2. H. A. Bethe, *Intermediate Quantum Mechanics* (Benjamin, New York, 1964).
3. R. O. Jones and O. Gunnarsson, Rev. Mod. Phys. **61**, 689 (1989).
4. R. M. Dreizler and E. K. U. Gross, *Density Functional Theory* (Springer, Berlin, 1990).
5. R. G. Parr and W. Yang, *Density-Functional Theory of Atoms and Molecules* (Oxford University Press, New York, 1989).
6. W. Kohn and L. J. Sham, Phys. Rev. **140**, A1133 (1965).
7. U. von Barth and L. Hedin, J. Phys. C **5**, 1629 (1972).
8. J. P. Perdew and Y. Wang, Phys. Rev. B **45**, 13244 (1992).
9. J. P. Perdew and A. Zunger, Phys. Rev. B **23**, 5048 (1981).
10. S. H. Vosko, L. Wilk, and M. Nusair, Can. J. Phys. **58**, 1200 (1980).
11. D. C. Langreth and M. J. Mehl, Phys. Rev. B **28**, 1809 (1983).
12. J. P. Perdew, Phys. Rev. Lett. **55**, 1665 (1985).
13. J. P. Perdew and Y. Wang, Phys. Rev. B **33**, 8800 (1986); ibid. **40**, 3399 (1989) (E).
14. J. P. Perdew, Phys. Rev. B **33**, 8822 (1986); ibid. **34**, 7406 (1986) (E).
15. A. D. Becke, J. Chem. Phys. **84**, 4524 (1986).
16. A. D. Becke, Phys. Rev. A **38**, 3098 (1988).
17. C. Lee, W. Yang, and R. G. Parr, Phys. Rev. B **37**, 785 (1988).
18. J. P. Perdew, in *Electronic Structure of Solids '91*, edited by P. Ziesche and H. Eschrig (Akademie Verlag, Berlin, 1991), p. 11.
19. J. P. Perdew, J. A. Chevary, S. H. Vosko, K. A. Jackson, M. R. Pederson, D. J. Singh, and C. Fiolhais, Phys. Rev. B **46**, 6671 (1992); ibid., **48**, 4978 (1993)(E).
20. J. P. Perdew, K. Burke, and M. Ernzerhof, Phys. Rev. Lett. **77**, 3865 (1996); ibid. **78**, 1396 (1997)(E).
21. J. P. Perdew, M. Ernzerhof, A. Zupan, and K. Burke, J. Chem. Phys. **108**, 1522 (1998).
22. J. Hafner, *From Hamiltonians to Phase Diagrams* (Dover, New York, 1988).
23. A. Szabo and N. S. Ostlund, *Modern Quantum Chemistry* (MacMillan, New York, 1982).
24. P. Fulde, *Electron Correlations in Molecules and Solids* (Springer, Berlin, 1993).
25. P. Hohenberg and W. Kohn, Phys. Rev. **136**, B864 (1964).
26. I. N. Levine, *Quantum Chemistry* (Allyn and Bacon, Boston, 1974).
27. D. P. Joubert, Int. J. Quantum Chem. **61**, 355 (1997).
28. M. Levy, Proc. Natl. Acad. Sci. USA **76**, 6062 (1979).

29. M. Levy, in *Recent Developments and Applications of Modern Density Functional Theory*, edited by J. M. Seminario (Elsevier, Amsterdam, 1996), p. 3.
30. J. Harris and R. O. Jones, J. Phys. F **4**, 1170 (1974).
31. D. C. Langreth and J. P. Perdew, Solid State Commun. **17**, 1425 (1975).
32. D. C. Langreth and J. P. Perdew, Phys. Rev. B **15**, 2884 (1977).
33. O. Gunnarsson and B. I. Lundqvist, Phys. Rev. B **13**, 4274 (1976).
34. E. R. Davidson, *Reduced Density Matrices in Quantum Chemistry* (Academic Press, New York, 1976).
35. J. C. Kimball, Phys. Rev. A **7**, 1648 (1973).
36. T. Ziegler, A. Rauk, and E. J. Baerends, Theoret. Chim. Acta **43**, 261 (1977).
37. K. Burke and J. P. Perdew, Int. J. Quantum Chem. **56**, 199 (1995).
38. M. Levy and J. P. Perdew, Phys. Rev. A **32**, 2010 (1985).
39. M. Levy, in *Single-Particle Density in Physics and Chemistry*, edited by N. H. March and B. M. Deb (Academic, London, 1987), p. 45.
40. A. Görling and M. Levy, Phys. Rev. B **47**, 13105 (1993).
41. A. Görling and M. Levy, Phys. Rev. A **50**, 196 (1994).
42. M. Levy, Phys. Rev. A **43**, 4637 (1991).
43. M. Levy and J. P. Perdew, Phys. Rev. B **48**, 11638 (1993); ibid., **55**, 13321 (1997)(E).
44. E. H. Lieb, Int. J. Quantum Chem. **24**, 243 (1983).
45. E. H. Lieb and S. Oxford, Int. J. Quantum Chem. **19**, 427 (1981).
46. G. L. Oliver and J. P. Perdew, Phys. Rev. A **20**, 397 (1979).
47. J. P. Perdew, in *Density Functional Theory of Many-Fermion Systems*, Vol. 21 of *Advances in Quantum Chemistry*, edited by S. B. Trickey (Academic, New York, 1990), p. 113.
48. E. Fermi and E. Amaldi, Accad. Ital. Rome **6**, 119 (1934).
49. J. P. Perdew, R. G. Parr, M. Levy, and J. L. Balduz, Phys. Rev. Lett. **49**, 1691 (1982).
50. J. P. Perdew, in *Density Functional Methods in Physics*, Vol. B123 of *NATO ASI Series*, edited by R. M. Dreizler and J. da Providencia (Plenum Press, New York, 1985), p. 265.
51. J. F. Janak, Phys. Rev. B **18**, 7165 (1978).
52. M. Levy, J. P. Perdew, and V. Sahni, Phys. Rev. A **30**, 2745 (1984).
53. N. W. Ashcroft and N. D. Mermin, *Solid State Physics* (Holt, Rinehart, and Winston, New York, 1976).
54. M. Gell-Mann and K. A. Brueckner, Phys. Rev. **106**, 364 (1957).
55. L. Onsager, L. Mittag, and M. J. Stephen, Ann. Phys. (Leipzig) **18**, 71 (1966).
56. R. A. Coldwell-Horsfall and A. A. Maradudin, J. Math. Phys. **1**, 395 (1960).
57. D. M. Ceperley and B. J. Alder, Phys. Rev. Lett. **45**, 566 (1980).
58. J. P. Perdew and Y. Wang, Phys. Rev. B **46**, 12947 (1992); ibid. **56**, 7018 (1997)(E). See also P. Gori-Giorgi and J. P. Perdew, Phys. Rev. B **66**, 165118 (2002).
59. J. P. Perdew and T. Datta, Phys. Stat. Sol. (b) **102**, 283 (1980).
60. C. Fiolhais, J. P. Perdew, S. Q. Armster, J. M. MacLaren, and M. Brajczewska, Phys. Rev. B **51**, 14001 (1995); ibid., **53**, 13193 (1996)(E).
61. E. Engel and S. H. Vosko, Phys. Rev. B **42**, 4940 (1990); ibid., **44**, 1446 (1991)(E).
62. S. Moroni, D. M. Ceperley, and G. Senatore, Phys. Rev. Lett. **75**, 689 (1995).
63. K. Burke, J. P. Perdew, and D. C. Langreth, Phys. Rev. Lett. **73**, 1283 (1994).

64. K. Burke, J. P. Perdew, and M. Ernzerhof, J. Chem. Phys. **109**, 3760 (1998).
65. M. Ernzerhof, in *Macmillan Encyclopedia of Physics* (Macmillan Publishing Company, New York, 1996), Vol. 2, p. 733.
66. J. P. Perdew, A. Savin, and K. Burke, Phys. Rev. A **51**, 4531 (1995).
67. J. P. Perdew, M. Ernzerhof, K. Burke, and A. Savin, Int. J. Quantum Chem. **61**, 197 (1997).
68. J. P. Perdew and M. Ernzerhof, in *Electronic Density Functional Theory: Recent Progress and New Directions*, edited by J. F. Dobson, G. Vignale, and M. Das (Plenum Press, New York, 1997).
69. S.-K. Ma and K. A. Brueckner, Phys. Rev. **165**, 18 (1968).
70. A. Zupan, K. Burke, M. Ernzerhof, and J. P. Perdew, J. Chem. Phys. **106**, 10184 (1997).
71. A. Zupan, J. P. Perdew, K. Burke, and M. Causá, Int. J. Quantum Chem. **61**, 287 (1997).
72. D. A. Kirzhnits, JETP **5**, 64 (1957).
73. L. J. Sham, in *Computational Methods in Band Theory*, edited by P. M. Marcus, J. F. Janak, and A. R. Williams (Plenum Press, New York, 1971), p. 458.
74. M. Rasolt and D. J. W. Geldart, Phys. Rev. B **34**, 1325 (1986).
75. M. Rasolt and H. L. Davis, Phys. Lett. A **86**, 45 (1981).
76. Y. Wang and J. P. Perdew, Phys. Rev. B **43**, 8911 (1991).
77. P. A. Antoniewicz and L. Kleinman, Phys. Rev. B **31**, 6779 (1985).
78. D. C. Langreth and S. H. Vosko, Phys. Rev. Lett. **59**, 497 (1987); ibid. **60**, 1984 (1988).
79. J. P. Perdew, K. Burke, and Y. Wang, Phys. Rev. B **54**, 16533 (1996); ibid. **57**, 14999 (1998) (E).
80. Y. Wang, J. P. Perdew, J. A. Chevary, L. D. MacDonald, and S. H. Vosko, Phys. Rev. A **41**, 78 (1990).
81. J. P. Perdew, Phys. Lett. A **165**, 79 (1992).
82. Z. Yan, J. P. Perdew, T. Korhonen, and P. Ziesche, Phys. Rev. A **55**, 4601 (1997).
83. D. C. Langreth and J. P. Perdew, Phys. Rev. B **21**, 5469 (1980).
84. R. Colle and O. Salvetti, Theoret. Chim. Acta **37**, 329 (1975).
85. J. P. Perdew, H. Q. Tran, and E. D. Smith, Phys. Rev. B **42**, 11627 (1990).
86. S. Ivanov and M. Levy, J. Phys. Chem. A **102**, 3151 (1998).
87. J. P. Perdew and K. Burke, Int. J. Quantum Chem. **57**, 309 (1996).
88. J. C. Slater, *The Self-Consistent Field for Molecules and Solids* (McGraw-Hill, New York, 1974).
89. H. Lee, C. Lee, and R. G. Parr, Phys. Rev. A **44**, 768 (1991).
90. K. Burke, J. P. Perdew, and M. Levy, in *Modern Density Functional Theory: A Tool for Chemistry*, Vol. 2 of *Theoretical and Computational Chemistry*, edited by J. M. Seminario and P. Politzer (Elsevier, Amsterdam, 1995), p. 29.
91. A. D. Becke, A. Savin, and H. Stoll, Theoret. Chim. Acta **91**, 147 (1995).
92. R. Merkle, A. Savin, and H. Preuss, J. Chem. Phys. **97**, 9216 (1992).
93. V. Tschinke and T. Ziegler, J. Chem. Phys. **93**, 8051 (1990).
94. A. D. Becke, J. Chem. Phys. **98**, 1372 (1993).
95. A. D. Becke, J. Chem. Phys. **98**, 5648 (1993).
96. J. P. Perdew, M. Ernzerhof, and K. Burke, J. Chem. Phys. **105**, 9982 (1996).
97. C. Adamo and V. Barone, J. Chem. Phys. **110**, 6158 (1999).

98. M. Ernzerhof and G. E. Scuseria, J. Chem. Phys. **110**, 5029 (1999).

99. N. C. Handy and A. J. Cohen, Mol. Phys. **99**, 403 (2001).

100. A. D. Becke, Int. J. Quantum Chem. **23**, 1915 (1983).

101. A. D. Becke, J. Chem. Phys. **109**, 2092 (1998).

102. A. D. Becke, J. Chem. Phys. **104**, 1040 (1996).

103. E. I. Proynov, S. Sirois, and D. R. Salahub, Int. J. Quantum Chem. **64**, 427 (1997).

104. T. Van Voorhis and G. E. Scuseria, J. Chem. Phys. **109**, 400 (1998).

105. M. Filatov and W. Thiel, Phys. Rev. A **57**, 189 (1998).

106. J. B. Krieger, J. Chen, G. J. Iafrate, and A. Savin, in *Electron Correlations and Materials Properties*, edited by A. Gonis and N. Kioussis (Plenum, New York, 1999).

107. J. P. Perdew, S. Kurth, A. Zupan, and P. Blaha, Phys. Rev. Lett. **82**, 2544 (1999); ibid. **82**, 5179 (1999)(E).

108. S. Kurth, J. P. Perdew, and P. Blaha, Int. J. Quantum Chem. **75**, 889 (1999).

109. M. Seidl, J. P. Perdew, and S. Kurth, Phys. Rev. A **62**, 012502 (2000).

110. C. Adamo, M. Ernzerhof, and G. E. Scuseria, J. Chem. Phys. **112**, 2643 (2000).

111. J. P. Perdew and J. Tao, work in progress.

112. J. P. Perdew and K. Schmidt, in *Density Functional Theory and Its Applications to Materials*, edited by V. E. van Doren, C. van Alsenoy, and P. Geerlings (American Institute of Physics, AIP Conference Proceedings Vol. **577**, 2001).

113. J. M. Pitarke and A. G. Eguiluz, Phys. Rev. B **63**, 045116 (2001).

114. F. Furche, Phys. Rev. B **64**, 195120 (2001).

115. M. Fuchs and X. Gonze, Phys. Rev. B **65**, 235109 (2002).

116. Z. Yan, J. P. Perdew, and S. Kurth, Phys. Rev. B **61**, 16430 (2000).

117. M. Lein, E. K. U. Gross, and J. P. Perdew, Phys. Rev. B **61**, 13431 (2000).

118. M. Seidl, J. P. Perdew, and S. Kurth, Phys. Rev. Lett. **84**, 5070 (2000).

119. T. Grabo, T. Kreibich, S. Kurth, and E. K. U. Gross, in *The Strong Coulomb Correlations and Electronic Structure Calculations: Beyond Local Density Approximations*, edited by V. Anisimov (Gordon and Breach, Amsterdam, 2000).

120. S. Kümmel and J. P. Perdew, Phys. Rev. Lett. **90**, 043004 (2003)

2 Orbital-Dependent Functionals for the Exchange-Correlation Energy: A Third Generation of Density Functionals

Eberhard Engel

Institut für Theoretische Physik,
J. W.Goethe - Universität Frankfurt,
Robert-Mayer-Straße 6-8,
60054 Frankfurt/Main, Germany
engel@th.physik.uni-frankfurt.de

Eberhard Engel

2.1 Introduction

This chapter is devoted to orbital-dependent exchange-correlation (xc) functionals, a concept that has attracted more and more attention during the last ten years. After a few preliminary remarks, which clarify the scope of this review and introduce the basic notation, some motivation will be given why such implicit density functionals are of definite interest, in spite of the fact that one has to cope with additional complications (compared to the standard xc-functionals). The basic idea of orbital-dependent xc-functionals is then illustrated by the simplest and, at the same time, most important functional of this type, the exact exchange of density functional theory (DFT – for a review see e.g. [1], or the chapter by J. Perdew and S. Kurth in this volume).

Given some orbital-dependent xc-functional E_{xc} the first question to be addressed is the evaluation of the corresponding multiplicative xc-potential v_{xc}. This is possible via the optimized potential method[1] (OPM) [2,3], which is described in Sect. 2.2. After an outline of three different strategies for the derivation of the crucial OPM integral equation, a few exact relations for the OPM xc-potential are summarized. In addition, the Krieger-Li-Iafrate (KLI) approximation [4] to the OPM integral equation is presented.

Once one has all basic ingredients of this third generation of DFT together, it is very instructive to analyze, in some detail, the exchange-only (x-only) limit, in which correlation is completely neglected (Sect. 2.3). On the

---

[1] The method is sometimes also termed optimized effective potential (OEP).

one hand, the simple functional form of the exact exchange and its universal applicability allows a quantitative examination of the KLI approximation for a variety of systems. In this way one can explicitly verify the high accuracy of the KLI approximation, which makes it an important tool for the application of orbital-dependent xc-functionals. On the other hand, the exact x-only results can be used to investigate the properties of the standard functionals like the local density approximation (LDA) [5] and the generalized gradient approximation (GGA) [6–15].

The systematic derivation of implicit correlation functionals is discussed in Sect. 2.4. In particular, perturbation theory based on the Kohn-Sham (KS) Hamiltonian [16–18] is used to derive an exact relation for E_{xc}. This expression is then expanded to second order in the electron-electron coupling constant e^2 in order to obtain the simplest first-principles correlation functional [18]. The corresponding OPM integral equation as well as extensions like the random phase approximation (RPA) [19,20] and the interaction strength interpolation (ISI) [21] are also introduced.

Two semi-empirical orbital-dependent xc-functionals are reviewed in Sect. 2.5. Both functionals had been in the literature for quite some time before it was realized that they should be understood as implicit density functionals in the same sense as the exact exchange. The first is the self-interaction corrected (SIC) form of the LDA [22], and the second is the Colle-Salvetti correlation functional [23].

Finally, the performance of the presently available implicit correlation functionals is studied in Sect. 2.6. In particular, the success of the first-principles perturbative correlation functional with the description of dispersion forces is demonstrated [24]. On the other hand, this functional leads to a divergent correlation potential in the case of finite systems [25]. This failure prompts an approximate handling of the associated OPM integral equation in the spirit of the KLI approximation, which avoids the asymptotic divergence and produces comparatively accurate atomic correlation potentials.

The status of implicit functionals is summarized in Sect. 2.7. In addition, it is shown that the concept of implicit functionals is not restricted to the xc-energy, but can equally well be applied to such quantities as the 2-particle density. In this way implicit functionals provide access to quantities which are beyond the traditional realm of DFT.

2.1.1 Preliminaries and Notation

First of all, a few words on the scope of this review seem to be appropriate. For simplicity, all explicit formulae in this chapter will be given for spin-saturated systems only. Of course, the complete formalism can be extended to spin-density functional theory (SDFT) and all numerical results for spin-polarized systems given in this paper were obtained by SDFT calculations. In addition, the discussion is restricted to the nonrelativistic formalism – for its relativistic form see Chap. 3. The concept of implicit functionals has also been extended to time-dependent phenomena to which Chap. 4 is devoted. The present discussion, on the other hand, focuses completely on ground-state problems, assuming this state to be non-degenerate (and using the Born-Oppenheimer approximation in the case of polyatomic systems). The presentation is furthermore restricted to zero temperature. However, the extension to finite temperature essentially requires the appropriate replacement of the occupation function Θ_k.

In order to introduce the notation, the basic relations of DFT are now summarized, starting with the KS equations[2] [5],

$$\left[-\frac{\nabla^2}{2m} + v_s(\mathbf{r}) \right] \phi_k(\mathbf{r}) = \epsilon_k \phi_k(\mathbf{r}) . \tag{2.1}$$

Throughout this chapter ϕ_k and ϵ_k always denote the KS orbitals and eigenvalues, respectively. As usual, the total KS potential v_s is given by the sum of the external (nuclear) potential v_{ext}, the Hartree potential v_H and the xc-potential v_{xc}:

$$v_s(\mathbf{r}) = v_{ext}(\mathbf{r}) + v_H(\mathbf{r}) + v_{xc}(\mathbf{r}) \tag{2.2}$$

$$v_H(\mathbf{r}) = e^2 \int d^3 r' \frac{n(\mathbf{r}')}{|\mathbf{r} - \mathbf{r}'|} \tag{2.3}$$

$$v_{xc}(\mathbf{r}) = \frac{\delta E_{xc}[n]}{\delta n(\mathbf{r})} . \tag{2.4}$$

The density is obtained by summing up the energetically lowest KS states,

$$n(\mathbf{r}) = \sum_k \Theta_k |\phi_k(\mathbf{r})|^2 , \tag{2.5}$$

which is implemented via the occupation function

$$\Theta_k = \begin{cases} 1 \text{ for } \epsilon_k \leq \epsilon_F \\ 0 \text{ for } \epsilon_F < \epsilon_k \end{cases} , \tag{2.6}$$

with ϵ_F being the Fermi energy (ϵ_F is always identified with the eigenvalue ϵ_{HOMO} of the highest occupied KS state). The total energy functional is given by

$$E_{tot}[n] = T_s[n] + E_{ext}[n] + E_H[n] + E_{xc}[n] \left(+ E_{ion} \right) . \tag{2.7}$$

Its components are the KS kinetic energy,

$$T_s[n] = -\frac{1}{2m} \sum_k \Theta_k \int d^3 r \, \phi_k^\dagger(\mathbf{r}) \nabla^2 \phi_k(\mathbf{r}) , \tag{2.8}$$

---

[2] In all formulae $\hbar = 1$, but $e \neq 1 \neq m$ is used.

the external potential energy,

$$E_{\text{ext}}[n] = \int d^3r \, v_{\text{ext}}(\boldsymbol{r}) n(\boldsymbol{r}) , \qquad (2.9)$$

the Hartree energy,

$$E_{\text{H}}[n] = \frac{e^2}{2} \int d^3r \int d^3r' \, \frac{n(\boldsymbol{r}) \, n(\boldsymbol{r}')}{|\boldsymbol{r} - \boldsymbol{r}'|} , \qquad (2.10)$$

and the xc-energy $E_{\text{xc}}[n]$ which is defined by (2.7). In the case of polyatomic systems one has to add the electrostatic repulsion of the nuclei (or the ions in a pseudopotential framework),

$$E_{\text{ion}} = \sum_{\alpha<\beta=1}^{N_{\text{ion}}} \frac{Z_\alpha Z_\beta e^2}{|\boldsymbol{R}_\alpha - \boldsymbol{R}_\beta|} \qquad \Longleftrightarrow \qquad v_{\text{ext}}(\boldsymbol{r}) = - \sum_{\alpha=1}^{N_{\text{ion}}} \frac{Z_\alpha e^2}{|\boldsymbol{r} - \boldsymbol{R}_\alpha|} , \quad (2.11)$$

with the $\boldsymbol{R}_\alpha$ and Z_α denoting the nuclear (ionic) positions and charges. However, for the DFT formalism this last energy contribution is irrelevant, so that it is omitted in the subsequent discussion.

2.1.2   Motivation for Orbital-Dependent Functionals

The first question to be addressed is: Why would one think about using orbital-dependent functionals, given the tremendous success of the GGA? The answer to this question necessarily consists of a list of situations in which the GGA, which is by now the standard workhorse of DFT, fails.

Heavy Elements. The first problem to be mentioned here is the least important. When comparing the quality of GGA results from different regions of the periodic table one finds that there is a tendency of the GGA to loose accuracy with increasing nuclear charge (note that increasing charge automatically implies the presence of higher angular momentum). GGAs are known to be very accurate for light molecules, involving constituents from the first and second row. For these systems the GGA, which consistently stretches bond lengths (R_e) and reduces bond energies (D_e) compared with the LDA[3], corrects the LDA's underestimation of R_e and the accompanying overestimation of D_e. However, in the case of heavy constituents the LDA results are often rather close to the experimental numbers, so that the GGA overcorrects the LDA values. One example for this behavior is shown in Table 2.1, where the LDA and GGA values for the cohesive properties of gold are listed. In particular, the LDA lattice constant is already very accurate. When the gradient corrections are switched on, the lattice constant is expanded as usual, leading to a significant error. The same effect is observed for a number of $5d$ metals [26] and also for molecules containing fifth row elements [27,28].

---

[3] In this chapter the parameterization of [29] has been utilized for all explicit LDA calculations.

Table 2.1. Cohesive properties (equilibrium lattice constant, a_0, and cohesive energy, E_{coh}) of gold: PW91-GGA versus LDA results on the basis of fully relativistic LAPW calculations [26]. The nonrelativistic forms of the functionals [29,30] are compared with their relativistic counterparts (RLDA, RGGA) [31,32]

Au	a_0 (bohr)	$-E_{coh}$ (eV)
LDA	7.68	4.12
RLDA	7.68	4.09
GGA	7.87	2.91
RGGA	7.88	2.89
expt.	7.67	3.78

It seems worthwhile to emphasize that this deficiency of the GGA can not be explained by relativistic effects: The inclusion of relativistic corrections in the GGA [32] does not improve the results (see Table 2.1 – the fully relativistic T_s and thus the fully relativistic KS equations have been applied in all calculations). This observation suggests that the GGA has some difficulties with the treatment of higher angular momentum (d and f), similarly to the LDA [33].

Negative Ions. In contrast to the loss of accuracy for heavy elements, the second problem of the GGA, its failure for negative ions, is of qualitative nature. It originates from the (semi-)local density-dependence of the LDA and GGA exchange potential. The situation is most easily analyzed in the case of the LDA, for which one has

$$v_x^{LDA}(\boldsymbol{r}) = -\frac{(3\pi^2)^{\frac{1}{3}}e^2}{\pi} \, n(\boldsymbol{r})^{\frac{1}{3}} \; . \qquad (2.12)$$

In the asymptotic regime of finite systems, in which the density decays exponentially, one thus finds an exponential decay of v_x^{LDA},

$$n(\boldsymbol{r}) \xrightarrow[r \to \infty]{} e^{-\alpha r} \qquad \Longrightarrow \qquad v_x^{LDA}(\boldsymbol{r}) \xrightarrow[r \to \infty]{} e^{-\alpha r/3} \; . \qquad (2.13)$$

The same is true for the LDA correlation potential. Moreover, for neutral atoms the electrostatic potential of the nucleus cancels with the monopole term in v_H, (2.3). Consequently, the total v_s also decays faster than $1/r$. This implies that, within the framework of the LDA, a neutral atom does not exhibit a Rydberg series of excited states and thus is not able to bind an additional electron, i.e. to form a negative ion.

This problem is also present for all GGAs, whose potential typically depends on the first two gradients of the density,

$$v_x^{GGA}[n] = v_x^{GGA}(n, (\nabla n)^2, \nabla^2 n, \nabla n \cdot \nabla(\nabla n)^2) \; .$$

While the inclusion of these gradients can affect the asymptotic form of the exchange potential (for details see Sect. 2.3), the standard form of the GGA is incompatible with the $1/r$ behavior which is required to obtain a Rydberg series [34]. As a consequence, GGAs do not predict the existence of negative ions either.

How should v_x really look like in the asymptotic regime? The basic behavior of v_x is most easily illustrated by a two-electron system like the helium atom. In this case the exchange energy just has to cancel the self-interaction included in the Hartree term,

$$E_x^{He}[n] = -\frac{e^2}{4} \int d^3r \int d^3r' \, \frac{n(\mathbf{r})n(\mathbf{r}')}{|\mathbf{r} - \mathbf{r}'|} \, . \tag{2.14}$$

The functional derivative of $E_x^{He}[n]$ is then trivially given by

$$v_x^{He}(\mathbf{r}) = -\frac{e^2}{2} \int d^3r' \, \frac{n(\mathbf{r}')}{|\mathbf{r} - \mathbf{r}'|} \, . \tag{2.15}$$

Although the density decays exponentially, v_x^{He} asymptotically goes like $-1/r$,

$$v_x^{He}(\mathbf{r}) \xrightarrow{r \to \infty} -\frac{e^2}{|\mathbf{r}|} \, . \tag{2.16}$$

This $-1/r$ behavior of the exact v_x is found quite generally for all finite systems (see Sect. 2.2). The same statement then applies to the total v_s, as long as the system is neutral. Physically the reason for this result is very simple: If one electron moves sufficiently far away from the other electrons bound by the nucleus, it must experience the remaining net charge of the system, which consists of $N-1$ electrons and N protons. However, v_H, as defined by (2.3), still contains the Coulomb repulsion of the far out electron, which has to be eliminated by v_x. As a consequence of (2.16), the exact v_s generates a Rydberg series and is thus able to bind an additional electron[4].

This argument as well as (2.14) and (2.15) indicate that one needs a rather nonlocal exchange functional to reproduce the $-1/r$ behavior: The component of v_x which cancels the self-interaction in the Hartree potential must be as nonlocal as v_H itself, which is a quite nonlocal Coulomb integral.

Dispersion Forces. The LDA and GGA also fail to reproduce dispersion forces (one type of van der Waals forces). In this case the problem is due to the short-ranged nature of the LDA/GGA correlation functional. In the LDA the correlation energy density is simply given by the energy density e_c^{HEG} of the homogeneous electron gas (HEG), evaluated with the local density,

$$E_c^{LDA}[n] = \int d^3r \, e_c^{HEG}(n(\mathbf{r})) \, . \tag{2.17}$$

---

[4] Note, however, that (2.16) is only a necessary but not a sufficient criterium for the stability of a negative ion: Ultimately, the stability depends on the relative value of the total energies of the N and the $N+1$ electron systems.

One immediately realizes that only regions in space with non-vanishing density contribute to the correlation energy. Now consider two neutral closed-subshell atoms which are so far apart that there exists no overlap between the their densities, as depicted in Fig. 2.1. The density of this system is identical

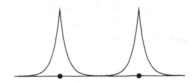

Fig. 2.1. Electronic density of two atoms at large separation

to the sum of the two atomic densities. This is the situation in which dispersion forces become important, as there is neither an electrostatic interaction between the two atoms nor can any bonding orbitals be formed. Only the attraction between virtual dipole excitations on the two atoms can lead to molecular bonding, i.e. the London dispersion force. In the LDA, however, any molecular bonding provided by E_c requires the atomic densities to overlap, as the binding energy must result from the nonlinear density dependence of E_c^{LDA}, $E_b = E_c^{LDA}[n_A + n_B] - E_c^{LDA}[n_A] - E_c^{LDA}[n_B]$. This means that dispersion forces can not be described by the LDA.

As in the case of negative ions this problem is not resolved by using the GGA,

$$E_c^{GGA}[n] = \int d^3r \, e_c^{GGA}(n, (\nabla n)^2, \nabla^2 n) . \tag{2.18}$$

Its correlation energy density $e_c^{GGA}(\mathbf{r})$ only takes into account the density in the immediate vicinity of $\mathbf{r}$. $e_c^{GGA}(\mathbf{r})$ thus vanishes wherever $n(\mathbf{r})$ vanishes. Neither the LDA nor the GGA can mediate the long-range force generated by virtual excitations. Not only the exact exchange functional is very nonlocal, but also the exact correlation functional.

Strongly Correlated Systems. The third class of systems for which the LDA and the GGA have fundamental problems are strongly correlated systems. The most prominent examples of such systems are the $3d$ transition metal monoxides MnO, FeO, CoO, and NiO. These systems, which crystallize in the rock salt structure, are insulating antiferromagnets of type II (Mott insulators). Both the LDA and the GGA, on the other hand, predict FeO and CoO to be metallic and by far underestimate the band gap in MnO and NiO [35–37]. This is illustrated in Fig. 2.2 in which the LDA band structure for FeO is plotted – the band structures obtained with the most frequently applied GGAs are rather similar to their LDA counterpart [36,37].

The origin of this problem is not yet really understood. There are only some indications where one might have to look: On the one hand, there exists

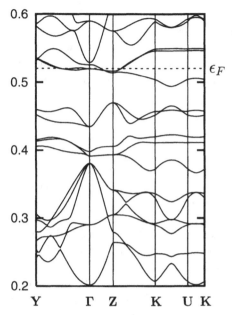

Fig. 2.2. Band structure of antiferromagnetic (type II) FeO within LDA obtained by plane-wave-pseudopotential calculation (the valence space of Fe includes the $3s$, $3p$, $3d$ and $4s$ states, $E_{\text{cut}} = 300$ Ry, 10 special k-points)

one variant of the GGA which predicts FeO and CoO to be antiferromagnetic insulators [37] (although with the size of the gaps being much too small). This GGA is the only functional of this type whose kernel has been optimized to reproduce the exact atomic exchange potentials as accurately as possible [38]. This points at the importance of an accurate exchange potential for describing Mott insulators. Furthermore, the explicitly self-interaction corrected form of the LDA also leads to the correct ground-states [39]. While no definitive conclusions are possible, these two results suggest that the inappropriate handling of the self-interaction is responsible for the failure of the LDA and the standard GGAs.

2.1.3 Basic Concept of Orbital-Dependent Functionals

This is the right point to clarify the meaning of the term orbital-dependent functional. The natural starting point for the discussion is the exact exchange E_{x} of DFT, which is the most simple functional of this type. The exact E_{x} is defined as the Fock expression written in terms of KS orbitals [40,8],

$$E_{\text{x}} := -\frac{e^2}{2} \sum_{kl} \Theta_k \Theta_l \int \mathrm{d}^3r \int \mathrm{d}^3r \, \frac{\phi_k^\dagger(\boldsymbol{r})\phi_l(\boldsymbol{r})\phi_l^\dagger(\boldsymbol{r}')\phi_k(\boldsymbol{r}')}{|\boldsymbol{r} - \boldsymbol{r}'|} \; . \tag{2.19}$$

This is the most appropriate definition as it guarantees the exact cancellation of the self-interaction energy contained in E_H, which has been identified as the origin of the problem of the LDA/GGA with negative ions (and also seems to be relevant for the description of Mott insulators). It automatically induces a corresponding definition of the correlation functional of DFT,

$$E_c := E_{xc} - E_x . \tag{2.20}$$

It must be emphasized that E_x and E_c are not identical with the exchange and correlation energies defined in conventional many-body theory. Although the functional form of E_x agrees with the exchange term of the Hartree-Fock (HF) approach, a difference originates from the orbitals inserted into the Fock expression: In (2.19) the KS orbitals are used, which are solutions of the KS equations (2.1) with their multiplicative potential v_s. The ϕ_k do not agree with the HF orbitals which satisfy the nonlocal HF equations. The difference between the resulting exchange energies as well as the difference between T_s and the full kinetic energy are absorbed into E_c.

The right-hand side of (2.19) is a density functional in the same sense as the kinetic energy T_s: The KS orbitals ϕ_k are uniquely determined by the density n, as n uniquely determines v_s (which is guaranteed by the Hohenberg-Kohn theorem [41] for noninteracting systems), which then allows the unambiguous calculation of the ϕ_k. E_x thus represents an implicit density functional, in contrast to the explicit density functionals LDA and GGA. This argument can be directly extended to the more general class of functionals $E_{xc}[\phi_k, \epsilon_k]$ which do not only depend on the occupied ϕ_k, but also on the unoccupied KS states and the KS eigenvalues, as v_s uniquely determines the complete KS spectrum.

The step from explicitly density-dependent to orbital-dependent xc-functionals is in some sense analogous to the transition from the Thomas-Fermi variational equation to the KS equations: In the latter transition the most important part of $E_{tot}[n]$, the kinetic energy, is recast in orbital-dependent form. The same concept is now applied to E_{xc}. In that sense, one can call orbital-dependent functionals a third generation of density functionals.

At this point, the idea of implicit xc-functionals might appear as a purely formal concept. In the following sections, however, it will be shown that implicit functionals can be used in practice.

2.2 Optimized Potential Method (OPM)

The most important question is how to calculate the multiplicative potential which corresponds to xc-functionals of the type (2.19). There are three distinct ways for the derivation of the basic equation which yields this potential. As all three are rather instructive, all of them will be gone through in the following, assuming the xc-functional to be of the general form $E_{xc}[\phi_k, \epsilon_k]$.

2.2.1 Direct Functional Derivative

The simplest way to derive the OPM equation is the transformation of the functional derivative (2.4) into derivatives with respect to ϕ_k and ϵ_k, using the chain rule for functional differentiation [18],

$$\frac{\delta E_{xc}[\phi_k, \epsilon_k]}{\delta n(r)} = \int d^3 r' \frac{\delta v_s(r')}{\delta n(r)} \sum_k \left\{ \int d^3 r'' \left[\frac{\delta \phi_k^\dagger(r'')}{\delta v_s(r')} \frac{\delta E_{xc}}{\delta \phi_k^\dagger(r'')} + c.c. \right] \right.$$
$$\left. + \frac{\delta \epsilon_k}{\delta v_s(r')} \frac{\partial E_{xc}}{\partial \epsilon_k} \right\} \tag{2.21}$$

(k is not restricted to the occupied states). Now one has expressed $\delta E_{xc}/\delta n$ in terms of quantities which can be evaluated: The functional derivatives of E_{xc} can be easily calculated for any explicit expression at hand. For instance, for E_x one finds

$$\frac{\delta E_x}{\delta \phi_k^\dagger(r')} = -e^2 \Theta_k \sum_l \Theta_l \phi_l(r') \int d^3 r \frac{\phi_l^\dagger(r) \phi_k(r)}{|r - r'|} \tag{2.22}$$

and $\partial E_x/\partial \epsilon_k = 0$. The functional derivatives $\delta \phi_k^\dagger/\delta v_s$ and $\delta \epsilon_k/\delta v_s$ are evaluated by varying v_s infinitesimally and looking how ϕ_k and ϵ_k react (via (2.1)). Using first order perturbation theory one obtains

$$\frac{\delta \phi_k^\dagger(r)}{\delta v_s(r')} = -\phi_k^\dagger(r') G_k(r', r) \tag{2.23}$$

$$\frac{\delta \epsilon_k}{\delta v_s(r)} = \phi_k^\dagger(r) \phi_k(r) , \tag{2.24}$$

with the Green's function

$$G_k(r, r') = \sum_{l \neq k} \frac{\phi_l(r) \phi_l^\dagger(r')}{\epsilon_l - \epsilon_k} . \tag{2.25}$$

It remains to deal with the factor $\delta v_s/\delta n$. The inverse of this quantity is the static response function of the KS auxiliary system, i.e. the KS response function,

$$\frac{\delta n(r)}{\delta v_s(r')} = \chi_s(r, r') = -\sum_k \Theta_k \phi_k^\dagger(r) G_k(r, r') \phi_k(r') + c.c. \tag{2.26}$$

So, if one multiplies (2.21) by χ_s, and integrates over r, one ends up with the OPM integral equation, a Fredholm equation of first kind,

$$\int d^3 r' \chi_s(r, r') v_{xc}(r') = \Lambda_{xc}(r) , \tag{2.27}$$

with the inhomogeneity given by

$$\Lambda_{xc}(\boldsymbol{r}) = \sum_k \left\{ - \int d^3r' \left[\phi_k^\dagger(\boldsymbol{r}) G_k(\boldsymbol{r}, \boldsymbol{r}') \frac{\delta E_{xc}}{\delta \phi_k^\dagger(\boldsymbol{r}')} + c.c. \right] + |\phi_k(\boldsymbol{r})|^2 \frac{\partial E_{xc}}{\partial \epsilon_k} \right\}.$$

(2.28)

Equation (2.27) is the central equation of the OPM. It allows the calculation of the multiplicative xc-potential for a given orbital- and eigenvalue-dependent functional E_{xc}. Note that (2.27) is linear in E_{xc}, so that each component of E_{xc} can be treated separately.

Any self-consistent KS calculation consists of the alternate solution of the KS equations (2.1) and the calculation of v_s from the resulting ϕ_k. Thus, at some point of this cycle, one has to evaluate v_{xc}. If this is an LDA or GGA potential, one just has to take the density and its derivatives and insert these quantities into some analytical formula. In the OPM, on the other hand, the solution of (2.27) replaces the insertion of n into the LDA or GGA functional.

2.2.2   Total Energy Minimization

The physics behind the OPM integral equation becomes more transparent in the second derivation of (2.27). This alternative derivation, which, in fact, represents the original approach [2,3], relies on energy minimization. Its starting point is a total energy functional given in terms of the KS orbitals and eigenvalues, $E_{tot}[\phi_k, \epsilon_k]$. As already pointed out, the Hohenberg-Kohn theorem for noninteracting particles guarantees that there is a unique relation between n and v_s. Thus, the standard minimization of E_{tot} with respect to n can be substituted by a minimization with respect to v_s,

$$\frac{\delta E_{tot}[\phi_k, \epsilon_k]}{\delta v_s(\boldsymbol{r})} = 0$$

(2.29)

(for fixed particle number). The derivative in (2.29) can be handled as in (2.21),

$$\frac{\delta E_{tot}[\phi_k, \epsilon_k]}{\delta v_s(\boldsymbol{r})} = \sum_k \left\{ \int d^3r' \left[\frac{\delta \phi_k^\dagger(\boldsymbol{r}')}{\delta v_s(\boldsymbol{r})} \frac{\delta E_{tot}}{\delta \phi_k^\dagger(\boldsymbol{r}')} + c.c. \right] + \frac{\delta \epsilon_k}{\delta v_s(\boldsymbol{r})} \frac{\partial E_{tot}}{\partial \epsilon_k} \right\}.$$

(2.30)

In addition to the ingredients which are already known, (2.30) contains the functional derivatives of E_{tot} with respect to ϕ_k and ϵ_k, which can be evaluated from (2.7)-(2.10),

$$\frac{\delta E_{tot}}{\delta \phi_k^\dagger(\boldsymbol{r})} = \Theta_k \left[-\frac{\nabla^2}{2m} + v_{ext}(\boldsymbol{r}) + v_H(\boldsymbol{r}) \right] \phi_k(\boldsymbol{r}) + \frac{\delta E_{xc}}{\delta \phi_k^\dagger(\boldsymbol{r})}$$

(2.31)

$$\frac{\partial E_{tot}}{\partial \epsilon_k} = \frac{\partial E_{xc}}{\partial \epsilon_k}.$$

(2.32)

One can then use the KS equations to rewrite $\delta E_{\text{tot}}/\delta\phi_k^\dagger$,

$$\frac{\delta E_{\text{tot}}}{\delta\phi_k^\dagger(\boldsymbol{r})} = \Theta_k\left[\epsilon_k - v_{\text{xc}}(\boldsymbol{r})\right]\phi_k(\boldsymbol{r}) + \frac{\delta E_{\text{xc}}}{\delta\phi_k^\dagger(\boldsymbol{r})} . \tag{2.33}$$

Insertion of (2.23) and (2.24) as well as (2.32) and (2.33) into (2.30) leads to

$$\sum_k \int \mathrm{d}^3r' \left\{ \phi_k^\dagger(\boldsymbol{r})\, G_k(\boldsymbol{r},\boldsymbol{r}')\left[\Theta_k\phi_k(\boldsymbol{r}')\left(v_{\text{xc}}(\boldsymbol{r}') - \epsilon_k\right) + \frac{\delta E_{\text{xc}}}{\delta\phi_k^\dagger(\boldsymbol{r}')}\right] + c.c.\right\}$$
$$+ \sum_k |\phi_k(\boldsymbol{r})|^2 \frac{\partial E_{\text{xc}}}{\partial\epsilon_k} = 0 . \tag{2.34}$$

After identification of the ingredients of $\chi_{\text{s}}(\boldsymbol{r},\boldsymbol{r}')$ and $\Lambda_{\text{xc}}(\boldsymbol{r})$, which show up in (2.34), and use of the orthogonality relation

$$\int \mathrm{d}^3r\, \phi_k^\dagger(\boldsymbol{r})\, G_k(\boldsymbol{r},\boldsymbol{r}') = \int \mathrm{d}^3r'\, G_k(\boldsymbol{r},\boldsymbol{r}')\, \phi_k(\boldsymbol{r}') = 0 \tag{2.35}$$

one again ends up with the OPM integral equation (2.27).

At first glance this derivation suggests that the x-only limit of the OPM is conceptually identical to the HF approach, as in this limit $E_{\text{tot}}[\phi_k,\epsilon_k]$ agrees with the HF energy functional. It thus seems worthwhile to emphasize the difference between the two schemes once again: The HF approach corresponds to a free minimization of the total energy functional with respect to the ϕ_k and ϵ_k. Equation (2.29), on the other hand, is not equivalent to a free minimization of E_{tot}: Rather the ϕ_k and ϵ_k have to satisfy the KS equations with their multiplicative total potential. This requirement represents a subsidiary condition to the minimization of E_{tot}. The subsidiary condition is actually implemented into the OPM equation via (2.33). In Sect. 2.3 this point will be investigated further from a quantitative point of view.

2.2.3 Invariance of the Density

The starting point of this third derivation of the OPM integral equation is the identity of the KS density n_{s} with the density n of the interacting system [42,43],

$$n_{\text{s}}(\boldsymbol{r}) - n(\boldsymbol{r}) = 0 . \tag{2.36}$$

Note that the relation (2.36) relies on the complete framework of the Hohenberg-Kohn and KS formalism. In particular, it implies the application of the minimum principle for the total energy, which underlies all of ground-state DFT. This common DFT background provides the link between (2.36) and the arguments of Sects. 2.2.1 and 2.2.2.

The KS density n_{s} can now be written in terms of the Green's function G_{s} of the KS system, while the interacting n can be expressed in terms of the 1-particle Green's function G of the interacting system,

$$-\mathrm{i}\,\mathrm{tr}\left\{G_{\text{s}}(\boldsymbol{r}t, \boldsymbol{r}t^+) - G(\boldsymbol{r}t, \boldsymbol{r}t^+)\right\} = 0 . \tag{2.37}$$

Here t^+ indicates an infinitesimal positive time-shift of t, i.e. $t^+ \equiv \lim_{\epsilon \to 0}(t + |\epsilon|)$. The KS and the full many-body Green's function are defined by the ground-state expectation values of the time-ordered product of the corresponding field operators, $\hat{\psi}_0$ and $\hat{\psi}$,

$$G_s(rt, r't') = -\mathrm{i}\,\langle \Phi_0 | T\hat{\psi}_0(rt)\hat{\psi}_0^\dagger(r't') | \Phi_0 \rangle \qquad (2.38)$$

$$G(rt, r't') = -\mathrm{i}\,\langle \Psi_0 | T\hat{\psi}(rt)\hat{\psi}^\dagger(r't') | \Psi_0 \rangle \,, \qquad (2.39)$$

with $|\Phi_0\rangle$ being the KS ground-state (i.e. a Slater determinant of the KS orbitals ϕ_k) and $|\Psi_0\rangle$ denoting the true ground-state of the interacting system.

The interacting Green's function obeys the well-known Dyson equation,

$$G(1,2) = G_0(1,2) + \int \mathrm{d}3\,\mathrm{d}4\, G_0(1,3)\Sigma(3,4)G(4,2)\,. \qquad (2.40)$$

Here G_0 represents the Green's function of electrons which just experience the external potential v_ext, Σ is the full self-energy of the interacting system,

$$\Sigma(3,4) = \Sigma_\mathrm{xc}(3,4) + \delta(3,4)v_\mathrm{H}(r_3)\,, \qquad (2.41)$$

and the Harvard notation $1 = (r_1 t_1)$, $\int \mathrm{d}3 = \int \mathrm{d}^3r_3 \int \mathrm{d}t_3$ and $\delta(3,4) = \delta^{(3)}(r_3 - r_4)\delta(t_3 - t_4)$ has been used. On the other hand, the KS Green's function satisfies a Dyson equation in which the self-energy is simply given by $v_\mathrm{H} + v_\mathrm{xc}$,

$$G_s(1,2) = G_0(1,2) + \int \mathrm{d}3\,\mathrm{d}4\, G_0(1,3)\delta(3,4)[v_\mathrm{H}(r_3) + v_\mathrm{xc}(r_3)]G_s(4,2) \quad (2.42)$$

($v_\mathrm{H} + v_\mathrm{xc}$ is the only self-energy insertion that shows up in the case of an effective single-particle system). If one now subtracts (2.40) and (2.42) from each other one ends up with a relation between G and G_s,

$$G(1,2) = G_s(1,2) + \int \mathrm{d}3\,\mathrm{d}4\, G_s(1,3)\Big[\Sigma_\mathrm{xc}(3,4) - \delta(3,4)v_\mathrm{xc}(r_3)\Big]G(4,2)\,. \qquad (2.43)$$

Equation (2.43) is a Dyson equation whose irreducible kernel is given by the difference between the full self-energy and the KS self-energy. Upon insertion of (2.43) into (2.37) one obtains

$$-\mathrm{i}\,\mathrm{tr} \int \mathrm{d}3\mathrm{d}4\, G_s(1,3)\Big[\Sigma_\mathrm{xc}(3,4) - \delta(3,4)v_\mathrm{xc}(r_3)\Big]G(4,1^+) = 0\,. \quad (2.44)$$

Equation (2.44) is a complicated integral equation connecting the KS Green's function, the xc-component of the full self-energy, the xc-potential and the full Green's function. Does this relation have anything to do with the OPM equation (2.27)? The first step towards an answer to this question is provided by a repeated use of the Dyson equation (2.43). After insertion of (2.43) the leading term in (2.44) contains the product of $G_s(1,3)$ with $G_s(4,1)$. Partial evaluation of the 4-integration then yields,

$$\int d^3 r_3 \int dt_3\, \chi_s(1,3) v_{xc}(\boldsymbol{r}_3) = -i \operatorname{tr} \int d3\, d4\, G_s(1,3) \Sigma_{xc}(3,4) G_s(4,1^+)$$

$$-i \operatorname{tr} \int d3\, d4\, G_s(1,3) \Big[\Sigma_{xc}(3,4) - \delta(3,4) v_{xc}(\boldsymbol{r}_3) \Big]$$

$$\times \int d5\, d6\, G_s(4,5) \Big[\Sigma_{xc}(5,6) - \delta(5,6) v_{xc}(\boldsymbol{r}_5) \Big] G(6,1^+) ,$$
$$(2.45)$$

where the KS response function

$$\chi_s(1,3) = -i \Big[\langle \Phi_0 | T \hat{\psi}_0^\dagger(\boldsymbol{r}_1 t_1) \hat{\psi}_0(\boldsymbol{r}_1 t_1) \hat{\psi}_0^\dagger(\boldsymbol{r}_3 t_3) \hat{\psi}_0(\boldsymbol{r}_3 t_3) | \Phi_0 \rangle - n(\boldsymbol{r}_1) n(\boldsymbol{r}_3) \Big]$$

$$= -i \operatorname{tr} \big[G_s(1,3) G_s(3,1) \big] \tag{2.46}$$

has been introduced in order to make the similarity of (2.45) with (2.27) more apparent. The left-hand side of (2.45) is identical with that of the OPM equation, if one performs the dt_3 integration and identifies the static response function (zero-frequency limit of the Fourier transform of $\chi_s(1,3)$),

$$\int d t_3\, \chi_s(1,3) = \int d t_3\, \chi_s(\boldsymbol{r}_1, \boldsymbol{r}_3, t_1 - t_3) = \chi_s(\boldsymbol{r}_1, \boldsymbol{r}_3, \omega = 0) \equiv \chi_s(\boldsymbol{r}_1, \boldsymbol{r}_3) .$$

On the other hand, the right-hand side of (2.45) is still quite different from the inhomogeneity (2.28). In fact, the right-hand side depends on v_{xc} itself, so that (2.45) represents a nonlinear integral equation for v_{xc}.

Where does this fundamental difference to (2.27) come from? To answer this question one has to remember that in the first two approaches some arbitrary orbital-dependent E_{xc} has been assumed, i.e. the form of E_{xc} has not been specified. On the other hand, in the present approach the use of the Dyson equation for both the KS and the interacting system automatically implies the use of the exact E_{xc}. In order to make closer contact between the first two and this third derivation, one thus has to study the exact E_{xc} in more detail. This will be the subject of Sect. 2.4. In the present section the comparison of (2.45) with (2.27) will for simplicity be restricted to the x-only limit, which corresponds to a lowest order expansion of E_{xc} in the coupling constant e^2. In this limit the right-hand side of (2.45) reduces to

$$-i \operatorname{tr} \int d3\, d4\, G_s(1,3) \Sigma_x(3,4) G_s(4,1^+) ,$$

as each factor of Σ_{xc} or v_{xc} introduces an additional factor of e^2. Insertion of the exchange contribution Σ_x to the full self-energy, i.e. the standard 1-loop self-energy diagram, then leads to the exchange component of (2.28), obtained by use of (2.22). One has thus explicitly verified that in the x-only limit (2.45) agrees with the standard OPM equation.

2.2.4   Exact Relations Related to OPM

Before investigating the correlation component of E_{xc} in more detail it seems worthwhile to list a few exact relations which emerge from an analysis of the OPM integral equation. One first recognizes that the OPM equation determines v_{xc} only up to an additive constant. In fact, as norm-conservation requires that

$$\int d^3r \, \chi_s(\boldsymbol{r}, \boldsymbol{r}') = \int d^3r' \, \chi_s(\boldsymbol{r}, \boldsymbol{r}') = 0 \,, \qquad (2.47)$$

one can add any constant to v_{xc} without altering the left-hand side of (2.27). In the process of solving the OPM equation one thus has to ensure the normalization of v_{xc} in some explicit form. For finite systems one usually requires v_{xc} to vanish far outside, $\lim_{|\boldsymbol{r}| \to \infty} v_{xc}(\boldsymbol{r}) = 0$. One way to implement this condition is the use of an identity for the highest occupied KS state [44] which results from this normalization. In the case of the exchange this identity reads

$$\int d^3r \, v_x(\boldsymbol{r}) |\phi_F(\boldsymbol{r})|^2 = \frac{1}{2} \int d^3r \, \phi_F^\dagger(\boldsymbol{r}) \frac{\delta E_x}{\delta \phi_F^\dagger(\boldsymbol{r})} + \text{c.c.}$$

$$= -e^2 \sum_l \Theta_l \int d^3r \int d^3r' \, \frac{\phi_F^\dagger(\boldsymbol{r}) \phi_l(\boldsymbol{r}) \phi_l^\dagger(\boldsymbol{r}') \phi_F(\boldsymbol{r}')}{|\boldsymbol{r} - \boldsymbol{r}'|} \,, \qquad (2.48)$$

where ϕ_F denotes the highest occupied orbital (the Fermi level is assumed to be non-degenerate). Equation (2.48) allows the unambiguous normalization of v_x in the case of finite systems. An analogous, though more complicated statement is available for v_c [45]. For solids, on the other hand, it is more convenient to fix the average of v_{xc} in the unit cell.

It has already been mentioned that for physical reasons the exact exchange potential of finite systems must asymptotically behave as

$$v_x(\boldsymbol{r}) \xrightarrow[r \to \infty]{} -\frac{e^2}{r} \,. \qquad (2.49)$$

It is thus very pleasing that one finds exactly this behavior for the solution of (2.27): Equation (2.49) can be verified by an examination of the x-only OPM integral equation for large r, requiring the standard normalization $\lim_{r \to \infty} v_x(\boldsymbol{r}) = 0$, i.e. the validity of (2.48) [3]. Equation (2.49) provides an alternative to (2.48) for the normalization of v_x.

It seems worthwhile to point out that the behavior (2.49) can not be obtained by differentiation of the asymptotic form of the exact exchange energy density e_x. The asymptotic form of e_x follows from (2.19) [46],

$$e_x(\boldsymbol{r}) \xrightarrow[r| \to \infty]{} -\frac{e^2 n(\boldsymbol{r})}{2r} \,. \qquad (2.50)$$

Given the relation between e_x and v_x in the case of the LDA, one might thus be tempted to expect

$$\frac{de_x(\boldsymbol{r})}{dn(\boldsymbol{r})} \xrightarrow[r \to \infty]{} -\frac{e^2}{2r}$$

to be the asymptotic behavior of v_x. However, E_x is a nonlocal functional, so that de_x/dn has nothing to do with $\delta E_x/\delta n(r)$. In fact, (2.49) and (2.50) directly reflect the Coulomb integral structure of the self-interaction part of E_x.

One can also establish a necessary condition for the eigenvalue-dependence of E_{xc} [47]. In fact, direct integration over (2.27) yields

$$\int d^3r \int d^3r' \, \chi_s(r, r') \, v_{xc}(r') = \sum_k \frac{\delta E_{xc}}{\delta \epsilon_k} = 0 \qquad (2.51)$$

(provided that the integral over r exists and that the integrations over r, r' and the summation over k can be interchanged).

The OPM leads back to the conventional functional derivative $v_{xc} = \delta E_{xc}/\delta n$ for explicitly density-dependent expressions. In this case E_{xc} depends on the ϕ_k only via n, so that (2.27) reduces to

$$\begin{aligned}
\int d^3r' \, \chi_s(r, r') \, v_{xc}(r') &= -\sum_k \int d^3r' \, \phi_k^\dagger(r) G_k(r, r') \frac{\delta E_{xc}[n]}{\delta \phi_k^\dagger(r')} + c.c. \\
&= -\sum_k \int d^3r' \, \phi_k^\dagger(r) G_k(r, r') \phi_k(r') \frac{\delta E_{xc}[n]}{\delta n(r')} + c.c. \\
&= \int d^3r' \, \chi_s(r, r') \frac{\delta E_{xc}[n]}{\delta n(r')} \ .
\end{aligned}$$

If one now multiplies both sides by χ_s^{-1} one recovers the original definition of v_{xc}.

One further limit of (2.27) appears to be worth a comment: If there is only one occupied orbital ($k = F$), the exchange component of (2.27) reads

$$\phi_F^\dagger(r) \int d^3r' \, G_F(r, r') \phi_F(r') \left[v_x(r') + e^2 \int d^3r'' \frac{\phi_F^\dagger(r'') \phi_F(r'')}{|r'' - r'|} \right] + c.c. = 0$$

(upon insertion of (2.22)). One thus easily identifies

$$v_x(r) = -e^2 \int d^3r' \frac{|\phi_F(r')|^2}{|r' - r|} = -\frac{e^2}{2} \int d^3r' \frac{n(r')}{|r' - r|} \qquad (2.52)$$

as solution of the OPM integral equation for spin-saturated two-electron systems, in perfect agreement with (2.15). For these systems the exchange potential just has to eliminate the self-interaction of the electrons, but does not include any Pauli repulsion among equal spins.

2.2.5 Krieger–Li–Iafrate Approximation

One has now reached the point at which it is clear that, as a matter of principle, one can handle orbital-dependent functionals in a fashion consistent with DFT. Moreover, the subsequent sections of this review will show that the OPM integral equation can also be solved in practice. However, in view of the complicated structure of (2.27), (2.28) and, in particular, of (2.45), the question quite naturally arises how efficient the OPM is? The answer obviously depends on the system under consideration and on the numerical implementation of the OPM. Nevertheless, as a rule of thumb, one might say that OPM calculations are essentially one or two orders of magnitude less efficient than the corresponding GGA calculations. Consequently, an approximate (semi-analytical) solution of the OPM integral equation is of definite interest.

The main reason for the inefficiency of the OPM is the presence of the Green's function (2.25) both in the response function (2.26) and in the inhomogeneity (2.28). This Green's function depends on the complete KS spectrum, not just the occupied states. A full solution of (2.27) thus requires the evaluation and, perhaps, the storage of all occupied and unoccupied KS states.

Is there a way to avoid this evaluation? Indeed, such a procedure has been suggested by Krieger, Li and Iafrate (KLI) [4]. The idea is to use a closure approximation for the Green's function, i.e. to approximate the eigenvalue difference in the denominator of (2.25) by some average $\Delta\bar{\epsilon}$ [2,4],

$$G_k(\boldsymbol{r},\boldsymbol{r}') \approx \sum_{l \neq k} \frac{\phi_l(\boldsymbol{r})\phi_l^\dagger(\boldsymbol{r}')}{\Delta\bar{\epsilon}} = \frac{\delta^{(3)}(\boldsymbol{r}-\boldsymbol{r}') - \phi_k(\boldsymbol{r})\phi_k^\dagger(\boldsymbol{r}')}{\Delta\bar{\epsilon}}. \tag{2.53}$$

Insertion into the OPM integral equation leads to

$$v_{\mathrm{xc}}(\boldsymbol{r}) = \frac{1}{2n(\boldsymbol{r})} \sum_k \left\{ \left[\phi_k^\dagger(\boldsymbol{r}) \frac{\delta E_{\mathrm{xc}}}{\delta\phi_k^\dagger(\boldsymbol{r})} + c.c. \right] + |\phi_k(\boldsymbol{r})|^2 \left[\Delta v_k - \Delta\bar{\epsilon}\frac{\partial E_{\mathrm{xc}}}{\partial\epsilon_k} \right] \right\}$$

$$\Delta v_k = \int d^3r \left\{ \Theta_k |\phi_k(\boldsymbol{r})|^2 v_{\mathrm{xc}}(\boldsymbol{r}) - \phi_k^\dagger(\boldsymbol{r})\frac{\delta E_{\mathrm{xc}}}{\delta\phi_k^\dagger(\boldsymbol{r})} \right\} + c.c. \,. \tag{2.54}$$

This approximation is completely unambiguous as soon as E_{xc} is independent of ϵ_k. On the other hand, for eigenvalue-dependent E_{xc} the presence of $\partial E_{\mathrm{xc}}/\partial\epsilon_k$ introduces a new energy scale via $\Delta\bar{\epsilon}$. Given the initial idea of the closure approximation, however, it is obvious that this term should be neglected. The only situation in which one can seriously investigate the consequences of this step is the relativistic exchange [47]. In this case neglect of the $\partial E_{\mathrm{xc}}/\partial\epsilon_k$ contribution represents an excellent approximation. The KLI approximation is thus always understood to imply the neglect of the $\partial E_{\mathrm{xc}}/\partial\epsilon_k$ term,

$$v_{\mathrm{xc}}^{\mathrm{KLI}}(\boldsymbol{r}) = \frac{1}{2n(\boldsymbol{r})} \sum_k \left\{ \left[\phi_k^\dagger(\boldsymbol{r}) \frac{\delta E_{\mathrm{xc}}}{\delta\phi_k^\dagger(\boldsymbol{r})} + c.c. \right] + |\phi_k(\boldsymbol{r})|^2 \Delta v_k^{\mathrm{KLI}} \right\}. \tag{2.55}$$

A careful look at (2.54) and (2.55) shows that one has not yet found a full solution of the problem, as $v_{\mathrm{xc}}^{\mathrm{KLI}}$ appears both on the left-hand and on the right-hand side of (2.55). Fortunately, one can recast (2.54) and (2.55) as a set of linear equations which allow the determination of Δv_k without prior knowledge of $v_{\mathrm{xc}}^{\mathrm{KLI}}$ [4], thus providing an analytical solution of the integral equation (2.55). Alternatively, one can iterate (2.54) and (2.55) until self-consistency, starting with some approximation for $\Delta v_k^{\mathrm{KLI}}$, e.g. obtained from the LDA.

When applied to the exact exchange, the KLI scheme is as efficient as a Hartree-Fock calculation, and often only slightly less efficient than a GGA calculation. At this point one should nevertheless keep in mind that the KLI approximation only speeds up the calculation of G_k, but not that of the other ingredients of the OPM equation. The most time-consuming step in a KLI calculation is usually the evaluation of $\delta E_{\mathrm{xc}}/\delta \phi_k^\dagger$: As soon as the exact exchange is used the evaluation of Slater integrals is required, which usually costs more time than the calculation of density gradients.

The KLI approximation preserves both the KLI identity (2.48) and the asymptotic behavior of v_{x}, (2.49) (for finite systems). It is exact for spin-saturated two-electron systems, i.e. it also satisfies (2.52). Moreover, all applications available so far point at the rather high accuracy of this approximation, at least in the case of the exact exchange (see Sect. 2.3).

2.3 Exchange-Only Results

Before addressing the issue of correlation in more detail, it is instructive to study the x-only limit from a quantitative point of view. This analysis serves two purposes: The first is to assert the accuracy of the KLI approximation. As is clear from the discussion of Sect. 2.2.5, any large-scale application of orbital-dependent functionals will have to rely on the efficiency of the KLI approximation. One thus has to make sure that this approach yields reasonable results at least for the simplest orbital-dependent functional, the exact E_{x}. The second aim of this section is to demonstrate that orbital-dependent functionals are worth the increased computational effort, i.e. that they in fact yield improvements over the standard functionals.

2.3.1 Accuracy of the KLI Approximation

In Table 2.2 the x-only ground-state energies of closed-subshell atoms resulting from different DFT methods are compared with the corresponding HF values (All calculations were performed fully numerically, relying on finite differences methods). In the first column the energies obtained by solution of the full OPM equation (2.27) for the exchange (2.19) are given. In the DFT context, this exact handling of the exact exchange functional provides the reference data for the x-only limit. For all other methods the energies are given relative to this reference standard. Among these methods the KLI approximation for the exact exchange is of primary interest here. The first

Table 2.2. Exchange-only ground-state energies of closed-subshell atoms: Self-consistent OPM results [48] versus KLI, LDA, PW91-GGA [30] and HF [49] energies (all energies in mhartree)

Atom	E_{tot} OPM	$E_{\text{tot}} - E_{\text{tot}}^{\text{OPM}}$			
		KLI	LDA	GGA	HF
He	−2861.7	0.0	138.0	6.5	0.0
Be	−14572.4	0.1	349.1	18.2	−0.6
Ne	−128545.4	0.6	1054.7	−23.5	−1.7
Mg	−199611.6	0.9	1362.8	−0.5	−3.1
Ar	−526812.2	1.7	2294.8	41.2	−5.3
Ca	−676751.9	2.2	2591.8	25.7	−6.3
Zn	−1777834.4	3.7	3924.5	−252.6	−13.8
Kr	−2752042.9	3.2	5176.8	−18.4	−12.0
Sr	−3131533.4	3.6	5535.4	−8.8	−12.2
Pd	−4937906.0	4.5	6896.0	−65.2	−15.0
Cd	−5465114.4	6.0	7292.6	−31.9	−18.7
Xe	−7232121.1	6.1	8463.8	54.9	−17.3
Ba	−7883526.6	6.5	8792.5	15.7	−17.3
Yb	−13391416.3	10.0	10505.6	−852.4	−39.9
Hg	−18408960.5	9.1	13040.4	−221.5	−31.0
Rn	−21866745.7	8.5	14424.3	8.3	−26.5
Ra	−23094277.9	8.7	14807.2	0.5	−25.8
No	−32789472.7	12.9	17202.9	−373.1	−39.5

observation is that the KLI energies are extremely close to the full OPM energies. For helium the KLI approximation is exact, as explained in Sect. 2.2.5. All other KLI energies are higher than the corresponding OPM values, consistent with the fact that the full OPM generates that potential which minimizes the energy expression at hand. The deviation of the KLI approximation systematically increases with the atomic size. Nevertheless, even for very heavy atoms it is still no larger than 15 mhartree. An idea of the relevance of this error is obtained by comparing with the corresponding LDA and GGA energies, which are also listed in Table 2.2. Even the GGA values, which drastically improve on the LDA energies, are more than an order of magnitude further away from the exact OPM data than the KLI numbers.

The next comparison to be made is that of OPM and Hartree-Fock results. In Sect. 2.2.2 it has been emphasized that the x-only OPM represents a restricted Hartree-Fock energy minimization: One minimizes the same energy expression, but under the subsidiary condition of having a multiplicative exchange potential. How important is this subsidiary condition? As Table 2.2 shows, the differences are rather small. For He the OPM energy is identical with the HF value, as in this case the HF equation can be trivially recast as a KS equation with the OPM exchange potential (2.52). Moreover, even for the heaviest elements the differences between OPM and HF energies are

Table 2.3. Exchange-only ground-state energies of diatomic molecules: Self-consistent OPM [50] results versus KLI [51], LDA and HF [52] energies at the experimental bond lengths, R_e (all energies in mhartree)

	state	R_e (bohr)	$-E_{tot}$ KLI	$E_{tot} - E_{tot}^{KLI}$ OPM	LDA	HF
H_2	$^1\Sigma$	1.400	1133.6	0.5	89.9	0.0
Li_2	$^1\Sigma$	5.046	14870.4		473.7	-1.2
Be_2	$^1\Sigma$	4.600	29127.4		666.2	-6.3
B_2	$^3\Sigma$	3.003	49085.2		823.6	
C_2	$^1\Sigma$	2.348	75394.0		956.3	
N_2	$^1\Sigma$	2.075	108985.1	5.6	1229.0	-8.5
O_2	$^3\Sigma$	2.281	149681.3	11.5	1447.0	
F_2	$^1\Sigma$	2.678	198760.2	16.3	1703.3	
LiH	$^1\Sigma$	3.014	7986.8		282.6	-0.5
BH	$^1\Sigma$	2.336	25129.0		499.1	-2.6
NH	$^3\Sigma$	2.047	54982.9	3.4	711.3	
FH	$^1\Sigma$	1.733	100067.5	11.0	916.3	-3.3
BF	$^1\Sigma$	2.386	124162.1		1312.1	-6.8
CO	$^1\Sigma$	2.132	112783.3	6.7	1252.5	-7.7
NO	$^2\Pi$	2.175	129295.5		1336.5	

below 40 mhartree. The additional variational freedom of the HF approach thus appears to be of very limited importance. The x-only OPM is in many respects physically equivalent to the Hartree-Fock approximation.

This statement is corroborated by Table 2.3, in which the x-only ground-state energies of a number of diatomic molecules are presented (evaluated at the experimental bond lengths). In Table 2.3 the KLI energies [51] are used as reference numbers, for a reason that will become clear in a moment. All other energies are given relative to the KLI values. If one compares the KLI and HF energies one finds, as expected, that the latter energies are always lower – with the exception of H_2, as for this spin-saturated two-electron system both energies must be identical.

On the other hand, the full OPM results [50] are energetically higher than the KLI data, although the OPM by construction produces the optimum exchange potential. How can that happen? The answer is hidden in the technical details of the calculations. The HF results were obtained fully numerically, using large real-space grids [52]. All DFT calculations rely on basis set expansions. In the case of the KLI (and LDA) calculations extremely large two-center basis sets have been used [51], so that the KLI numbers are essentially converged with respect to the basis set size. On the other hand, the OPM results were obtained with standard Gaussian basis sets of more modest size, so that the basis set limit is not yet reached. Clearly, converged OPM energies must lie between the KLI and the HF numbers. One can therefore conclude that the error of the KLI approximation is smaller than the impact

Table 2.4. Exchange-only ionization potentials of atoms: Self-consistent OPM results versus KLI, LDA, PW91-GGA and HF data. Also given is the highest occupied eigenvalue ϵ_{HOMO} obtained within the OPM (all energies in mhartree)

	$-\epsilon_{HOMO}$ OPM	IP OPM	KLI	IP-IP^{OPM} LDA	GGA	HF
He	918	862	0	−51	4	0
Be	309	295	0	−14	6	1
Mg	253	242	0	−4	12	1
Ca	196	188	0	1	12	0
Sr	179	171	0	3	13	0
Cu	240	231	−2	47	54	5
Ag	222	215	−1	36	41	3
Au	223	216	−2	38	42	2
Li	196	196	0	−11	4	0
Na	182	181	0	−2	10	1
K	148	147	0	2	10	0
Rb	138	137	0	4	12	1
Cs	124	123	0	5	11	0
Zn	293	276	0	34	44	5

of different basis sets. In other words: The appropriate choice of the basis set is more important than the handling of the OPM integral equation. Compared with the full OPM, the KLI approximation either allows to speed up molecular calculations (keeping the basis set fixed) or to gain higher accuracy by enlarging the basis set.

Until now total energies have been considered. However, the physical and chemical properties usually depend on energy differences. In Table 2.4 the most simple energy difference, namely the ionization potential (IP), is studied for atoms. Again the KLI results are extremely close to the OPM data, which agree very well with the Hartree-Fock IPs. On the other hand, one finds the (well-known) errors in the case of the LDA and the GGA.

The most critical energy difference one can look at for atoms is the electron affinity (EA). In Table 2.5 the EAs of F^- and Na^- as prototype negative ions are listed together with the highest occupied eigenvalues for the full OPM and the KLI approximation. One first should note the mere existence of these systems within the OPM [53], as negative ions can not be handled by the LDA and the GGA. This deficiency of the conventional density functionals, which was an important motivation for studying implicit functionals (Sect. 2.1.2), is automatically resolved by use of the exact exchange. The existence of negative ions is a direct consequence of the $-1/r$ behavior of the exact exchange potential. As the KLI potential is particularly close to the full OPM potential in the valence regime, the KLI EAs are almost identical to the corresponding OPM values.

Table 2.5. Exchange-only electron affinities of atoms: Self-consistent KLI versus OPM results. Also given is the highest occupied eigenvalue ϵ_{HOMO} (all energies in mhartree)

Atom	Method	$-\epsilon_{HOMO}$	EA
F$^-$	OPM	181.0	48.5
	KLI	180.4	48.5
Na$^-$	OPM	13.3	58.4
	KLI	13.2	58.3

One next observes the huge difference between the EA and the highest occupied eigenvalue (ϵ_{HOMO}). This discrepancy is somewhat surprising given the fact that the IPs of neutral atoms are in reasonable agreement with the corresponding ϵ_{HOMO} (see Table 2.4) and that one can prove that the exact EA is identical with the exact ϵ_{HOMO} [54] (including correlation). However, one has to keep in mind that the data in Table 2.5 correspond to the x-only limit. The difference between the EA and ϵ_{HOMO} simply reflects the important role which correlation plays for negative ions. For the same reason the x-only EAs should not be expected to be close to the experimental EAs.

In the quantum chemical context the most interesting quantities are the spectroscopic constants. Corresponding data for some diatomic molecules are given in Table 2.6. As full OPM results for these quantities are not yet available, the KLI numbers can only be compared with Hartree-Fock data. However, for each individual molecular geometry the exact OPM energy must be somewhere in between the KLI and the HF energy (for fixed basis set). Consequently, the OPM energy surface lies in between the KLI and the HF surface. As long as the latter two surfaces are very close, one can also be sure that the KLI and OPM results agree very well. This is exactly what one finds: The KLI and HF spectroscopic constants (as a measure for the energy surface) show very good agreement, in particular if one takes into account that not all HF results in Table 2.6 might be fully converged with respect to the basis set size. In conclusion, one can state that, in the x-only limit, the KLI results are essentially identical with the OPM values, which, in turn, are identical to the HF data.

Until now only global quantities like total energies and energy differences have been considered. However, the OPM also offers the possibility to analyze local quantities like the exchange potential. In Fig. 2.3 the exchange potential of neon is shown. The only difference between the full OPM result and the KLI potential is found in the transition region from the K- to the L-shell, where the shell oscillation of v_x^{KLI} is not as pronounced as that of the exact potential. However, this oscillation has little impact on total energies. For large r both potentials go like $-1/r$, i.e. both potentials are self-interaction free. On the other hand, the LDA and GGA curves, which are rather close to each other in the relevant regime, differ substantially from the exact v_x: Consistent with the

Table 2.6. Exchange-only spectroscopic constants of diatomic molecules: Self-consistent KLI [51] versus HF [55,56] results

	method	R_e (bohr)	D_e (eV)	ω (cm^{-1})
H_2	KLI	1.386	3.638	4603
	HF	1.386	3.631	4583
Li_2	KLI	5.266	0.168	338
	HF	5.224	0.160	316
B_2	KLI	3.068	0.608	972
	HF	3.096	0.75	939
C_2	KLI	2.332	0.281	1933
	HF		0.38	1912
N_2	KLI	2.011	4.972	2736
	HF	2.04	4.952	2738
O_2	KLI	2.184	1.441	1981
	HF	2.21	1.455	2002
F_2	KLI	2.496	-1.607	1283
	HF	2.508	-1.627	1257
LiH	KLI	3.037	1.483	1427
	HF	3.038	1.462	1406
FH	KLI	1.694	4.203	4501
	HF	1.695	4.197	4472
CO	KLI	2.080	7.530	2444
	HF	2.105	7.534	2439
Cl_2	KLI	3.727	(1.083)	613
	HF	3.732	(1.23)	614

argument in Sect. 2.1.2, they decay much more rapidly for large r. Moreover, they show no shell oscillation at all (see also Sect. 2.3.2). Figure 2.3 explains the findings in Tables 2.2–2.6 from a microscopic perspective.

At this point a side remark seems appropriate. All potentials shown in Fig. 2.3 originate from self-consistent calculations within the corresponding schemes. One might then ask how these curves change if the same density (and thus the same orbitals) are used for the evaluation of the different functionals? This issue is addressed in Fig. 2.4 in which the solution of the OPM integral equation on the basis of three different sets of orbitals is plotted.

In addition to the exact x-only orbitals used for Fig. 2.3 also the exact KS orbitals [57] and the LDA orbitals are inserted into (2.22)–(2.28). It turns out that the three solutions are almost indistinguishable. The origin of the orbitals (and thus of the density) is much less important for the structure of atomic v_{xc} than the functional form of E_{xc}. In other words: Fig. 2.3 would look very similar if all functionals were evaluated with the same density.

Returning to the accuracy of the KLI approximation, Fig. 2.5 provides a comparison analogous to Fig. 2.3 for the case of a solid.

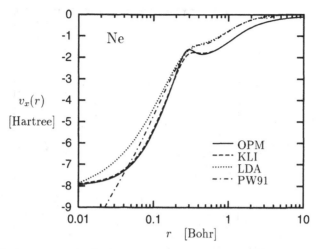

Fig. 2.3. Exchange potential of Ne: Self-consistent OPM, KLI, LDA and PW91-GGA results

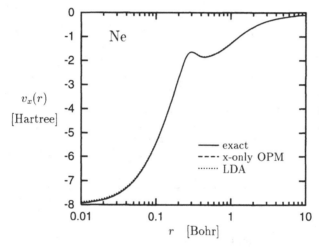

Fig. 2.4. Exchange potential of Ne: Importance of self-consistency. Solution of OPM equation with exact KS, x-only OPM and LDA orbitals

The exchange potential of bulk aluminum is plotted along the [110] direction. As all potentials originate from pseudopotential calculations, the attractive part of v_x associated with the core electrons is missing in Fig. 2.5 – the comparison completely focuses on the delocalized valence states of the metal. Again the KLI approximation is rather close to the OPM potential. The agreement is particularly convincing in view of the GGA result: The gradient corrections to the LDA even go into the wrong direction.

Finally, Fig. 2.6 shows the most significant deviation of a KLI result from the full OPM solution observed so far. In Fig. 2.6 the Colle-Salvetti corre-

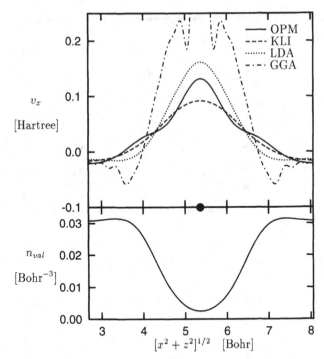

Fig. 2.5. Exchange potential of fcc aluminum in the [110] direction: Full OPM versus KLI approximation, LDA and PW91-GGA (● indicates the position of atom). All potentials have been evaluated with the KS states/density resulting from a self-consistent x-only KLI calculation within the plane-wave pseudopotential scheme ($E_{\mathrm{cut}} = 100\,\mathrm{Ry}$, 44 special k-points for integration over Brillouin zone, 750 states per k-point in G_k in the case of the OPM). The corresponding valence density is also given

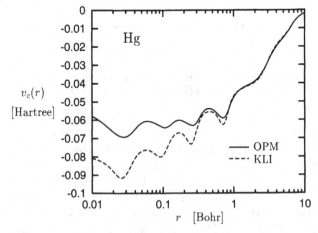

Fig. 2.6. KLI approximation versus full OPM solution: Colle-Salvetti correlation potential for neutral Hg

lation functional [23] (for details see Sect. 2.5.2) is examined for Hg. One finds excellent agreement of the KLI approximation with the exact solution of (2.27) for the outermost shells and in the asymptotic regime. On the other hand, for the inner shells the deviations are clearly larger than the differences visible in Figs. 2.3 and 2.5. Nevertheless, even in this regime the shell oscillations are reproduced correctly. Given the limited importance of v_c in the inner shell region, the enhanced inaccuracy of the KLI approximation for this particular functional should thus not be overemphasized.

In any case, one can conclude that the KLI scheme provides an excellent approximation to the OPM integral equation in the case of the exact exchange.

2.3.2 Properties of the Exact Exchange: Comparison with Explicit Density Functionals

The exact OPM results do not only allow an analysis of the KLI approximation, but also of conventional density functionals, as is already clear from the discussion of Sect. 2.3.1. In this section the comparison of LDA and GGA data with the corresponding OPM reference results will be extended in order to highlight some properties of the exact exchange. In turn, the limitations of the standard functionals, which become obvious in this comparison, provide additional motivation for resorting to implicit functionals.

One important property that has been emphasized a number of times now is the $-1/r$ decay of the exact exchange potential of finite systems. In view of the long-range character of the underlying self-interaction integral, one might ask whether it is possible to reproduce this behavior by some explicit density functional? In Sect. 2.1.2 it has been demonstrated that the LDA potential decays exponentially. In the case of the GGA, the second explicit density functional of interest,

$$E_x^{GGA}[n] = \int d^3r\, e_x^{GGA}(n, (\nabla n)^2)\,,$$

the situation is somewhat more complicated. For the GGA the asymptotic behavior depends on the detailed structure of the functional's kernel e_x^{GGA}. While most GGAs lead to exponentially decaying potentials, Becke made an attempt [15] to incorporate the $-1/r$ behavior into the GGA form by requiring e_x^{GGA} to satisfy (2.50) (B88-GGA). However, as emphasized in Sect. 2.2.4, the asymptotic form of the energy density is not related to that of the potential. In fact, one can show that GGAs can not satisfy (2.49) and (2.50) simultaneously [34]. Nevertheless, the potential resulting from the B88-GGA goes like $1/r^2$ for large r, so that one might expect an improvement over the exponential decay of all other GGAs.

However, it is not only the ultimate asymptotic form that matters, but also the point beyond which it sets in. This is illustrated in Fig. 2.7, which

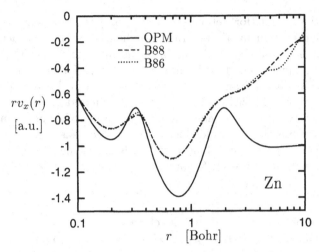

Fig. 2.7. Asymptotic form of v_x: B86- [14] and B88-GGA [15] versus OPM result for Zn. The GGA potentials have been evaluated with the exact x-only density

shows the exchange potential of Zn. The B88-GGA is compared with its predecessor B86 [14] and the exact v_x. The two GGAs are chosen as they have been constructed in the same fashion, i.e. in both cases the ansatz for e_x^{GGA} has been optimized to reproduce exact atomic exchange energies as well as possible, with similar success. On the other hand, the analytic form of the kernel and the resulting asymptotic behavior differ. As Fig. 2.7 shows, this difference has no effect in the physically relevant part of the asymptotic regime. Therefore, the HOMOs of the two GGAs are almost identical. The two GGA potentials are also very close in the inner shell regime. As a consequence, one finds little difference between the performance of the B88- and that of the B86-GGA, in spite of the formal improvement of v_x by the B88 ansatz. As a matter of principle, the semi-local form of the GGA kernel does not allow the reproduction of the exact v_x in the asymptotic regime.

How close can GGAs come to the exact v_x in the inner shell regime? Figure 2.3 already indicated that the conventional density functionals do not follow the shell oscillations of the exact v_x. This point is investigated further in Fig. 2.8 which focuses on the nonlocal contribution to v_x. As nonlocal contribution one understands the difference between a given v_x and the corresponding LDA potential, both evaluated with the same density. In this way one separates the fine structure in v_x from its smooth average behavior, which is well reproduced by the LDA. In order to ease the comparison of the GGAs with the exact result, $v_x - v_x^{LDA}$ is further corrected for the global shift resulting from the asymptotic behavior: Compared with the exponential decay of v_x^{LDA} the $-1/r$ behavior of the OPM potential essentially leads to a global attractive shift in v_x that is irrelevant for the inner shell features. This shift is well approximated by the difference between the highest occupied eigen-

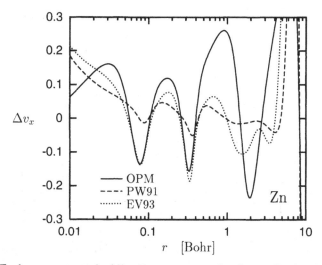

Fig. 2.8. Exchange potential of Zn: Percentage nonlocal contribution in the exact v_x and two GGAs [30,38]. All potentials have been evaluated with the exact x-only density

Table 2.7. Exchange energy of Zn: PW91- [30] and EV93-GGA [38] versus exact OPM result.

	OPM	PW91	EV93
$-E_x$	69.619	69.686	69.805

values of the two schemes which is therefore subtracted. In order to obtain the percentage nonlocal contribution, the difference is finally normalized with respect to the exact v_x,

$$\Delta v_x = \frac{v_x(r) - v_x^{LDA}(r) - \epsilon_{HOMO} + \epsilon_{HOMO}^{LDA}}{v_x^{OPM}(r)}. \qquad (2.56)$$

Figure 2.8 shows that the size of the shell oscillations in the OPM potential is of the order of 10–20%. Looking at the PW91-GGA it is obvious that its deviation on the local level is much larger than the corresponding global error [38]: The 10–15% deviation of v_x^{PW91} from v_x^{OPM} has to be compared with the 0.1% error in E_x (see Table 2.7). This picture is not specific to Zn, but quite characteristic for all atoms.

This imbalance prompts the question whether one can improve the accuracy of the GGA potential? An attempt in this direction led to the EV93-GGA [38] whose kernel was optimized to reproduce the exact atomic exchange potentials as accurately as possible. Figure 2.8 demonstrates the (limited) success of this strategy. The EV93-GGA reproduces the amplitudes of the shell oscillations in v_x^{OPM} much better than other GGAs. However, there is a

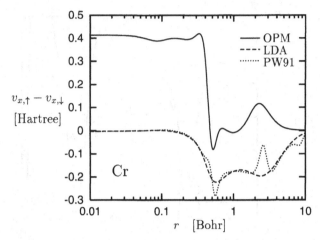

Fig. 2.9. Exchange potential of Cr: Spin balance of the exact result versus LDA and PW91-GGA

price to be paid for this improvement, which becomes obvious from Table 2.7: Any improvement of v_x is accompanied by some loss of accuracy for E_x. Insisting on the optimum local accuracy precludes the subtle error cancellation required to achieve optimum accuracy for integrated quantities like E_x. This again points at the limitations of the GGA form.

There is one further consequence of the nonlocality of the exact E_x worth to be noted. As soon as one goes to open-shell systems, the difference between the highest occupied eigenvalues of the majority spin (spin-up) and the minority spin (spin-down) channels comes into play: It determines the relative stability of different spin states, i.e. the magnetization in the case of solids. As the highest occupied eigenvalues depend strongly on v_x, the balance between the spin-up and the spin-down exchange potential has a major impact on the local magnetic moments. In Fig. 2.9 the difference between the spin-up and the spin-down exchange potential of Cr is shown. The deviations of the LDA and the GGA from the exact result are obvious [48]. For large r the exact $v_{x,\sigma}$ is dominated by the $-1/r$ tail for both spins σ ($\sigma = \uparrow, \downarrow$). In the valence regime the difference $v_{x,\uparrow}^{OPM} - v_{x,\downarrow}^{OPM}$ is thus less affected by the actual positions of the spin-aligned $4s$ and $3d$ electrons than the LDA and GGA potentials, which directly reflect the structures of the valence spin-densities: $v_{x,\uparrow}^{OPM} - v_{x,\downarrow}^{OPM}$ is closer to zero and repulsive, while the LDA and GGA results necessarily must be attractive. In the L- and K-shell regime the difference between spin-up and spin-down densities essentially vanishes, so that in the LDA and GGA the difference $v_{x,\uparrow} - v_{x,\downarrow}$ approaches zero. The exact functional, on the other hand, leads to an almost constant shift between $v_{x,\uparrow}$ and $v_{x,\downarrow}$. The nonlocality of the exact E_x propagates the differences between spin-up and spin-down in the valence regime into the inner shell region.

Once one has studied atomic exchange potentials, the next step must be an analysis of the bonding regime of molecules and solids. Corresponding results are given in Figs. 2.10–2.12. Figure 2.10 shows the exchange potential of the simplest molecule, H_2, for which the exact v_x reduces to a pure self-interaction correction (as H_2 is a spin-saturated two-electron system). Both the LDA and the B88-GGA [15] potential are compared to the exact v_x. In addition to the shift resulting from the $-1/r$ asymptotics of the exact v_x, one observes a minimum of the exact potential in the bonding regime, while the LDA potential peaks at the nuclear sites. One again notices the difference between the nonlocal Coulomb integral and the direct $n^{1/3}$ dependence of v_x^{LDA}. The peak structure is even more pronounced in the case of the GGA whose potential diverges weakly at the positions of the protons[5]. Nevertheless one finds an overall improvement by the GGA which, on average, generates a more attractive v_x in the bonding regime.

Figure 2.11 provides the corresponding comparison for N_2. In addition to the features already observed for H_2, one can now see the shell structure in the molecular v_x. Note that in the case of the exact exchange the KLI approximation is used for the evaluation of v_x. In analogy to the situation for atoms (Fig. 2.3) one expects the shell structure in the exact v_x to be even more pronounced than that obtained with the KLI approximation. The GGA potential also exhibits an indication of the shell structure, while this feature is completely absent in the LDA.

As a complement to Fig. 2.5, the exchange potential of bulk Si is plotted in Fig. 2.12 (along the [111]-direction). Two GGAs are compared to the LDA and the exact v_x. One can see that the GGA overcorrects the error of the LDA. An improvement is only observed in high density regions, i.e. in the bonding regime between the nearest neighbor atoms. On the other hand, the GGA's dependence on the local density gradients introduces some artificial structure in the low density region.

As a final example of the role of the self-interaction component in v_x some band gaps obtained with the exact exchange are listed in Table 2.8. All-electron OPM results [58] based on the Korringa-Kohn-Rostoker (KKR) method and the atomic sphere approximation (ASA) are compared with full potential plane-wave pseudopotential (PWPP) data [59] for C, Si and Ge. In addition to the x-only data, also the values resulting from the combination of the exact E_x with either LDA or GGA correlation are listed. The single-particle contribution Δ_s, i.e. the direct band gap, is separated from the xc-contribution Δ_{xc} to the band gap E_g,

$$E_g = \Delta_s + \Delta_{xc} . \qquad (2.57)$$

While the former is given by the standard difference between the highest eigenvalue of the valence band and the lowest eigenvalue of the conduction

---

[5] This divergence is much weaker than that of the nuclear potential, so that it does not have any adverse effect in practical calculations.

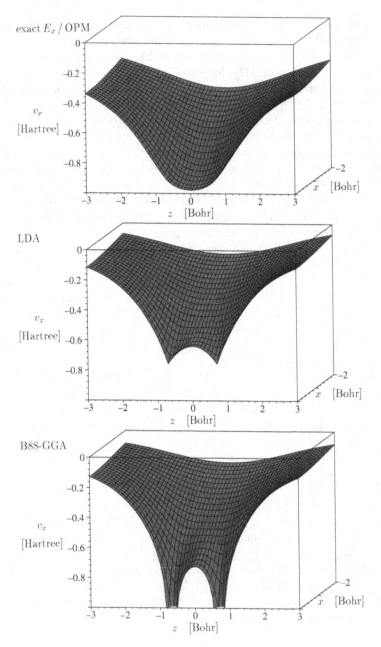

Fig. 2.10. Exchange potential of H_2: LDA and B88-GGA versus exact v_x

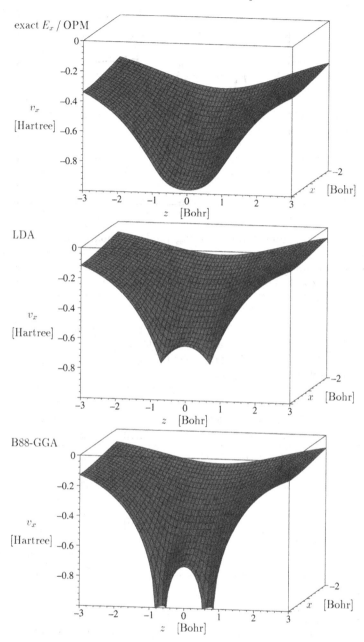

Fig. 2.11. Exchange potential of N_2: LDA and B88-GGA versus KLI approximation to the exact v_x

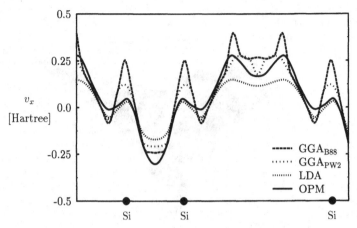

Fig. 2.12. Exchange potential of Si along [111] direction of diamond structure: Exact exchange potential (OPM) versus LDA as well as B88- and PW91-GGA data (• indicates the position of atom). All results were obtained by plane-wave pseudopotential calculations with a local pseudopotential ($E_{cut} = 25$ Ry, 19 special k-points)

Table 2.8. Band gap E_g of semiconductors: KKR-ASA [58] and plane-wave pseudopotential [59] OPM results versus LDA, HF [60] and experimental data (all energies in eV). The direct gap Δ_s and the contribution Δ_{xc} of the derivative discontinuity of E_{xc} are given separately

	E_x	E_c	method	C	Si	Ge
Δ_s	LDA	LDA	KKR-ASA	4.15	0.54	0.40
	LDA	LDA	PW-PP	4.16	0.49	
Δ_s	exact	LDA	KKR-ASA	4.58	1.12	1.03
	exact	LDA	PW-PP	5.06	1.44	
	exact	GGA	PW-PP		0.97	0.72
	exact	—	PW-PP		1.23	0.94
Δ_{xc}	exact	—	PW-PP	8.70	5.62	
E_g	HF	—	LAPW		7.4	6.4
E_g		expt.		5.48	1.17	0.87

band, the latter originates from the derivative discontinuity of the exact E_{xc} (see e.g. [1]). For comparison, also the corresponding LDA and HF band gaps are listed (in the case of the LDA the direct gap is identical to the total gap, as the LDA does not explicitly depend on the particle number).

The LDA results serve two purposes: On the one hand, they document the well-known underestimation of band gaps by the LDA. On the other hand, they show how close converged KKR-ASA and PWPP results are for the solids under consideration. When replacing the LDA exchange by the exact E_x, the direct gap is consistently enlarged. In fact, the direct gaps obtained with the exact E_x are in much better agreement with the experimental data than the LDA gaps, irrespectively of the correlation functional applied. Surprisingly, one finds that for Si the inclusion of correlation on the LDA level increases Δ_s compared with the x-only result, while the inclusion of a GGA for E_c leads to a reduced gap. One also notices the enhanced deviations between the KKR-ASA and the PWPP data for the direct gap, which either indicates some limitations of the ASA in the case of the exact E_x, or points at convergence problems.

The picture becomes even less clear as soon as the derivative discontinuity of E_x is taken into account. The corresponding contribution Δ_x is much larger than Δ_s [59], so that the agreement with experiment is completely lost. In fact, the sum of Δ_s and Δ_x obtained in the x-only OPM calculation is almost equal to the very large band gap that one finds in the Hartree-Fock approximation [60]. In that sense, the x-only OPM and the HF scheme are again equivalent. Obviously, the correlation contribution to Δ_{xc} must cancel the large Δ_x. As long as no adequate correlation functional with derivative discontinuity is available, Δ_x has to be ignored.

In Sect. 2.1.2 the insufficient handling of the self-interaction was suggested as one possible reason for the failure of the LDA and the GGA for Mott insulators. Can one solve this problem by using the exact E_x? A preliminary answer to this question is given in Fig. 2.13, in which the band structure of FeO is again considered. Within the PWPP approach, the exact E_x is combined with the LDA for E_c. While the resulting band structure is quite different from its LDA counterpart (compare with Fig. 2.2), FeO is again predicted to have a metallic ground-state, as by the LDA and the GGA. For the interpretation of this negative result it is important to realize the technical limitations of the calculation. First of all, the KLI approximation is utilized. Given the results of Sect. 2.3.1, this should not be a serious point. More important might be the fact that the $3s$ state of Fe has neither been included in the valence space of the Fe pseudopotential nor has it been taken into account via some nonlinear core-correction. This neglect leads to an incorrect ground-state in the case of diatomic FeO [61]. Furthermore, only 3 special k-points have been used for integrations over the Brillouin zone, and the plane-wave cut-off of 250 Ry is not particularly large in view of the localized $3p$ and $3d$ states. It remains to be investigated to which extent these technical limitations affect the bands shown in Fig. 2.13, so that it seems too early to draw definitive conclusions concerning the description of Mott insulators by using the exact exchange.

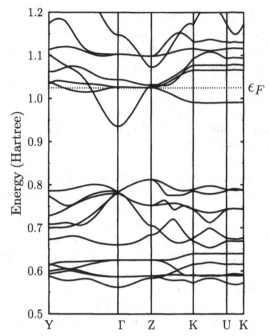

Fig. 2.13. Band structure of antiferromagnetic (type II) FeO obtained by plane-wave-pseudopotential calculation with exact E_x and LDA correlation on the basis of the KLI approximation (the valence space of Fe includes the $3p$, $3d$ and $4s$ states, $E_{cut} = 250\,\mathrm{Ry}$, 3 special k-points)

2.4   First-Principles Implicit Correlation Functionals

Some of the examples considered in the previous section already indicated that the exact exchange, while providing obvious progress compared with the LDA and the GGA, has to be combined with an appropriate orbital-dependent correlation functional in order to be useful in practice. Given the first-principles nature of the exact E_x, it is natural to derive such a correlation functional in a systematic fashion. The first task is to establish a suitable expression for the exact E_{xc} which can serve as starting point for the subsequent discussion of different approximations. This exact formula for E_{xc} at the same time resolves the discrepancy which was found between the original OPM equation (2.27) and the Sham-Schlüter equation (2.45).

2.4.1   Many-Body Theory on the Basis of the Kohn–Sham System: Exact Expression for E_{xc}

Let us assume for a moment that v_s, the total Kohn-Sham potential, is known [16,17,62]. This allows the definition of a noninteracting N-particle

Hamiltonian $\hat{H}_s$, which is the sum of the kinetic energy and an external potential term based on v_s,

$$\hat{H}_s = \hat{T} + \int \mathrm{d}^3r \, \hat{n}(\boldsymbol{r}) v_s(\boldsymbol{r}) \,. \tag{2.58}$$

The ground-state $|\Phi_0\rangle$ (assumed to be nondegenerate) corresponding to $\hat{H}_s$ is obtained by solution of the Schrödinger equation,

$$\hat{H}_s|\Phi_0\rangle = E_s|\Phi_0\rangle \,. \tag{2.59}$$

$|\Phi_0\rangle$ is a Slater determinant of the KS orbitals,

$$|\Phi_0\rangle = \prod_{\epsilon_k \leq \epsilon_F} \hat{b}_k^\dagger \, |0\rangle \,, \tag{2.60}$$

where $\hat{b}_k$ $(\hat{b}_k^\dagger)$ denotes the annihilation (creation) operator for the single-particle KS state ϕ_k and $|0\rangle$ is the corresponding KS vacuum, $\hat{b}_k|0\rangle = 0$. The ground-state energy E_s and density are given by

$$E_s = T_s + \int \mathrm{d}^3r \, n(\boldsymbol{r}) v_s(\boldsymbol{r}) = \sum_k \Theta_k \epsilon_k \tag{2.61}$$

$$n(\boldsymbol{r}) = \langle \Phi_0 | \hat{n}(\boldsymbol{r}) | \Phi_0 \rangle = \sum_k \Theta_k |\phi_k(\boldsymbol{r})|^2 \,. \tag{2.62}$$

By construction (2.62) is identical with the density of the interacting system.

The KS system, characterized by $\hat{H}_s$, provides all the ingredients required for standard many-body theory. In particular, the explicit form of the KS field operator in Heisenberg representation,

$$\hat{\psi}_0(\boldsymbol{r}t) = e^{\mathrm{i}\hat{H}_s t} \, \hat{\psi}(\boldsymbol{r}) \, e^{-\mathrm{i}\hat{H}_s t} = \sum_k \hat{b}_k \phi_k(\boldsymbol{r}) e^{-\mathrm{i}\epsilon_k t} \,, \tag{2.63}$$

allows the evaluation of the KS Green's function (2.38),

$$G_s(\boldsymbol{r}t, \boldsymbol{r}'t') = -\mathrm{i}\Theta(t - t') \sum_{\epsilon_F < \epsilon_k} \phi_k(\boldsymbol{r})\phi_k^\dagger(\boldsymbol{r}') \, e^{-\mathrm{i}\epsilon_k(t-t')}$$
$$+ \mathrm{i}\Theta(t' - t) \sum_{\epsilon_k \leq \epsilon_F} \phi_k(\boldsymbol{r})\phi_k^\dagger(\boldsymbol{r}') \, e^{-\mathrm{i}\epsilon_k(t-t')} \,. \tag{2.64}$$

In order to derive an exact relation for E_{xc} one now uses $\hat{H}_s$ as the noninteracting reference Hamiltonian. First, the total $\hat{H}$ of the interacting system is decomposed into $\hat{H}_s$ and a remainder $\hat{H}_1$, whose main part is the electron-electron interaction $\hat{W}$. In addition, $\hat{H}_1$ has to compensate those parts of $\hat{H}_s$ which are not present in $\hat{H}$,

$$\hat{H}_1 = \hat{W} - \int \mathrm{d}^3r \, \hat{n}(\boldsymbol{r}) v_{\mathrm{Hxc}}(\boldsymbol{r}) \,. \tag{2.65}$$

Here, v_{Hxc} represents the electron-electron interaction components in v_s,

$$v_{Hxc}(\boldsymbol{r}) = v_s(\boldsymbol{r}) - v_{ext}(\boldsymbol{r}) = v_H(\boldsymbol{r}) + v_{xc}(\boldsymbol{r}) . \qquad (2.66)$$

In the second step, a coupling constant g is introduced into the total Hamiltonian,

$$\hat{H}(g) = \hat{H}_s + g\hat{H}_1 , \qquad (2.67)$$

which allows the use of the coupling constant integration trick. The ground-state $|\Psi_0(g)\rangle$ corresponding to $\hat{H}(g)$ (also assumed to be nondegenerate) is obtained from the interacting Schrödinger equation,

$$\hat{H}(g)|\Psi_0(g)\rangle = E(g)|\Psi_0(g)\rangle . \qquad (2.68)$$

One can now apply the coupling constant integration scheme to the g-dependent ground-state energy,

$$E(g) = \langle\Psi_0(g)|\hat{H}(g)|\Psi_0(g)\rangle . \qquad (2.69)$$

One starts by differentiating $E(g)$ with respect to g,

$$\frac{\partial}{\partial g}E(g) = \langle\Psi_0(g)|\hat{H}_1|\Psi_0(g)\rangle , \qquad (2.70)$$

using the fact that $|\Psi_0(g)\rangle$ is normalized for all g,

$$\langle\Psi_0(g)|\Psi_0(g)\rangle = 1 .$$

One can then integrate (2.70) with respect to g. The integration starts at $g = 0$, for which $\hat{H}(g)$ agrees with the KS Hamiltonian, and ends at $g = 1$, where $\hat{H}(g)$ is identical with the true interacting Hamiltonian. On the left-hand side, the integration thus leads to the difference between the energy $E(1) = E_{tot}$ of the interacting system (which is the energy one is interested in) and the KS energy $E(0) = E_s$,

$$E(1) - E(0) = E_{tot} - E_s \equiv E_1 = \int_0^1 dg \, \langle\Psi_0(g)|\hat{H}_1|\Psi_0(g)\rangle . \qquad (2.71)$$

For the evaluation of E_1 the concept of adiabatic switching is applied to $\hat{H}_1$, i.e. $\hat{H}_1$ is switched off for large positive and negative times, using some exponential switching factor. The standard machinery of many-body perturbation theory then leads to an expression which connects the interacting ground-state $|\Psi_0\rangle$ with the KS ground-state,

$$|\Psi_0\rangle = A \lim_{\epsilon\to 0} \frac{\hat{U}_{I,\epsilon}(0,\mp\infty)|\Phi_0\rangle}{\langle\Phi_0|\hat{U}_{I,\epsilon}(0,\mp\infty)|\Phi_0\rangle} \qquad (2.72)$$

$$A = \lim_{\epsilon_1,\epsilon_2\to 0} \left[\frac{\langle\Phi_0|\hat{U}_{I,\epsilon_1}(+\infty,0)|\Phi_0\rangle \, \langle\Phi_0|\hat{U}_{I,\epsilon_2}(0,-\infty)|\Phi_0\rangle}{\langle\Phi_0|\hat{U}_{I,\epsilon_1}(+\infty,0) \, \hat{U}_{I,\epsilon_2}(0,-\infty)|\Phi_0\rangle} \right]^{1/2} \qquad (2.73)$$

(A ensures the correct normalization of $|\Psi_0\rangle$). The main ingredient of (2.72) is the interaction picture time-evolution operator,

$$
\hat{U}_{I,\epsilon}(t,t') = \sum_{n=0}^{\infty} \frac{(-ig)^n}{n!} \int_{t'}^{t} dt_1 \cdots \int_{t'}^{t} dt_n \, e^{-\epsilon(|t_1|+\cdots+|t_n|)}
$$

$$
\times T\big[\hat{H}_{1,I}(t_1) \cdots \hat{H}_{1,I}(t_n)\big] , \qquad (2.74)
$$

which is given by a power series of $\hat{H}_1$ in the interaction picture,

$$
\hat{H}_{1,I}(t) = e^{i\hat{H}_s t} \, \hat{H}_1 \, e^{-i\hat{H}_s t} = \hat{W}_I(t) - \int d^3r \, \hat{\psi}_0^\dagger(rt)\hat{\psi}_0(rt)v_{\mathrm{Hxc}}(r) . \quad (2.75)
$$

Insertion of (2.72) into the coupling constant integral (2.71) leads to the standard energy correction which results from switching on some perturbation to a noninteracting reference system,

$$
E_1 = \lim_{\epsilon \to 0} \int_0^1 dg \sum_{n=0}^{\infty} \frac{(-ig)^n}{n!} \int_{-\infty}^{\infty} dt_1 \cdots \int_{-\infty}^{\infty} dt_n \, e^{-\epsilon(|t_1|+\cdots+|t_n|)}
$$

$$
\times \frac{\langle\Phi_0|T\hat{H}_{1,I}(0)\hat{H}_{1,I}(t_1) \cdots \hat{H}_{1,I}(t_n)|\Phi_0\rangle}{\langle\Phi_0|\hat{U}_{I,\epsilon}(+\infty,-\infty)|\Phi_0\rangle} . \qquad (2.76)
$$

Using (2.71) and (2.7) one can finally extract E_{xc},

$$
E_{\mathrm{xc}} = E_1 - E_{\mathrm{H}} + \int d^3r \, n(r)v_{\mathrm{Hxc}}(r)
$$

$$
= \frac{1}{2} \int d^3r \int d^3r' \frac{e^2}{|r-r'|} \Big[\langle\Phi_0|\hat{\psi}^\dagger(r)\hat{\psi}^\dagger(r')\hat{\psi}(r')\hat{\psi}(r)|\Phi_0\rangle - n(r)n(r')\Big]
$$

$$
+ \lim_{\epsilon \to 0} \sum_{n=1}^{\infty} \frac{(-i)^n}{(n+1)!} \int_{-\infty}^{\infty} dt_1 \cdots \int_{-\infty}^{\infty} dt_n \, e^{-\epsilon(|t_1|\cdots+|t_n|)}
$$

$$
\times \langle\Phi_0|T\hat{H}_{1,I}(0)\hat{H}_{1,I}(t_1) \cdots \hat{H}_{1,I}(t_n)|\Phi_0\rangle_l , \quad (2.77)
$$

where the index l indicates that only linked diagrams are to be included in the evaluation of (2.77) via Wick's theorem (this restriction corresponds to the cancellation of the denominator of (2.76)). The first term on the right-hand side (second line) represents the first order contribution with respect to the perturbation $\hat{H}_1$ and is easily identified as the exchange energy (2.19). The second term, which absorbs all higher orders in $\hat{H}_1$, provides an exact expression for the correlation energy E_c. The ingredients required for the evaluation of this expression via Wick's theorem are the KS Green's function (2.64), the Coulomb interaction, and v_{Hxc} (as $\hat{H}_1$ depends on this potential). The Hartree component of v_{Hxc} is readily calculated from the KS orbitals, so that (2.77) depends on three basic ingredients, ϕ_k, ϵ_k and v_{xc}.

One thus ends up with an exact representation of E_{xc} in terms of the KS orbitals and eigenvalues as well as the xc-potential [47]. Consequently, (2.77) is not an algebraic assignment of some well-defined expression to E_{xc}, but rather represents a highly nonlinear functional equation, as E_{xc} is given in terms of it's own functional derivative[6]. This now explains the nonlinear character of the Sham–Schlüter equation (2.45). By analyzing the exact E_{xc} which determines the right-hand side (2.28) of the OPM equation one finds that exactly the same nonlinearity exists in the conventional OPM.

How can one deal with this nonlinearity? There are two possible strategies: Either one tries to solve the nonlinear OPM equation, which is a highly nontrivial task, that has not yet been attempted; or, as an alternative, one can try to linearize the xc-energy functional and therefore the OPM equation. This route is pursued in the next section.

2.4.2 Perturbative Approach to the Sham–Schlüter Equation: Second Order Correlation Functional

Given the origin of (2.77), an expansion of E_{xc} (and thus v_{xc}) in powers of e^2 is the natural approach to the linearization of the OPM equation [18]. The lowest order term in this expansion is the exchange energy. All higher order terms correspond to E_{c},

$$E_{\mathrm{xc}} = \sum_{l=1}^{\infty} e^{2l}\, E_{\mathrm{xc}}^{(l)}[n] = E_{\mathrm{x}} + E_{\mathrm{c}}^{(2)} + \dots \tag{2.78}$$

$$v_{\mathrm{xc}} = \sum_{l=1}^{\infty} e^{2l}\, v_{\mathrm{xc}}^{(l)}[n] = v_{\mathrm{x}} + v_{\mathrm{c}}^{(2)} + \dots \; . \tag{2.79}$$

After insertion of (2.78) and (2.79) into the OPM equation both its right-hand and its left-hand side are given as power series with respect to e^2. The identity of both sides is now required order by order. In the lowest order (e^2) the left-hand side of the OPM equation just contains v_{x}, while the right-hand side is determined by (2.22). This simply reflects the fact that E_{x} is a well-defined functional of the ϕ_k only. To lowest order one ends up with the standard linear OPM equation for the exact exchange,

$$\int \mathrm{d}^3 r'\, \chi_{\mathrm{s}}(\boldsymbol{r}, \boldsymbol{r}') v_{\mathrm{x}}(\boldsymbol{r}') = \Lambda_{\mathrm{x}}(\boldsymbol{r}) \tag{2.80}$$

$$\Lambda_{\mathrm{x}}(\boldsymbol{r}) = -\sum_{k} \int \mathrm{d}^3 r'\, \phi_k^\dagger(\boldsymbol{r}) G_k(\boldsymbol{r}, \boldsymbol{r}') \frac{\delta E_{\mathrm{x}}}{\delta \phi_k^\dagger(\boldsymbol{r}')} + c.c. \tag{2.81}$$

---

[6] Note, however, that this result is consistent with the basic statements of DFT: As v_{xc} is a density functional itself, the right-hand side of (2.77) is an implicit density functional.

The first time that the nonlinearity shows up is in the order e^4. This lowest order correlation functional can be written as

$$E_c^{(2)} = E_c^{MP2} + E_c^{\Delta HF} . \qquad (2.82)$$

The first of these terms is an expression which basically looks like the standard second order Møller-Plesset (MP2) correction to the HF energy,

$$E_c^{MP2} = \frac{e^4}{2} \sum_{ijkl} \Theta_i \Theta_j (1 - \Theta_k)(1 - \Theta_l) \frac{(ij\|kl)[(kl\|ij) - (kl\|ji)]}{\epsilon_i + \epsilon_j - \epsilon_k - \epsilon_l} . \qquad (2.83)$$

However, the Slater integrals $(ij\|kl)$,

$$(ij\|kl) = \int d^3 r_1 \int d^3 r_2 \frac{\phi_i^\dagger(\boldsymbol{r}_1)\phi_k(\boldsymbol{r}_1)\phi_j^\dagger(\boldsymbol{r}_2)\phi_l(\boldsymbol{r}_2)}{|\boldsymbol{r}_1 - \boldsymbol{r}_2|} , \qquad (2.84)$$

are calculated with the KS orbitals ϕ_k, and the denominator of (2.83) relies on the KS eigenvalues ϵ_k, so that E_c^{MP2} can give results which are quite different from standard MP2 data (see below). The second contribution to (2.82) takes into account that the present perturbation expansion is not based on the HF Hamiltonian, but on the KS Hamiltonian,

$$E_c^{\Delta HF} = \sum_{il} \frac{\Theta_i (1 - \Theta_l)}{\epsilon_i - \epsilon_l} \left| \langle i|v_x|l\rangle + e^2 \sum_j \Theta_j (ij\|jl) \right|^2 . \qquad (2.85)$$

It involves the difference between the orbital expectation value of the nonlocal HF-type exchange potential $\sum_j \Theta_j \, j\|j$ and that of v_x,

$$\langle i|v_x|l\rangle = \int d^3 r \, \phi_i^\dagger(\boldsymbol{r})v_x(\boldsymbol{r})\phi_l(\boldsymbol{r}) . \qquad (2.86)$$

At this point, it seems worthwhile to emphasize the difference between an expansion with respect to the perturbing Hamiltonian $\hat{H}_1$ and the present expansion in powers of e^2. The former leads to the same basic expressions (2.83) and (2.85), but with v_x replaced by the full v_{xc}. Consequently, the nonlinearity is not resolved in this type of expansion. On the other hand, as soon as one expands v_{xc} with respect to e^2, only the lowest order contribution v_x is relevant in (2.85) due to the quadratic structure of this expression. As net result one finds a well-defined linear relation for the correlation functional $E_c^{(2)}$.

The contribution $E_c^{\Delta HF}$ once again illustrates the relation between the x-only OPM and the standard HF approach. In the x-only limit, the OPM corresponds to a minimization of the HF energy expression under the subsidiary condition that the orbitals satisfy the KS equations (see Sect. 2.2.2). The OPM energy is thus slightly higher than the HF value, which results from a free minimization of the same energy expression. The difference between the two energies can be evaluated order by order, using the difference between the HF and the x-only OPM Hamiltonian as perturbation. To lowest order this procedure leads to the energy (2.85). This expression is always negative, consistent with the fact that the HF energy must be below the x-only OPM value. On the other hand, if one examines (2.85) quantitatively, one usually finds it to be rather small (see Sects. 2.6).

As already emphasized, $E_c^{(2)}$ is well-defined as soon as v_x is known. Thus, the first step of a self-consistent application of (2.82) is the solution of the x-only OPM equation (2.80). Once v_x is available, it remains to evaluate

$$v_c^{(2)}(\boldsymbol{r}) = \frac{\delta E_c^{(2)}[\phi_k, \epsilon_k, v_x]}{\delta n(\boldsymbol{r})} . \tag{2.87}$$

How can this be done? Clearly, the functional derivative of E_c^{MP2} with respect to n can be handled as in (2.21), as this term does not depend on v_x. The same is true for the ϕ_k and ϵ_k dependence of $E_c^{\Delta\mathrm{HF}}$. The subsequent discussion focuses on the v_x dependence of $E_c^{\Delta\mathrm{HF}}$.

One starts by realizing that the explicit v_x dependence of $E_c^{\Delta\mathrm{HF}}$ is not fundamentally different from the ϕ_k and ϵ_k dependence. One thus has to include a functional derivative with respect to this additional variable, when eliminating the original derivative $\delta/\delta n$ via the chain rule (as in (2.21)). This leads to an additional contribution to the inhomogeneity (2.28),

$$\Delta\Lambda_c^{(2)}(\boldsymbol{r}) = \int \mathrm{d}^3 r' \frac{\delta v_x(\boldsymbol{r}')}{\delta v_s(\boldsymbol{r})} \frac{\delta E_c^{\Delta\mathrm{HF}}}{\delta v_x(\boldsymbol{r}')} . \tag{2.88}$$

The first ingredient, the functional derivative of v_x with respect to v_s, is accessible via the x-only OPM equation. If one differentiates (2.80) with respect to v_s and isolates the desired derivative one ends up with

$$\frac{\delta v_x(\boldsymbol{r}_2)}{\delta v_s(\boldsymbol{r}_1)} = \int \mathrm{d}^3 r_3 \, \chi_s^{-1}(\boldsymbol{r}_2, \boldsymbol{r}_3) \left[\frac{\delta\Lambda_x(\boldsymbol{r}_3)}{\delta v_s(\boldsymbol{r}_1)} - \int \mathrm{d}^3 r_4 \frac{\delta\chi_s(\boldsymbol{r}_3, \boldsymbol{r}_4)}{\delta v_s(\boldsymbol{r}_1)} v_x(\boldsymbol{r}_4) \right] . \tag{2.89}$$

The functional derivative of Λ_x with respect to the KS potential can again be obtained by using the chain rule,

$$\begin{aligned}
\frac{\delta\Lambda_x(\boldsymbol{r}_3)}{\delta v_s(\boldsymbol{r}_1)} &= -\sum_k \int \mathrm{d}^3 r_4 \left[\phi_k^\dagger(\boldsymbol{r}_1) G_k(\boldsymbol{r}_1, \boldsymbol{r}_4) \frac{\delta\Lambda_x(\boldsymbol{r}_3)}{\delta\phi_k^\dagger(\boldsymbol{r}_4)} + c.c. \right] \\
&\quad + \sum_k |\phi_k(\boldsymbol{r}_1)|^2 \frac{\partial\Lambda_x(\boldsymbol{r}_3)}{\partial\epsilon_k} .
\end{aligned} \tag{2.90}$$

The second expression in (2.89) involves the derivative of the linear response function with respect to the KS potential. Using the definition (2.26) of the linear KS response function, one can rewrite this quantity as the second functional derivative of n with respect to v_s, i.e. the quadratic response function of the KS system,

$$\frac{\delta \chi_s(r_3, r_4)}{\delta v_s(r_1)} = \frac{\delta^2 n(r_3)}{\delta v_s(r_4)\delta v_s(r_1)} . \tag{2.91}$$

This function can be evaluated in the same fashion as χ_s (compare with [47]). As $\delta E_c^{\Delta HF}/\delta v_x$ can be evaluated directly, all ingredients of (2.88) are known, and $v_c^{(2)}$ can be calculated.

In principle, one could now turn to the third order terms. The third order functional depends not only on v_x, but also on $v_c^{(2)}$. However $v_c^{(2)}$ is well-defined by now, so that also $E_c^{(3)}$ is well-defined. Moreover, following the route sketched above the corresponding third order potential can also be handled. The scheme presented here thus establishes a recursive procedure for calculating E_{xc} and the corresponding v_{xc} order by order in e^2.

On the other hand, it is obvious that this approach sooner or later will become excessively cumbersome. In addition, there are quite a number of systems for which perturbation theory fails, most notably metals. In the next step one thus has to consider extensions of $E_c^{(2)}$ which allow the treatment of metallic systems.

2.4.3 Extensions of the Second Order Functional

Random Phase Approximation. The most straightforward approach to go beyond the second order functional is a resummation of certain classes of diagrams. Starting from the exact representation (2.77) of E_{xc}, one could, for example, resum the ring diagrams [19,20] (random phase approximation – RPA), which are known to form the most important class for the description of metals,

$$E_{xc}^{RPA} = \frac{i}{2} \sum_{n=1}^{\infty} \frac{1}{n} \int d1\, d1' \cdots \int dn\, dn'\, \delta(t_1) \frac{e^2 \delta(t_1 - t_1')}{|r_1 - r_1'|} \chi_s(1', 2) \cdots$$
$$\times \frac{e^2 \delta(t_n - t_n')}{|r_n - r_n'|} \chi_s(n', 1) , \tag{2.92}$$

with $\chi_s(1, 2)$ being given by (2.46). Note that it is not trivial to identify approximations like the RPA in the present context, because in (2.77) the standard diagrammatic contributions are intertwined with diagrams depending on v_{xc}. The lowest order ring diagram studied in Sect. 2.4.2 is the most obvious example. Contributions similar to $E_c^{\Delta HF}$ are found to all orders. It is thus not clear *a priori* what it means to restrict E_{xc}^{RPA} to the KS ring diagrams only. However, the quantitative dominance of the second order ring diagram over $E_c^{\Delta HF}$ seems to legitimize this approach. It seems worthwhile to mention that, for non-metallic systems, the second order exchange diagram included in (2.82) can be directly added to (2.92).

Interaction Strength Interpolation (ISI). The evaluation of expressions like (2.92) and, in particular, its derivatives is quite challenging. Consequently, one might ask whether it is possible to account for the higher order contributions in a more efficient way? This is the basic idea behind the interaction strength interpolation (ISI) [21], which makes an attempt to recast the higher order terms in the form of an explicit density functional. The starting point for the derivation of the ISI is the adiabatic connection, which makes the transition from the non-interacting KS system to the fully interacting system in such a way that the density remains the same all along the way [46,63,64]. The coupling constant integration trick then leads to a formula analogous to (2.71),

$$E_{\mathrm{xc}} = \int_0^1 \mathrm{d}g\, W_g[n] , \qquad (2.93)$$

with[7]

$$W_g[n] = \langle \Psi_0(g)[n]|\hat{W}|\Psi_0(g)[n]\rangle - E_{\mathrm{H}}[n] . \qquad (2.94)$$

The basic idea of the ISI is to obtain W_g in the interesting regime $g \approx 1$ from an interpolation between the weak ($g \ll 1$) and the strong ($g \gg 1$) interaction limit. The former limit is well-known by now: For a weak interaction one can expand E_{xc} in powers of the coupling constant g which automatically yields a corresponding expansion for W_g,

$$E_{\mathrm{xc}} = \int_0^1 \mathrm{d}g \left(E_{\mathrm{x}} + \sum_{l=2}^{\infty} l g^{l-1} E_{\mathrm{c}}^{(l)} \right) . \qquad (2.95)$$

The lowest two orders of W_g are thus determined by E_{x} and $E_{\mathrm{c}}^{(2)}$.

The strong interaction limit requires a new concept. However, for $g \to \infty$ the electrostatic forces completely dominate over kinetic effects, so that a simple model system ("point charge plus continuum model") is sufficient to extract W_g [65]. In this way one finds the two leading orders of the expansion of W_g in powers of $1/\sqrt{g}$,

$$\lim_{g \to \infty} W_g[n] = W_\infty[n] + W_\infty'[n]g^{-1/2} + \dots , \qquad (2.96)$$

with

$$W_\infty[n] = \int \mathrm{d}^3 r \left[An^{4/3} + B\frac{(\nabla n)^2}{n^{4/3}} \right] \qquad (2.97)$$

$$W_\infty'[n] = \int \mathrm{d}^3 r \left[Cn^{3/2} + D\frac{(\nabla n)^2}{n^{7/6}} \right] \qquad (2.98)$$

---

[7] $|\Psi_0(g)\rangle$ is not identical with the ground-state introduced in Sect. 2.4.1, as the perturbation in the adiabatic connection is different from (2.65).

(for the values of the coefficients $A - D$ see [21]). Interpolation between E_x, $E_c^{(2)}$ on the one hand and W_∞, W_∞' on the other hand then leads to

$$E_{xc}^{ISI} = W_\infty + \frac{2X}{Y}\left[(1+Y)^{1/2} - 1 - Z\ln\left(\frac{(1+Y)^{1/2} + Z}{1+Z}\right)\right], \quad (2.99)$$

with the abbreviations

$$X = \frac{xy^2}{z^2} \qquad Y = \frac{x^2y^2}{z^4} \qquad Z = \frac{xy^2}{z^3}$$
$$x = -4E_c^{(2)} \qquad y = W_\infty' \qquad z = W_\infty - E_x. \quad (2.100)$$

Equations (2.99) and (2.100) should be understood as an effective resummation of the KS perturbation series. The correlation part of E_{xc}^{ISI} is obtained by subtraction of the exact E_x from (2.99).

2.5   Semi-empirical Orbital-Dependent Exchange-Correlation Functionals

Given the complexity of the first-principles implicit correlation functionals of Sect. 2.4, one is automatically led to look for simpler and thus more efficient semi-empirical alternatives. Two functionals of this type have been suggested for use within the OPM.

2.5.1   Self-interaction Corrected LDA

The first of these functionals is the self-interaction corrected LDA (SIC-LDA) [22]. It has been emphasized that the self-interaction error of the LDA and the GGA is a major source of problems. It is thus tempting to try to eliminate this self-interaction in a semi-empirical form. This is the main idea behind the SIC-LDA. The starting point is the spin-density dependent version of the standard LDA, $E_{xc}^{LDA}[n_\uparrow, n_\downarrow]$. In contrast to the exact $E_{xc}[n_\uparrow, n_\downarrow]$, this functional does not reduce to a pure Coulomb self-interaction integral if only one single particle with spin up and density $n_\uparrow = |\phi_{1,\uparrow}(r)|^2$ is present,

$$E_{xc}^{LDA}[n_\uparrow, 0] \neq -\frac{e^2}{2}\int d^3r\int d^3r' \frac{|\phi_{1,\uparrow}(r)|^2|\phi_{1,\uparrow}(r')|^2}{|r - r'|} = E_{xc}^{exact}[|\phi_{1,\uparrow}|^2, 0].$$
$$(2.101)$$

In a many-particle system this self-interaction error is present for all particles in the system. In the SIC-LDA one eliminates the self-interaction component *a posteriori* by explicitly subtracting the erroneous terms for the individual

KS states of both spins,

$$
\begin{aligned}
E_{\text{xc}}^{\text{SIC–LDA}} &= E_{\text{xc}}^{\text{LDA}}[n_\uparrow, n_\downarrow] - \sum_{k,\sigma} \Theta_{k,\sigma} \left\{ E_{\text{H}}\left[|\phi_{k,\sigma}|^2\right] + E_{\text{xc}}^{\text{LDA}}\left[|\phi_{k,\sigma}|^2, 0\right] \right\} \\
&= E_{\text{xc}}^{\text{LDA}}[n_\uparrow, n_\downarrow] - \sum_{k,\sigma} \Theta_{k,\sigma} E_{\text{xc}}^{\text{LDA}}\left[|\phi_{k,\sigma}|^2, 0\right] \\
&\quad - \frac{e^2}{2} \sum_{k,\sigma} \Theta_{k,\sigma} \int d^3 r \int d^3 r' \frac{|\phi_{k,\sigma}(\boldsymbol{r})|^2 |\phi_{k,\sigma}(\boldsymbol{r}')|^2}{|\boldsymbol{r} - \boldsymbol{r}'|} .
\end{aligned}
\tag{2.102}
$$

By construction, $E_{\text{xc}}^{\text{SIC–LDA}}$ satisfies (2.101).

The standard scheme for the application of this functional implies the use of orbital-dependent Kohn-Sham potentials: A separate KS equation has to be solved for each individual KS state. This procedure leads, in general, to non-orthogonal KS orbitals, so that an *a posteriori* orthogonalization is required [66]. However, it has been realized very early [67] that the SIC-LDA should be applied within the framework of the OPM. For any orbital-dependent functional, the OPM produces the corresponding KS-type multiplicative potential, which automatically avoids the problem of non-orthogonality.

On the other hand, the use of the OPM does not resolve the unitarity problem which is inherent to this functional [68–70]: If one performs a unitary transformation among the KS orbitals, the individual orbital densities will change, even if the transformation only couples degenerate KS states. Consequently, also the value of $E_{\text{xc}}^{\text{SIC–LDA}}$ changes. An additional prescription which defines a suitable representation of the KS orbitals (which usually implies a localization) is necessary for practical calculations [71]. While the results for atoms are not very sensitive to this unitarity problem, it becomes more important in extended systems. For that reason the SIC-LDA has rarely been applied to molecules [72–74].

2.5.2   Colle–Salvetti Functional

A second orbital-dependent expression, originally introduced for use with the Hartree-Fock scheme, is the Colle-Salvetti (CS) correlation functional [23]. The starting point for the derivation of the CS functional is an approximation for the correlated wavefunction $\Psi(\boldsymbol{r}_1\sigma_1, \dots \boldsymbol{r}_N\sigma_N)$. The ansatz for $\Psi(\boldsymbol{r}_1\sigma_1, \dots \boldsymbol{r}_N\sigma_N)$ consists of a product of the HF Slater determinant and Jastrow factors,

$$
\Psi(\boldsymbol{r}_1\sigma_1, \dots \boldsymbol{r}_N\sigma_N) = \Phi^{\text{HF}}(\boldsymbol{r}_1\sigma_1, \dots \boldsymbol{r}_N\sigma_N) \prod_{i<j} \left[1 - \varphi(\boldsymbol{r}_i, \boldsymbol{r}_j)\right] . \tag{2.103}
$$

CS then used a model for the correlation functions $\varphi(\boldsymbol{r}_i, \boldsymbol{r}_j)$ which satisfies the electron-electron cusp condition at $\boldsymbol{r}_i = \boldsymbol{r}_j$ [75]. The free parameter in the model is adjusted to the correlation energy of the He atom [23]. The final functional reads

$$E_c^{CS} = -\frac{ab}{4} \int d^3r \, \gamma \, \xi \left[4 \sum_{k\sigma} \Theta_k n_\sigma |\nabla \phi_{k\sigma}|^2 - |\nabla n|^2 - \sum_\sigma n_\sigma \nabla^2 n_\sigma + n \nabla^2 n \right]$$

$$-a \int d^3r \, \gamma \, \frac{n}{\eta} \, , \qquad (2.104)$$

with the abbreviations

$$\gamma(\boldsymbol{r}) = 4 \frac{n_\uparrow(\boldsymbol{r}) n_\downarrow(\boldsymbol{r})}{n(\boldsymbol{r})^2} \qquad (2.105)$$

$$\eta(\boldsymbol{r}) = 1 + dn(\boldsymbol{r})^{-1/3} \qquad (2.106)$$

$$\xi(\boldsymbol{r}) = \frac{n(\boldsymbol{r})^{-5/3} \exp[-cn(\boldsymbol{r})^{-1/3}]}{\eta(\boldsymbol{r})} \, . \qquad (2.107)$$

This functional depends on the spin-density n_σ and the kinetic energy of spin σ, $\sum_k \Theta_k |\nabla \phi_{k\sigma}|^2$. In the DFT context, this latter dependence makes E_c^{CS} an implicit functional for which the OPM has to be utilized. Therefore, the CS correlation functional has been suggested as a first candidate for going beyond the exact exchange [76].

However, like the correlation part of the SIC-LDA, this CS functional is rather local. Its nonlocality is restricted to the first gradients of the KS orbitals. In this respect it is very similar to the Meta-GGA [77]. In the case of the Meta-GGA higher gradients of the density, i.e. its Laplacian, are effectively included into the GGA via the kinetic energy expression $\sum_k \Theta_k |\nabla \phi_{k\sigma}|^2$. However, in contrast to the CS-functional, the Meta-GGA has not yet been applied within the OPM (although, in principle, it should be treated within the OPM). In any case, neither the SIC-LDA nor the CS and Meta-GGA functionals are sufficiently nonlocal to deal with dispersion forces: The argument given in Sect. 2.1.2 for the LDA applies equally well to these types of functionals.

2.6 Analysis of the Orbital-Dependent Correlation

2.6.1 Description of Dispersion Forces by Second Order Correlation Functional

Given the motivation for implicit correlation functionals the first question to be addressed is that of dispersion forces. As none of the semi-empirical functionals of Sect. 2.5 can deal with these long-range forces, the present discussion focuses on the second order correlation functional (2.82) as the simplest first-principles functional.

Consider two neutral atoms A and B, separated by a sufficiently large distance R, so that no molecular orbitals are formed. For this system, the overlap between the atomic orbitals $\phi_{k,A}$ centered on atom A and $\phi_{l,B}$ centered on atom B vanishes exponentially with increasing R. The sums over all KS states in (2.83) and (2.85) then split up into sums over the atomic states,

$$\sum_i \longrightarrow \sum_{i_A} + \sum_{i_B} .$$

Thus, if one takes the atoms apart, only those Slater integrals in $E_c^{(2)}$ survive which do not link orbitals from different atomic centers at the same point $\boldsymbol{r}$. Consequently, neither $E_c^{\Delta\mathrm{HF}}$ nor the exchange component of E_c^{MP2} contribute to the interaction between the two atoms for large R, as for these terms all sums over states must belong to the same atom. Moreover, in the case of the ring diagram, which provides the direct matrix elements in E_c^{MP2}, each individual ring must only involve states of one of the atoms, e.g.

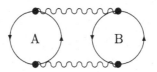

The complete ring diagram decomposes into four terms, as for both rings either atom A or atom B can be chosen. The combinations AA and BB contribute to the atomic correlation energies of A and B, as is the case of $E_c^{\Delta\mathrm{HF}}$ and the exchange component of E_c^{MP2}. Only the combinations AB and BA, which represent the interaction of virtual particle-hole excitations on the two centers, lead to some molecular binding energy. Moreover, no other component of the total energy functional contributes to the interaction between the two atoms, provided that A and B are closed-subshell atoms (so that no static multipole moments are present). The interaction energy between A and B thus reduces to

$$E_{\mathrm{c,int}}^{(2)} = e^4 \sum_{i_A k_A} \Theta_{i_A}(1-\Theta_{k_A}) \sum_{j_B l_B} \Theta_{j_B}(1-\Theta_{l_B}) \frac{(i_A j_B \| k_A l_B)(k_A l_B \| i_A j_B)}{\epsilon_{i_A} + \epsilon_{j_B} - \epsilon_{k_A} - \epsilon_{l_B}} .$$

$$(2.108)$$

If one expands (2.108) in powers of $1/R$ and re-introduces the frequency integration inherent in the ring diagram, one ends up with an expression [47] which is much more familiar,

$$E_{\mathrm{c,int}}^{(2)} = -\frac{C_6}{R^6} = -3\frac{e^4}{R^6} \int_0^\infty \frac{\mathrm{d}u}{\pi} \, \alpha_A(iu)\alpha_B(iu) . \qquad (2.109)$$

Here $\alpha_A(iu)$ denotes the atomic polarizability (for the case of closed subshells), evaluated at imaginary frequency,

$$\alpha_A(\omega) = \int \mathrm{d}^3 r_1 \int \mathrm{d}^3 r_2 \, z_1 \, z_2 \, \chi_{\mathrm{s},A}^{\mathrm{R}}(\boldsymbol{r}_1, \boldsymbol{r}_2, \omega) , \qquad (2.110)$$

whose basic ingredient is the frequency-dependent, retarded KS response function,

$$\chi_s^R(\boldsymbol{r}_1, \boldsymbol{r}_2, \omega) = \sum_{i,k} [\Theta_i - \Theta_k] \frac{\phi_i^\dagger(\boldsymbol{r}_1)\phi_k(\boldsymbol{r}_1)\phi_k^\dagger(\boldsymbol{r}_2)\phi_i(\boldsymbol{r}_2)}{\omega - \epsilon_k + \epsilon_i + i\eta} \ . \qquad (2.111)$$

Equation (2.109) has the standard form of the dispersion force. Obviously, $E_c^{(2)}$ is able to reproduce the correct long-range behavior proportional to $1/R^6$. However, the exact result for the coefficient C_6 involves the full atomic polarizabilities, while the present DFT-variant of C_6 is determined by the KS polarizabilities (as a consequence of second order perturbation theory). So, the next question is how close the KS coefficients come to the exact C_6? First calculations [78] show that, for light atoms, (2.109) yields reasonably accurate coefficients: They underestimate the full C_6 by 10–20%. On the other hand, for heavier atoms higher order correlation becomes important, and (2.109) is substantially off.

At this point it is has been verified that the functional (2.82) reproduces the long-range behavior of the dispersion force, but it is not yet clear how it performs in the intermediate (bonding) regime. However, in order to predict the equilibrium geometry of a van der Waals bond molecule, it is not just sufficient to have the correct asymptotic $1/R^6$ attraction. Rather, the complete energy surface must be accurate. This point is investigated in Fig. 2.14, in which the energy surface (E_b) of the He dimer is shown [24]. He$_2$ is a particularly critical system, which manifests itself in the scale used for Fig.2.14: It is meV, rather than the standard scale of eV. The He$_2$ bond is extremely

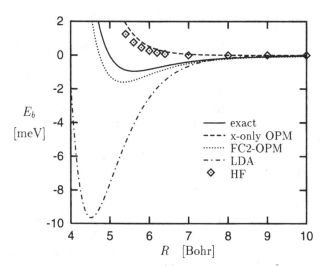

Fig. 2.14. Energy surface E_b of He$_2$: $E_c^{(2)}$ in combination with exact E_x (FC2) versus LDA, x-only OPM and HF [79] data as well as the exact result [80]

weak, which leads to a very delocalized ground-state wavefunction [81]. It thus provides an ideal testing ground for approximate correlation functionals. In Fig. 2.14 results from three DFT calculations are compared with HF data [79] and the exact E_b, obtained by variational calculations with correlated wavefunctions [80] (all E_b are strictly nonrelativistic). Focusing first on the x-only results, one immediately notices that dispersive bonding is a pure correlation effect. Both the exact x-only OPM calculation and its HF counterpart predict a repulsive energy surface. Moreover, as in so many other situations (see Sect. 2.3) the x-only OPM data are very close to the HF numbers. Two "correlated" DFT results are also shown in Fig. 2.14. On the one hand, the energy surface resulting from a standard LDA calculation is plotted. The LDA predicts the minimum of E_b to be too far in by 1 bohr and the corresponding well depth to be too large by an order of magnitude: As pointed out in Sect. 2.1.2, the LDA requires the densities of the two atoms to overlap, in order to produce binding. It therefore contracts the dimer far too much. The LDA is not suitable to describe such systems, in spite of the fact that it generates an attractive E_b.

The second "correlated" DFT calculation is based on the combination of the exact E_x with $E_c^{(2)}$ (FC2) [24] (the latter energy is added perturbatively to a self-consistent x-only calculation within the KLI approximation). The agreement of the resulting E_b with the exact surface is not perfect, but the resulting $E_b(R)$ is at least qualitatively correct. Thus $E_c^{(2)}$ does not only give the desired $1/R^6$ behavior, but also provides a realistic description of the rest of $E_b(R)$.

Given the fact that $E_c^{(2)}$ comes from second order perturbation theory, one might be tempted to consider this a trivial result: As the arguments given in the beginning of this section apply to any second order perturbative energy functional, the $1/R^6$ behavior is common to all of them. However, as pointed out earlier the correct long-range behavior does not imply that the complete $E_b(R)$ is accurate. This is demonstrated explicitly in Fig. 2.15, in which three second order results for $E_b(R)$ are compared. In addition to the FC2-result already shown in Fig. 2.14 the surface obtained with the conventional MP2 approach (second order perturbation theory on the basis of HF) is given. The FC2-result overestimates the exact well depth to roughly the same extent as the MP2 surface underestimates it (30%). The location of the minimum of $E_b(R)$ is predicted to be too large by 0.2 bohr by the MP2 approach, while the corresponding FC2-number is too small by 0.2 bohr.

In the third second order approach the FC2 energy functional is evaluated with LDA orbitals, i.e. the difference between the LDA and the FC2 functional is added perturbatively to the LDA surface. By construction, this functional yields the $1/R^6$ asymptotics. However, the C_6 coefficient obtained from (2.109) with LDA orbitals [78] is much larger than the C_6 resulting from x-only OPM orbitals (which is already too large). This is reflected by the large-R behavior of the corresponding energy surface. Even more impor-

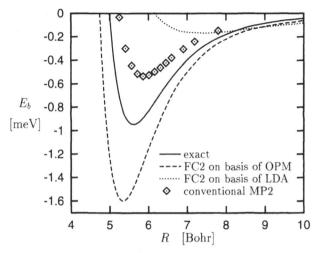

Fig. 2.15. Energy surface $E_b(R)$ of He$_2$: Different perturbation expansions to second order. $E_c^{(2)}$ in combination with exact E_x (FC2) versus conventional MP2 [82] and second order expansion on the basis of LDA orbitals as well as exact [80] result

tant is the complete failure of this third second order expansion in the vicinity of the minimum of E_b. The minimum is located roughly 2 bohr too far out and its depth is too small by an order of magnitude. This demonstrates that the application of second order perturbation theory does not automatically guarantee a realistic energy surface for dispersive bonds: A suitable noninteracting reference Hamiltonian, which provides the starting point for the expansion, is required. The energy surface obtained with the FC2-functional on the basis of the OPM should thus be understood as an encouraging result.

One final point related to the energy surface of He$_2$ seems worth a remark. In Fig. 2.16 the FC2 result is split into its individual components. In addition to the x-only part one has two contributions from $E_c^{(2)}$,

$$\Delta^{\mathrm{MP2}}(R) = E_c^{\mathrm{MP2}}[\mathrm{He_2}](R) - 2E_c^{\mathrm{MP2}}[\mathrm{He}](R) \tag{2.112}$$

$$\Delta^{\Delta\mathrm{HF}}(R) = E_c^{\Delta\mathrm{HF}}[\mathrm{He_2}](R) - 2E_c^{\Delta\mathrm{HF}}[\mathrm{He}](R) , \tag{2.113}$$

corresponding to the two ingredients of (2.82). One can explicitly see that $\Delta^{\Delta\mathrm{HF}}$ vanishes exponentially for large R, while Δ^{MP2} gives the dominant, attractive component of E_b. However, in the vicinity of the minimum of E_b the contribution of $\Delta^{\Delta\mathrm{HF}}$ is no longer negligible. The energy surface of He$_2$ is one of the rare cases in which the $E_c^{\Delta\mathrm{HF}}$ contribution is large.

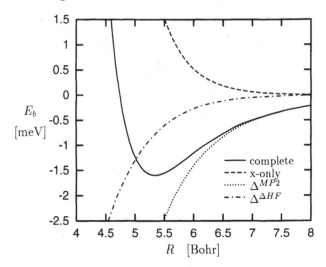

Fig. 2.16. Energy surface of He$_2$: Decomposition of FC2-OPM result into components

Table 2.9. Correlation energies $(-E_c)$ of closed-subshell atoms: LDA [29], PW91-GGA [30], CS [23], $E_c^{(2)}$ [18] and ISI [21] results (all energies obtained by insertion of x-only densities) in comparison with MP2 [84,85] and exact [83] energies (in mhartree). The contribution (2.85) to $E_c^{(2)}$ is also listed separately

	MP2	exact	ISI	$E_c^{(2)}$	$-E_c^{\Delta HF}$	LDA	GGA	CS
He	37	42	39	48	0.0	113	46	42
Be	76	94		124	0.6	225	94	93
Ne	388	391	410	477	1.7	746	382	375
Mg	428	438		522	3.2	892	450	451
Ar	709	722		866	5.4	1431	771	743
Ca	798			996	6.4	1581	847	824
Zn	1678			2016	14.9	2668	1526	1426
Cd	2618			3104	19.5	4571	2739	2412
Xe	3088			3487	17.7	5199	3149	2732

2.6.2 Comparison of Available Orbital-Dependent Approximations for E_c

After establishing the basic ability of $E_c^{(2)}$ to deal with dispersion forces, the next step is a quantitative study of more conventional systems. In Table 2.9 the correlation energies obtained with this functional for closed-subshell atoms are compared with various other approximations and the exact correlation energies [83] (which have been extracted by combining variational results for two- and three-electron systems with experimental data for the ionization energies of the remaining electrons). The LDA energies show the

Table 2.10. Correlation energies $(-E_c)$ of the He isoelectronic series: LDA [29], PW91-GGA [30], CS [23], $E_c^{(2)}$ [18] and ISI [21] results (all energies obtained by insertion of x-only densities) in comparison with MP2 [86] and exact [87] energies (in mhartree).

Ion	LDA	GGA	CS	$E_c^{(2)}$	ISI	exact	MP2
He	112.8	45.9	41.6	48.21	39.4	42.04	37.1
Ne^{8+}	203.0	61.7	40.6	46.81	45.0	45.69	44.4
Ca^{18+}	243.3	67.7	35.9	46.69	45.8	46.18	45.4
Zn^{28+}	267.2	71.3	33.2	46.67		46.34	45.7
Zr^{38+}	284.4	74.0	31.4	46.66	46.3	46.42	45.9
Sn^{48+}	297.7	76.0	30.0	46.65		46.47	46.0
Nd^{58+}	308.7	77.8	29.0	46.64	46.3	46.51	
Yb^{68+}	318.0	79.3	28.2	46.63		46.53	
Hg^{78+}	326.1	80.6	27.6	46.62	46.4	46.55	
Th^{88+}	333.2	81.7	27.0	46.62		46.56	
Fm^{98+}	339.6	82.8	26.0	46.62	46.4	46.57	

well-known overestimation of the correct atomic E_c by a factor of 2. The GGA [30] impressively improves on that. The CS-functional [23] also leads to rather accurate values for light atoms, but substantially underestimates the correct E_c of the heavier species. $E_c^{(2)}$ clearly overestimates atomic correlation energies, consistent with the result for the energy surface of He_2. The deviations are much larger than those observed for the GGA. Moreover, the accuracy of $E_c^{(2)}$ is obviously lower than that of the conventional MP2 scheme, in particular for atoms with many electrons. On the other hand, the ISI extension of $E_c^{(2)}$ seems to eliminate most of the error of the second order functional.

In Table 2.9 the $E_c^{\Delta HF}$ component of the complete $E_c^{(2)}$ is also listed separately. $E_c^{\Delta HF}$ vanishes for two-electron systems and is more than 2 orders of magnitude smaller than E_c^{MP2} for all other atoms. This result suggests that $E_c^{\Delta HF}$ can be neglected in most situations, which definitively simplifies the application of $E_c^{(2)}$.

Analogous data for the He isoelectronic series are given in Table 2.10 [62]. These numbers verify that $E_c^{(2)}$ is the only available density functional which satisfies the correct scaling law with respect to the nuclear charge Z. It becomes exact in the limit of large Z, in which the correlation energy of two-electron ions approaches a constant. The GGA energies, on the other hand, show a systematic increase with increasing Z. The opposite behavior is found for the CS functional, leading to an error of 50% for heavy ions. The ISI functional, whose dominating ingredient is $E_c^{(2)}$, also approaches a constant for large Z. However, the contributions to E_c^{ISI} beyond its second order basis

Table 2.11. Correlation energy and electron affinity of H^-: Results obtained by combination of the exact exchange with different correlation functionals (LDA [29], PW91-GGA [30], CS [23], C2 ($E_c^{(2)}$) [18] and ISI [21]) in comparison with MP2 [88] and exact [89] energies (in mhartree).

Method	$-E_c$	EA
F	—	−12.1
F+LDA	75.7	62.6
F+GGA	35.5	22.8
F+CS	31.2	18.9
F+C2	54.6	42.6
F+ISI	34.3	22.2
exact	39.8	27.8
MP2	27.3	15.2

$E_c^{(2)}$ do not completely vanish in this limit, so that the ISI correlation energy is slightly smaller than the exact result.

A more sensitive test for correlation functionals than total atomic correlation energies is provided by atomic EAs. In Table 2.11 the EAs for H^- obtained with various functionals are listed [62]. In all cases the exact exchange is used, only the correlation part of E_{xc} varies. As to be expected, the x-only calculation predicts H^- to be unbound, while LDA correlation produces an EA which is more than a factor of 2 too large. The CS result substantially underestimates the exact EA. $E_c^{(2)}$ clearly overestimates the correlation energy of H^- and thus also the EA: As for the correlation energies of neutral atoms and positive ions, $E_c^{(2)}$ and the conventional MP2 energies bracket the exact EA. The inclusion of higher order terms in $E_c^{(2)}$ via the ISI improves the agreement, although one notices a tendency to overcorrect the error of the straight second order expansion. Moreover, the GGA correlation functional yields a similarly accurate number, thus questioning the usefulness of the implicit correlation functionals.

The next step of the analysis of orbital-dependent correlation functionals consists of a look at covalently bond molecules. Table 2.12 lists the spectroscopic constants of N_2 obtained with a variety of approximations – the data for N_2 are quite characteristic for many diatomic molecules. First of all, one observes that the x-only approach yields a reasonably accurate bond length (R_e), but substantially underestimates the experimental dissociation energy (D_e), which emphasizes the importance of correlation. In Table 2.12 five different correlation functionals are added to the exact exchange. Use of GGA correlation improves D_e, although the remaining error is quite large. At the same time GGA correlation leads to a reduction of R_e, thus worsening the agreement with experiment. The correction also goes into the wrong direction in the case of the vibrational frequency. The same statements apply

Table 2.12. Spectroscopic constants of N_2: Results obtained by combination of the exact exchange with different correlation functionals (LDA [29], PW91-GGA [30], CS [23], SIC-LDA [22], C2 ($E_c^{(2)}$) [18] and ISI [21]) in comparison with HF, MP2, quadratic configuration interaction with single and double excitations (QCISD) [55], complete SIC-LDA, conventional LDA and PW91-GGA, as well as experimental [90] results (all OPM values rely on the KLI approximation, F+CS data from [91])

method	R_e (Bohr)	$E_b = D_e + \hbar\omega_e/2$ (eV)	ω_e (cm^{-1})
expt.	2.075	9.908	2360
HF	2.037	4.952	2738
MP2	2.135	9.333	2180
QCISD	2.105	8.488	2400
F	2.011	4.972	2736
F+GGA	1.997	7.574	2801
F+CS	1.998	7.818	
F+SIC-LDA	2.003	7.880	2770
F+C2		unbound	
F+ISI	2.235	12.225	1401
SIC-LDA(x+c)	1.876	−49.490	3245
LDA	2.068	11.601	2396
GGA	2.079	10.545	2352

to combinations of the exact exchange with the CS functional and with the SIC-LDA for correlation.

Turning to the first-principles orbital-dependent correlation functionals, one realizes that $E_c^{(2)}$ does not predict N_2 to be bound at all. To understand this result one has to go back to (2.83) and examine the structure of this expression. E_c^{MP2} represents the interaction of two simultaneous particle-hole excitations: The probability for these transitions is determined by the Slater integrals in the numerator of (2.83), while their lifetime depends on the energy gap in the denominator. If the separation of the two N atoms is increased, the highest occupied and the lowest unoccupied KS levels in the molecule approach each other further and further. So, with increasing R the energy gap in the denominator of (2.83) shrinks more and more, i.e. the lifetime of the excitations becomes longer and longer. As this divergence is not fully compensated by the numerator, E_c^{MP2} becomes larger and larger when the atoms are taken apart. This effect does not only show up for large R, but already in the vicinity of the equilibrium distance. As a consequence, one does not even find a local minimum in the energy surface.

The problem is intrinsically related to the existence of the Rydberg series in the OPM spectrum, which originates from the $-1/r$ behavior of the exact exchange potential. The conventional MP2 calculation gives quite reasonable results for N_2, as the underlying HF Hamiltonian does not yield a Rydberg series. The same basic effect shows up in various places: One finds, for example, that the correlation energy of Be is particularly overestimated by $E_c^{(2)}$, which is due to the presence of the low-lying unoccupied $2p$ states (compare Table 2.9). The presence of the Rydberg series is useful for the calculation of many atomic properties, most notably for the description of negative ions or of excited states. On the other hand, it is not very helpful if the treatment of correlation is based on some kind of perturbation theory.

This immediately raises the question whether the effective resummation of the perturbation series via the ISI can resolve this fundamental problem? The ISI functional indeed leads to a bound N_2. However, it does not perform particularly well from a quantitative point of view. It overestimates R_e much more than the x-only calculation underestimates it. At the same time, the dissociation energy is much too large and the vibrational frequency reflects the very shallow form of the energy surface.

For completeness, Table 2.12 also contains the spectroscopic constants obtained with the SIC-LDA for both exchange and correlation. In this calculation, the standard molecular orbitals were used for the evaluation of the xc-energy and potential, without further localization prescription. The importance of the unitarity problem discussed in Sect. 2.5.2 is obvious. In particular, the dissociation energy is completely off. This problem can be traced to the SIC energies of the molecular core states [72]: The Coulomb contribution to (2.102) resulting from the two-center molecular core states $(1\pi_g, 1\pi_u)$ differs substantially from that obtained with the one-center atomic core states $(1s)$, as

$$\phi_{1\pi_{g/u}} \approx \frac{1}{\sqrt{2}} \Big[\phi_{1s,a} \pm \phi_{1s,b} \Big].$$

Realistic SIC-LDA results for molecules or solids can only be obtained on the basis of some localization prescription for quasi-degenerate states [72]. Such a scheme essentially consists of using the localized linear combinations of $\phi_{1\pi_{g/u}}$ for the evaluation of the SIC functional and has to be applied to all core and semi-core states. It is obvious that such a prescription becomes rather difficult to handle in more complicated molecules involving several types of atoms. One thus has to realize that none of the presently available implicit functionals can compete with the standard LDA or GGA for covalently bond molecules.

In summary, one can say that, while KS perturbation theory offers a DFT description of dispersion forces, most of the other results obtained within this approach are somewhat disappointing. This is particularly obvious when one takes into account that the computational demands of FC2-calculations are not different from those of conventional MP2-calculations, as essentially the same energy expressions have to be evaluated. So, why should one apply the second order OPM approach, rather than the well-established MP2 scheme? The initial hope was that the perturbation series on the basis of the KS reference system converges faster than the MP perturbation series, so that the second order term is sufficient for most purposes. However, all results in this section indicate that to second order the KS series is slightly less accurate than the MP expansion. In addition, one has the degeneracy problem at the Fermi surface if $E_c^{(2)}$ is combined with the exact exchange potential. On the other hand, if the KS perturbation expansion is not based on the exact exchange, it does not converge at all [92]. It thus seems that an appropriate implicit correlation functional, which could be used with the exact exchange, remains to be found.

2.6.3 Analysis of the Second Order Correlation Potential

Asymptotic Divergence for Finite Systems. The conclusions of Sect. 2.6.2 are further corroborated by an analysis of the potential corresponding to $E_c^{(2)}$. In the applications discussed in Sects. 2.6.1 and 2.6.2 this functional has always been applied perturbatively, on the basis of self-consistent calculations with the exact E_x. However, the ultimate goal of KS perturbation theory is the derivation of a correlation functional which can be used self-consistently, rather than just be added *a posteriori*.

The most complicated step of a self-consistent treatment of $E_c^{(2)}$ is the evaluation of the potential corresponding to its $E_c^{\Delta HF}$ component: Especially the calculation of the quadratic response function (2.91) is very demanding in the case of molecular or solid state systems. Fortunately, $E_c^{\Delta HF}$ turns out to be much smaller than E_c^{MP2} in most situations, which suggests its neglect. Moreover, $E_c^{\Delta HF}$ vanishes identically for two-electron systems, as in this case the relevant matrix elements of the multiplicative OPM exchange potential coincide with those of the nonlocal HF-type potential. Thus, in order to avoid the ambiguity associated with the neglect of $E_c^{\Delta HF}$, the discussion in this section completely focuses on two-electron systems.

Before examining the full OPM equation for E_c^{MP2}, it is instructive to take a closer look at the KLI approximation, as it reveals the ingredients of $v_c^{MP2} = \delta E_c^{MP2}/\delta n$ more clearly [25]. The KLI results for v_c^{MP2} obtained with the exact x-only OPM states for H^-, He, Be^{2+} and Ne^{8+} are plotted in Fig. 2.17. One finds that $v_c^{MP2,KLI}$ diverges asymptotically for all systems (in the case of H^- the divergence sets in somewhat further out). What is the reason for this unphysical behavior? Reduced to the most essential ingredients the KLI approximation has the structure

$$v_c^{KLI}(\boldsymbol{r}) \sim \frac{1}{n(\boldsymbol{r})} \sum_k \phi_k^\dagger(\boldsymbol{r}) \frac{\delta E_c}{\delta \phi_k^\dagger(\boldsymbol{r})} + \cdots . \tag{2.114}$$

The expression on the right-hand side now has to be analyzed for large $\boldsymbol{r}$. For E_c^{MP2} the sum over k runs over all levels of the KS spectrum. In order to extract the behavior of the numerator of (2.114) for large $\boldsymbol{r}$, two basic situations must be distinguished: If k corresponds to an unoccupied level,

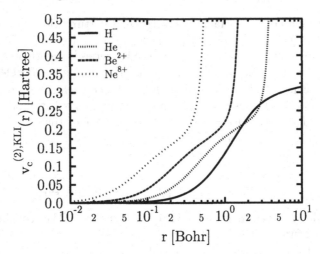

Fig. 2.17. Second order correlation potential of two-electron systems: Perturbative evaluation of KLI-approximation on the basis of exact x-only orbitals. (for two-electron systems $v_c^{(2)} = v_c^{MP2}$)

$\delta E_c^{MP2}/\delta \phi_k^\dagger(r)$ decays as fast as the highest occupied KS state ϕ_F (The Slater integrals in (2.83) always link an unoccupied with an occupied state, so that differentiation with respect to the unoccupied state breaks the integral and leaves an occupied state). If k denotes an occupied level $\delta E_c^{MP2}/\delta \phi_k^\dagger(r)$ decays slower than ϕ_F. On the other hand, the density in the denominator of (2.114) behaves like $|\phi_F(r)|^2$. Thus qualitatively one finds

$$v_c^{MP2,KLI}(r) \sim \frac{|\phi_F(r)\,\phi_{unocc}(r)|}{|\phi_F(r)|^2} . \qquad (2.115)$$

As all unoccupied states decay more slowly than ϕ_F, $v_c^{MP2,KLI}$ diverges for large r.

One might first hope that this divergence is a consequence of the KLI approximation. Unfortunately, this is not the case. One can explicitly verify that the divergence is present within the full OPM [25]. For the closed-subshell atoms considered in this section the OPM equation reduces to a radial integral equation [3],

$$\int_0^\infty dr'\, K(r,r')\, v_{xc}(r') = Q_{xc}(r) , \qquad (2.116)$$

with a radial response function K and a corresponding inhomogeneity Q_{xc},

$$K(r,r') = \frac{\delta[4\pi r^2 n(r)]}{\delta v_s(r')} \qquad (2.117)$$

$$Q_{xc}(r) = \sum_k \left\{ \int dr'\, \frac{\delta E_{xc}}{\delta \varphi_k(r')}\frac{\delta \varphi_k(r')}{\delta v_s(r)} + \frac{\partial E_{xc}}{\partial \epsilon_k}\frac{\delta \epsilon_k}{\delta v_s(r)} \right\} . \qquad (2.118)$$

Here φ_k denotes the radial part of the KS orbital, $\phi_{klm}(r) = \varphi_k(r)Y_{lm}(\Omega)/r$ (φ_k is chosen to be real). Both $K(r, r')$ and $Q_{\mathrm{xc}}(r)$ can be explicitly written in terms of the $\varphi_k(r)$ and the corresponding second (non-normalizable) solutions $\chi_k(r)$ of the radial KS equations for the KS eigenvalues [3].

On this basis now a *reductio ad absurdum* will be used to show that $v_{\mathrm{c}}^{\mathrm{MP2}}(r)$ diverges for large r. Let us assume that $v_{\mathrm{c}}^{\mathrm{MP2}}$ does not diverge, so that $v_{\mathrm{c}}^{\mathrm{MP2}}$ can be chosen to satisfy

$$v_{\mathrm{c}}^{\mathrm{MP2}}(r) \xrightarrow[r\to\infty]{} 0 \quad\Longrightarrow\quad v_{\mathrm{xc}}(r) = v_{\mathrm{x}}(r) + v_{\mathrm{c}}^{\mathrm{MP2}}(r) \xrightarrow[r\to\infty]{} 0 \,. \ (2.119)$$

With this assumption one finds [47]

$$\int_0^\infty dr'\, K(r, r')v_{xc}(r') \xrightarrow[r\to\infty]{} \Theta_F |\varphi_F(r)|^2 \int_0^r dr'\, \varphi_F(r')\chi_F(r')$$

$$\times \int_0^\infty dr''\, |\varphi_F(r'')|^2 v_{xc}(r'') \,. \tag{2.120}$$

The asymptotic behavior of the left-hand side of the OPM equation (2.116) is thus controlled by the product of the density of the highest occupied state and an indefinite integral over $\varphi_F(r)\chi_F(r)$, which is only weakly r-dependent. These r-dependent functions are multiplied by a constant, which is the orbital expectation value of v_{xc}. Equation (2.120) is separately valid for exchange and correlation. If one divides the OPM equation for $E_{\mathrm{c}}^{\mathrm{MP2}}$ by that for the exact exchange one ends up with

$$\lim_{r\to\infty} \frac{Q_{\mathrm{c}}^{\mathrm{MP2}}(r)}{Q_{\mathrm{x}}(r)} = \lim_{r\to\infty} \frac{\int_0^\infty dr'\, K(r, r')v_{\mathrm{c}}^{\mathrm{MP2}}(r')}{\int_0^\infty dr''\, K(r, r'')v_{\mathrm{x}}(r'')}$$

$$= \frac{\int_0^\infty dr'\, |\varphi_F(r')|^2 v_{\mathrm{c}}^{\mathrm{MP2}}(r')}{\int_0^\infty dr''\, |\varphi_F(r'')|^2 v_{\mathrm{x}}(r'')} = C \,. \tag{2.121}$$

Provided that the assumption (2.119) is correct, the ratio $Q_{\mathrm{c}}^{\mathrm{MP2}}(r)/Q_{\mathrm{x}}(r)$ must approach a constant for large r.

This ratio can be examined numerically, relying on standard finite differences methods for the evaluation of the radial inhomogeneities. Results for He are shown in Figs. 2.18 and 2.19.

For simplicity, only the contribution of the unoccupied Rydberg states to $Q_{\mathrm{c}}^{\mathrm{MP2}}$ is included in these plots, while the contribution of the continuum states to the corresponding sums in $E_{\mathrm{c}}^{\mathrm{MP2}}$ and (2.118) is neglected (compare the result including continuum states in [25]). Fig. 2.18 demonstrates the convergence of $Q_{\mathrm{c}}^{\mathrm{MP2}}$ with increasing number of Rydberg levels. As in the case of $E_{\mathrm{c}}^{\mathrm{MP2}}$ itself, summation up to the shell with $n = 10$ provides essentially the complete result. In addition, Fig. 2.18 illustrates the oscillatory nature of $Q_{\mathrm{xc}}(r)$: In fact, as a consequence of (2.47), the radial integral over Q_{xc} must vanish, as long as only discrete states are involved,

$$\int_0^\infty dr\, Q_{\mathrm{xc}}(r) = 0 \,.$$

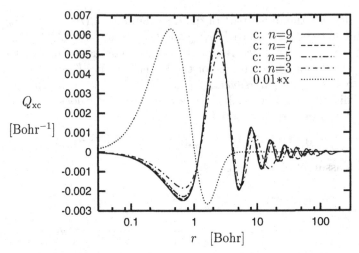

Fig. 2.18. Inhomogeneity of the radial OPM integral equation for He: Contribution of Rydberg states to Q_c^{MP2} versus exact exchange (Q_x is scaled by 10^{-2}). Q_c^{MP2} is given for four different sets of unoccupied states in the corresponding sums in E_c^{MP2} and (2.118). In order to cover the wide range of r logarithmic grids of up to 6400 points have been used

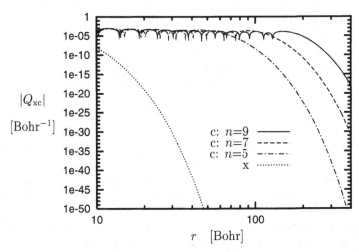

Fig. 2.19. Asymptotic behavior of Q_{xc} for He: Contribution of Rydberg states to Q_c^{MP2} versus exact Q_x. To allow for a logarithmic scale the absolute values are plotted (the zeros of Q_c^{MP2} are suppressed)

The asymptotic behavior of Q_c^{MP2}, which is of interest in the present context, is plotted in Fig. 2.19. For large r one finds that $Q_x(r)$ goes go to zero exponentially (as is immediately clear from (2.22) and (2.28)). On the other hand, the behavior of $Q_c^{\mathrm{MP2}}(r)$ depends on the highest unoccupied state included: The more Rydberg levels are taken into account, the further out the exponential decay sets in. In fact, Fig. 2.19 indicates that a complete resummation of all Rydberg states leads to a power law decay of $Q_c^{\mathrm{MP2}}(r)$. In any case, even if only a restricted number of unoccupied states is included, $Q_c^{\mathrm{MP2}}(r)$ vanishes much more slowly than $Q_x(r)$. The ratio between Q_c^{MP2} and $Q_x(r)$ diverges exponentially, in contradiction to (2.121). This result can also be verified analytically, as long as only a finite number of discrete unoccupied states is considered. Consequently, the assumption (2.119) must be wrong, which implies that v_c^{MP2} diverges[8] for large r.

It is tempting to associate this divergence with the use of second order perturbation theory: One could basically argue that the asymptotic region is a low density regime, while the perturbation expansion is typically a high density expansion. Within this physically motivated picture it is not surprising that for large r the second order expansion yields a non-physical potential. However, one should be aware that the basic mechanism which leads to the divergence of v_c^{MP2} is also present for any other functional which links occupied with unoccupied KS states at the same point r, including resummed forms of the perturbation series.

Approximate Second Order Correlation Potential. How can one avoid the asymptotic divergence of $v_c^{\mathrm{MP2}} = \delta E_c^{\mathrm{MP2}}/\delta n$ and related potentials? It is obvious from the discussion of the preceding section that a modification of the OPM procedure is required if one wants to keep the energy functional itself unchanged. A suitable modification is most easily introduced on the level of the individual pair correlation energies e_{ij}^{MP2},

$$e_{ij}^{\mathrm{MP2}} = \frac{e^4}{2} \sum_{k,l} (1 - \Theta_k)(1 - \Theta_l) \frac{(ij\|kl)[(kl\|ij) - (kl\|ji)]}{\epsilon_i + \epsilon_j - \epsilon_k - \epsilon_l} \quad (2.122)$$

$$E_c^{\mathrm{MP2}} = \sum_{i,j} \Theta_i \Theta_j e_{ij}^{\mathrm{MP2}} .$$

The basic structure of e_{ij}^{MP2} is somewhat similar to that of the Green's function (2.25) which suggests to apply a closure approximation (CA) in analogy to the KLI approximation (2.53), i.e. to approximate the denominator in (2.122) by some average eigenvalue difference,

$$\epsilon_i + \epsilon_j - \epsilon_k - \epsilon_l \approx \Delta\epsilon .$$

---

[8] The onset of this divergence is even visible in the basis set results for v_c^{MP2} shown in Fig. 3 of [93], in spite of the fact that a basis set representation of v_c^{MP2} ultimately truncates the divergent asymptotic behavior.

After this replacement one can use the completeness of the KS spectrum to eliminate all unoccupied states from (2.122),

$$e_{ij}^{CA} = \frac{e^4}{2} \frac{1}{\Delta\epsilon} \left\{ (ij|||ij) - (ij|||ji) - \sum_{k,l} \Theta_k \Theta_l (ij||kl)[(kl||ij) - (kl||ji)] \right\}$$
(2.123)

with the matrix element

$$(ij|||kl) = \int d^3r_1 \int d^3r_2 \, \frac{\phi_i^\dagger(\mathbf{r}_1)\phi_k(\mathbf{r}_1)\phi_j^\dagger(\mathbf{r}_2)\phi_l(\mathbf{r}_2)}{|\mathbf{r}_1 - \mathbf{r}_2|^2} \qquad (2.124)$$

(these higher order Coulomb integrals can be evaluated with the same methods as used for the standard Slater integrals). With the closure approximated pair-correlation energies one can then rewrite E_c^{MP2} in the form

$$E_c^{MP2} \equiv \sum_{i,j} \Theta_i \Theta_j w_{ij} \, e_{ij}^{CA}, \qquad (2.125)$$

with the weights w_{ij} given by

$$w_{ij} = e_{ij}^{MP2}/e_{ij}^{CA}. \qquad (2.126)$$

Until now E_c^{MP2} has only been recast, but not modified, as e_{ij}^{CA} and thus $\Delta\epsilon$ drop out of (2.125). The crucial step is the handling of the functional (2.125) in the OPM procedure. The form (2.125) suggests to restrict the OPM variation to the orbital-dependence of e_{ij}^{CA}, while keeping the weights w_{ij} fixed throughout the complete solution of (2.27) [25]. In this way the unoccupied KS states do no longer contribute to the r dependence of the inhomogeneity (2.28), they are only required for the evaluation of w_{ij}.

The potential obtained with this scheme for Ne is shown in Fig. 2.20 (this result is quite characteristic for all atoms considered so far). The closure approximated v_c^{MP2} is compared with the corresponding PW91-GGA and CS potentials as well as the exact v_c, which was extracted from Monte-Carlo calculations [57]. It is obvious from Fig. 2.20 that neither the GGA nor the CS potential have much in common with the exact v_c, while the modified second order potential $v_c^{MP2,CA}$ reproduces its main features: $v_c^{MP2,CA}$ is positive in the valence region and follows the shell structure of the exact v_c. Unfortunately, it clearly overestimates the shell oscillations and decays too slowly for large r. In particular, this latter deficiency prevents the direct use of $v_c^{MP2,CA}$ in applications. Nevertheless, $v_c^{MP2,CA}$ is the first DFT correlation potential which shows at least qualitative agreement with the exact result, reflecting the first-principles origin of $E_c^{(2)}$. In addition, one can hope that the quantitative agreement can be improved by a suitable resummation of the perturbation series (and, perhaps, a refined choice of w_{ij}).

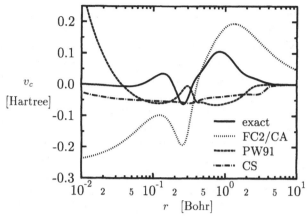

Fig. 2.20. Correlation potential of Neon: Closure approximated v_c^{MP2} (FC2/CA) [25] versus PW91-GGA [30], CS [23] and exact [57] result (all functionals have been evaluated with the exact KS orbitals)

2.7   Final Remarks

If one combines the results of Sect. 2.3 with those of Sect. 2.6 the emerging picture is somewhat ambiguous. On the one hand, the exact treatment of exchange is well established by now: Quite a number of practical realizations of the underlying x-only OPM formalism are available [3,50,51,58,59,94–98] and all applications indicate that DFT with the exact exchange is essentially equivalent to the HF approach (at least, in the case of ground-state problems). The use of the exact exchange most notably resolves the difficulties of the conventional xc-functionals with the description of negative ions. Furthermore, the KLI approximation [4] provides a very efficient and accurate tool for practical calculations with the exact exchange.

On the other hand, the orbital-dependent treatment of correlation represents a much more serious challenge than that of exchange: The systematic derivation of such functionals via standard many-body theory leads to rather complicated expressions. Their rigorous application within the OPM not only requires the evaluation of Coulomb matrix elements between the complete set of KS states, but, in principle, also relies on the knowledge of higher order response functions. In practical calculations, these first-principles functionals necessarily turn out to be rather inefficient, even if they are only treated perturbatively. In addition, the potential resulting from a large class of such functionals is non-physical for finite systems. Both problems are related to the presence of unoccupied states in the functionals which seems inevitable as soon as some variant of standard many-body theory is used for the derivation.

One thus has to find approximations that avoid the presence of Slater integrals connecting occupied with unoccupied states. Unfortunately, the available semi-empirical functionals, i.e. the SIC-LDA and the Colle-Salvetti functional, which both satisfy this requirement, do not give results comparable to what one obtains with standard GGAs. It seems that the derivation of suitable approximations must start from some first-principles expression like $E_c^{(2)}$ and then simplify this expression or, at least, the corresponding potential. One possible strategy of this type is the closure approximation illustrated in Sect. 2.6.3.

However, it may not be the construction of implicit correlation functionals which is the real domain of KS perturbation theory, but rather the density functional representation of the underlying correlated motion of the electrons. The crucial point is that the perturbation expansion on the basis of the KS Hamiltonian can be utilized for all kinds of many-body properties. This is true in particular for the many-body wavefunction $(r_1\sigma_1,\dots r_N\sigma_N|\Psi)$ itself, but also for the simplest quantity which reflects the correlated motion, the 2-particle density $\gamma(r_1, r_2)$,

$$\gamma(r_1, r_2) = \frac{N(N-1)}{2} \int d^3r_3 \dots d^3r_N \sum_{\sigma_1 \dots \sigma_N} |(r_1\sigma_1, r_2\sigma_2, r_3\sigma_3, \dots r_N\sigma_N|\Psi)|^2.$$

(2.127)

Using the perturbative approach of Sect. 2.4 to first order, one ends up with

$$\gamma^{(1)}(r_1, r_2)$$
$$= -\frac{1}{2} \sum_i \Theta_i \sum_k (1 - \Theta_k) \frac{\sum_j \Theta_j (kj||ji) + \langle k|v_x|i\rangle}{\epsilon_i - \epsilon_k}$$
$$\times \Big\{ \phi_i^\dagger(r_1)\phi_k(r_1)n(r_2) + n(r_1)\phi_i^\dagger(r_2)\phi_k(r_2)$$
$$- \sum_l \Theta_l \Big[\phi_l^\dagger(r_1)\phi_k(r_1)\phi_i^\dagger(r_2)\phi_l(r_2) + \phi_i^\dagger(r_1)\phi_l(r_1)\phi_l^\dagger(r_2)\phi_k(r_2) \Big] \Big\}$$
$$+ \frac{1}{2} \sum_{i,j} \Theta_i\Theta_j \sum_{k,l} (1 - \Theta_k)(1 - \Theta_l) \frac{(ij||kl) - (ij||lk)}{\epsilon_i + \epsilon_j - \epsilon_k - \epsilon_l}$$
$$\times \phi_i^\dagger(r_1)\phi_k(r_1)\phi_j^\dagger(r_2)\phi_l(r_2) + c.c.$$

(2.128)

This expression represents an implicit density functional for $\gamma(r_1, r_2)$ in the same sense as (2.19) is an implicit density functional for the exchange energy.

How realisticly does $\gamma^{(1)}$ describe the Coulomb correlation between the electrons? This question is answered in Fig. 2.21 which shows the 2-particle density of the He ground-state. For He one has only three relevant coordinates which fix the positions of the two electrons relative to the nucleus in the plane spanned by the three particles (the other three coordinates correspond to rotations of the triangle defined by the nucleus and the two electrons around the position of the nucleus). With the nucleus defining the origin of the coordinate system, the most suitable coordinates are the radial distances

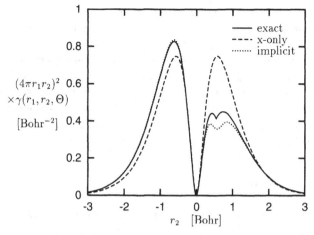

Fig. 2.21. 2-particle-density of helium: Radial structure for r_1=0.543 bohr, Θ = $\pi, 0$. First order perturbative implicit functional versus x-only and exact result [99]

between the two electrons and the nucleus, r_1, r_2, and the angle Θ between their positions, $\boldsymbol{r}_1 \cdot \boldsymbol{r}_2 = r_1 r_2 \cos \Theta$. As in the case of the radial single-particle density, the probability to find an electron in a certain radial range $[r_1, r_1 + \delta]$ is obtained by multiplication of γ by r_1^2. The appropriate quantity to plot is thus $(4\pi)^2 r_1^2 r_2^2 \gamma(r_1, r_2, \Theta)$, so that the radial density $4\pi r_1^2 n(r_1)$ is obtained by integration over r_2 and Θ without further manipulation,

$$4\pi r_1^2 n(r_1) = (4\pi)^2 r_1^2 \int_0^\infty r_2^2 \, dr_2 \int_0^\pi d\Theta \, \gamma(r_1, r_2, \Theta) \,.$$

In Fig. 2.21 r_1 has been set to 0.543 bohr, which is the radius at which $r_1^2 n(r_1)$ has its maximum. Together with the nucleus at the origin the position of the first electron defines a straight line along which the second electron is moved, i.e. Fig. 2.21 shows the r_2-dependence for $\Theta = 0$ (positive r_2-values) and $\Theta = \pi$ (negative r_2-values). For $r_2 = +0.543$ bohr the two electrons sit on top of each other.

The solid line represents the exact result, obtained from the most accurate variational wavefunction of Kinoshita [99]. One can see that the two electrons preferably move on opposite sides of the nucleus: The likelihood to come close to each other is only half as large as that of remaining on opposite sides. Moreover, the 2-particle density clearly shows the electron-electron cusp at $\boldsymbol{r}_1 = \boldsymbol{r}_2$. Figure 2.21 also provides the x-only result, which corresponds to the ground-state KS Slater determinant. As this determinant only contains Pauli, but not Coulomb correlation, the electrons move independently in the x-only approximation,

$$\gamma_s(r_1, r_2, \Theta) = |\phi_{1s}(r_1)|^2 |\phi_{1s}(r_2)|^2 \,.$$

The probability for the electrons to sit on top of each other is as high as that for sitting at the same radial distance on opposite sides of the nucleus. Finally, the implicit functional (2.128) is plotted. It incorporates most of the Coulomb correlation in γ. It slightly overestimates the suppression of the on-top position, but, overall, is rather realistic (the somewhat less pronounced cusp is a basis set, i.e. numerical, effect). This result demonstrates explicitly that the concept of implicit functionals can not only be used for xc-energies, but also for the description of local correlation, which might be of interest in the context of scattering and multiple excitation processes.

Acknowledgments

I would like to thank my colleagues R. M. Dreizler, A. Facco Bonetti, A. Höck, R. N. Schmid and S. H. Vosko for many valuable discussions and their contributions to the work presented in this review. It was a pleasure for the author to participate in the Coimbra school on DFT organized by C. Fiolhais, F. Nogueira and M. Marques in the Caramulo mountains. There can be no doubt that the inspiring environment and the excellent food at the conference site provided the right momentum for the joint effort which led to this book.

References

1. R. M. Dreizler and E. K. U. Gross, *Density Functional Theory* (Springer, Berlin, 1990).
2. R. T. Sharp and G. K. Horton, Phys. Rev. **90**, 317 (1953).
3. J. D. Talman and W. F. Shadwick, Phys. Rev. A **14**, 36 (1976).
4. J. B. Krieger, Y. Li, and G. J. Iafrate, Phys. Lett. A **146**, 256 (1990).
5. W. Kohn and L. J. Sham, Phys. Rev. **140 A**, 1133 (1965).
6. S.-K. Ma and K. A. Brueckner, Phys. Rev. **165**, 18 (1968).
7. D. C. Langreth and M. J. Mehl, Phys. Rev. Lett. **47**, 446 (1981).
8. D. C. Langreth and M. J. Mehl, Phys. Rev. B **28**, 1809 (1983).
9. C. D. Hu and D. C. Langreth, Phys. Rev. B **33**, 943 (1986).
10. J. P. Perdew and Y. Wang, Phys. Rev. B **33**, 8800 (1986).
11. J. P. Perdew, Phys. Rev. B **33**, 8822 (1986).
12. J. P. Perdew, in *Electronic Structure of Solids 1991*, edited by P. Ziesche and H. Eschrig (Akademie Verlag, Berlin, 1991), p. 11.
13. J. P. Perdew, K. Burke, and M. Ernzerhof, Phys. Rev. Lett. **77**, 3865 (1996).
14. A. D. Becke, J. Chem. Phys. **84**, 4524 (1986).
15. A. D. Becke, Phys. Rev. A **38**, 3098 (1988).
16. J. Harris and R. O. Jones, J. Phys. F **4**, 1170 (1974).
17. L. J. Sham, Phys. Rev. B **32**, 3876 (1985).
18. A. Görling and M. Levy, Phys. Rev. A **50**, 196 (1994).
19. T. Kotani, J. Phys.: Condens. Matter **10**, 9241 (1998).

20. E. Engel and A. Facco Bonetti, in *Quantum Systems in Theoretical Chemistry and Physics, Vol.1: Basic Problems and Model Systems*, edited by A. Hernández-Laguna, J. Maruani, R. McWeeny, and S. Wilson (Kluwer, Doordrecht, 2000), p. 227.
21. M. Seidl, J. P. Perdew, and S. Kurth, Phys. Rev. Lett. **84**, 5070 (2000).
22. J. P. Perdew and A. Zunger, Phys. Rev. B **23**, 5048 (1981).
23. R. Colle and O. Salvetti, Theoret. Chim. Acta (Berl.) **37**, 329 (1975).
24. E. Engel, A. Höck, and R. M. Dreizler, Phys. Rev. A **61**, 032502 (2000).
25. A. Facco Bonetti, E. Engel, R. N. Schmid, and R. M. Dreizler, Phys. Rev. Lett. **86**, 2241 (2001).
26. R. N. Schmid *et al.*, Adv. Quant. Chem. **33**, 209 (1998).
27. S. Varga *et al.*, J. Phys. Chem. A **104**, 6495 (2000).
28. W. Liu and C. van Wüllen, J. Chem. Phys. **113**, 2506 (2000).
29. S. H. Vosko, L. Wilk, and M. Nusair, Can. J. Phys. **58**, 1200 (1980).
30. J. P. Perdew *et al.*, Phys. Rev. B **46**, 6671 (1992).
31. E. Engel *et al.*, Phys. Rev. A **52**, 2750 (1995).
32. E. Engel, S. Keller, and R. M. Dreizler, Phys. Rev. A **53**, 1367 (1996).
33. R. O. Jones and O. Gunnarsson, Rev. Mod. Phys. **61**, 689 (1989).
34. E. Engel, J. A. Chevary, L. D. Macdonald, and S. H. Vosko, Z. Phys. D **23**, 7 (1992).
35. K. Terakura, T. Oguchi, A. R. Williams, and J. Kübler, Phys. Rev. B **30**, 4734 (1984).
36. T. C. Leung, C. T. Chan, and B. N. Harmon, Phys. Rev. B **44**, 2923 (1991).
37. P. Dufek, P. Blaha, and K. Schwarz, Phys. Rev. B **50**, 7279 (1994).
38. E. Engel and S. H. Vosko, Phys. Rev. B **47**, 13164 (1993).
39. A. Svane and O. Gunnarsson, Phys. Rev. Lett. **65**, 1148 (1990).
40. V. Sahni, J. Gruenebaum, and J. P. Perdew, Phys. Rev. B **26**, 4371 (1982).
41. P. Hohenberg and W. Kohn, Phys. Rev. **136 B**, 864 (1964).
42. L. J. Sham and M. Schlüter, Phys. Rev. Lett. **51**, 1888 (1983).
43. M. E. Casida, Phys. Rev. A **51**, 2505 (1995).
44. J. B. Krieger, Y. Li, and G. J. Iafrate, Phys. Lett. A **148**, 470 (1990).
45. T. Grabo, T. Kreibich, S. Kurth, and E. K. U. Gross, in *Strong Coulomb Correlations in Electronic Structure Calculations: Beyond the Local Density Approximation*, edited by V. I. Anisimov (Gordon and Breach, Amsterdam, 1999), p. 203.
46. O. Gunnarsson and B. I. Lundqvist, Phys. Rev. B **13**, 4274 (1976).
47. E. Engel *et al.*, Phys. Rev. A **58**, 964 (1998).
48. E. Engel and S. H. Vosko, Phys. Rev. A **47**, 2800 (1993).
49. C. F. Fischer, *The Hartree-Fock Method for Atoms* (Wiley, New York, 1977).
50. S. Ivanov, S. Hirata, and R. J. Bartlett, Phys. Rev. Lett. **83**, 5455 (1999).
51. E. Engel, A. Höck, and R. M. Dreizler, Phys. Rev. A **62**, 042502 (2000).
52. L. Laaksonen, P. Pyykkö, and D. Sundholm, Comput. Phys. Rep. **4**, 313 (1986).
53. Y. Li, J. B. Krieger, and G. J. Iafrate, Chem. Phys. Lett. **191**, 38 (1992).
54. C.-O. Almbladh and U. von Barth, Phys. Rev. B **31**, 3231 (1985).
55. B. G. Johnson, P. M. Gill, and J. A. Pople, J. Chem. Phys. **98**, 5612 (1993).
56. A. D. McLean, B. Liu, and G. S. Chandler, *AIP Conference Proceedings* (AIP, Woodbury, 1982), Vol. 86, p. 90.
57. C. J. Umrigar and X. Gonze, Phys. Rev. A **50**, 3827 (1994).
58. T. Kotani and H. Akai, Phys. Rev. B **54**, 16502 (1996).

59. M. Städele, J. A. Majewski, P. Vogl, and A. Görling, Phys. Rev. Lett. **79**, 2089 (1997).
60. S. Massidda, M. Posternak, and A. Baldereschi, Phys. Rev. B **48**, 5058 (1993).
61. E. Engel, A. Höck, and S. Varga, Phys. Rev. B **63**, 125121 (2001).
62. E. Engel and R. M. Dreizler, J. Comput. Chem. **20**, 31 (1999).
63. D. C. Langreth and J. P. Perdew, Phys. Rev. B **15**, 2884 (1977).
64. J. Harris, Phys. Rev. A **29**, 1648 (1984).
65. M. Seidl, J. P. Perdew, and M. Levy, Phys. Rev. A **59**, 51 (1999).
66. I. Lindgren, Int. J. Quantum Chem. Symp. **5**, 411 (1971).
67. M. R. Norman and D. D. Koelling, Phys. Rev. B **30**, 5530 (1984).
68. J. G. Harrison, J. Chem. Phys. **78**, 4562 (1983).
69. J. G. Harrison, J. Chem. Phys. **79**, 2265 (1983).
70. J. G. Harrison, J. Chem. Phys. **86**, 2849 (1987).
71. M. R. Pederson and C. C. Lin, J. Chem. Phys. **88**, 1807 (1988).
72. M. R. Pederson, R. A. Heaton, and C. C. Lin, J. Chem. Phys. **80**, 1972 (1984).
73. M. R. Pederson, R. A. Heaton, and C. C. Lin, J. Chem. Phys. **82**, 2688 (1985).
74. J. Chen, J. B. Krieger, and G. J. Iafrate, private communication (2001).
75. T. Kato, Pure Appl. Math. **10**, 151 (1957).
76. T. Grabo and E. K. U. Gross, Chem. Phys. Lett. **240**, 141 (1995).
77. J. P. Perdew, S. Kurth, A. Zupan, and P. Blaha, Phys. Rev. Lett. **82**, 2544 (1999).
78. M. Lein, J. F. Dobson, and E. K. U. Gross, J. Comput. Chem. **20**, 12 (1999).
79. D. M. Silver, Phys. Rev. A **21**, 1106 (1980).
80. R. A. Aziz and M. J. Slaman, J. Chem. Phys. **94**, 8047 (1991).
81. F. Luo *et al.*, J. Chem. Phys. **98**, 3564 (1993).
82. D. E. Woon, J. Chem. Phys. **100**, 2838 (1994).
83. S. J. Chakravorty *et al.*, Phys. Rev. A **47**, 3649 (1993).
84. J. R. Flores, J. Chem. Phys. **98**, 5642 (1993).
85. V. Termath, W. Klopper, and W. Kutzelnigg, J. Chem. Phys. **94**, 2002 (1991).
86. Y. Ishikawa and K. Koc, Phys. Rev. A **50**, 4733 (1994).
87. E. R. Davidson *et al.*, Phys. Rev. A **44**, 7071 (1991).
88. J. J. A. Montgomery, J. W. Ochterski, and G. A. Petersson, J. Chem. Phys. **101**, 5900 (1994).
89. D. E. Freund, B. D. Huxtable, and J. D. Morgan III, Phys. Rev. A **29**, 980 (1984).
90. K. P. Huber and G. L. Herzberg, *Molecular Spectra and Molecular Structure. IV. Constants of Diatomic Molecules* (Van Nostrand Reinhold, New York, 1979).
91. T. Grabo, T. Kreibich, and E. K. U. Gross, Mol. Engineering **7**, 27 (1997).
92. M. Warken, Chem. Phys. Lett. **237**, 256 (1995).
93. I. Grabowski, S. Hirata, S. Ivanov, and R. J. Bartlett, J. Chem. Phys. **116**, 4415 (2002).
94. T. Kotani, Phys. Rev. B **50**, 14816 (1994).
95. D. M. Bylander and L. Kleinman, Phys. Rev. Lett. **74**, 3660 (1995).
96. L. Fritsche and J. Yuan, Phys. Rev. A **57**, 3425 (1998).
97. A. Görling, Phys. Rev. Lett. **83**, 5459 (1999).
98. R. A. Hyman, M. D. Stiles, and A. Zangwill, Phys. Rev. B **62**, 15521 (2000).
99. T. Kinoshita, Phys. Rev. **105**, 1490 (1957).

3 Relativistic Density Functional Theory

Reiner Dreizler

Institut für Theoretische Physik,
J.W. Goethe - Universität Frankfurt,
Robert-Mayer-Straße 6-8,
60054 Frankfurt/Main, Germany
dreizler@th.physik.uni-frankfurt.de

Reiner Dreizler

3.1 Summary

In these lectures on relativistic density functional theory I had the choice to
provide a kind of survey, or to concentrate on a few specific aspects in greater
detail. I chose the first option. In order to give you the opportunity to fill in
the (often gory) details, I will distribute a list of references, augmented by
suitable comments on the contents of the papers cited.

In my lectures I will cover the topics:

1. *Introduction*, giving a brief summary of why one should work with quantum electrodynamics (QED) if one is interested in the density functional
   theory of relativistic Coulomb systems.
2. *Foundation*, containing some comments on the relativistic Hohenberg-
   Kohn theorem and indicating how the exact (but not easily solvable)
   relativistic Kohn-Sham equations (containing radiative corrections and
   all that) can be reduced to the standard approximate variant.
3. *Functionals*, with a mini-survey of the relativistic functionals that have
   been considered. The headings are well known from the non-relativistic
   case: LDA, GGA, OPM, etc.
4. *Results*, giving an indication of the performance of these functionals for
   a number of systems (mainly atoms) and a brief statement on some particular systems (molecules, solids).

A few words are also necessary concerning notation. Relativistic units
with
$$\hbar = c = 1 \; ; \quad m_0 = m \; ; \quad e \; .$$
will be used. This choice is debatable (e.g., if one considers expansions in
$1/c$), but in general use. I shall use the standard conventions of relativistic

theory, as found in most textbooks on relativistic quantum mechanics, e.g.,

$$\sum_{\mu=0}^{3} a_\mu b^\mu \longrightarrow a_\mu b^\mu \,,$$

with the metric

$$g_{\mu\nu} = \begin{pmatrix} 1 & 0 & 0 & 0 \\ 0 & -1 & 0 & 0 \\ 0 & 0 & -1 & 0 \\ 0 & 0 & 0 & -1 \end{pmatrix}$$

for time (0) and space (1–3) coordinates, or the Feynman dagger notation

$$\not{a} = \gamma_\mu a^\mu \,.$$

Relativistic corrections to the ground state energies of many particle systems are expected in two places. First there is the kinematic correction. The non-relativistic kinetic energy has to be replaced by its relativistic equivalent

$$-\frac{\nabla^2}{2m} \longrightarrow -i\nabla \cdot \gamma \,. \tag{3.1}$$

The second correction is a modification of the interaction energies. On the level of relativistic density functional theory for Coulomb systems this means, for instance, the replacement of the standard Hartree energy by its covariant form involving electron four-currents, j^μ and the photon propagator, $D^{(0)}_{\mu\nu}$,

$$\frac{e^2}{2} \int d^3x \int d^3y \, \frac{n(\boldsymbol{x})n(\boldsymbol{y})}{|\boldsymbol{x}-\boldsymbol{y}|} \longrightarrow \frac{1}{2} \int d^3x \int d^4y \, j^\mu(\boldsymbol{x}) D^{(0)}_{\mu\nu}(x-y) j^\nu(\boldsymbol{y}) \,. \tag{3.2}$$

A corresponding change applies to the other interaction terms.

The appropriate starting point for the discussion of relativistic Coulomb systems is QED [1–3]. The reason for using the full quantum field theory rather than just the Dirac equation is twofold: (i) With a quantum field theory the anti-particle sector is sorted correctly. This statement is illustrated by the small table comparing the free particle versions of the two options:

	charge	energy
Dirac	negative definite	negative / positive
QED	negative / positive	positive definite

The experimental situation with positive energies and oppositely charged particles and anti-particles is obviously described correctly by field theory. (ii) Possible questions of renormalization are quite apparent. I illustrate this remark by one example. The four-current of an electron in an external field is given by

$$j^\mu(x) = -i \lim_{y \to x} \text{Tr}\left[S_F(x,y)\gamma^\mu\right] \,. \tag{3.3}$$

This is the analogue of the non-relativistic case where, e.g., the density is the equal time, equal space limit of the Green function. The slightly more complicated symmetric limit

$$\lim_{\substack{S \\ y \to x}} = \frac{1}{2} \left[\lim_{y \to x, \, y^0 > x^0} + \lim_{x \to y, \, y^0 < x^0} \right]_{(x-y)^2 \geq 0} \tag{3.4}$$

is due to the requirement of charge conjugation invariance of the relativistic theory. In diagrammar the Green function is given by

(3.5)

The fermion (say electron) interacts in the sense of the Born approximation with an external source. Taking the symmetric limit corresponds to closing the lines on themselves

(3.6)

One immediately recognizes that the second diagram contains the lowest order vacuum polarization

$$-\mathrm{i}\Pi^{(0)}_{\mu\nu}(q) \quad = \quad$$, (3.7)

which is one of the three basic divergence contributions of QED. It has to, and can, be renormalized. It remains to say that attempts to set up extended Thomas-Fermi-type models for relativistic systems have been thwarted for quite some years by not recognizing this feature.

I will not dwell on the field theoretical details in the following, but we have to take note of the starting point, the QED Hamiltonian

$$\hat{H} = \int \mathrm{d}^3x \, \bar{\hat{\psi}}(x) \left(-\mathrm{i}\boldsymbol{\gamma} \cdot \boldsymbol{\nabla} + m \right) \hat{\psi}(x)$$

$$+ e \int \mathrm{d}^3x \, \hat{\jmath}^\mu(x) \hat{A}_\mu(x) + \int \mathrm{d}^3x \, \hat{\jmath}^\mu(x) v_{\mathrm{ext}\,\mu}(x)$$

$$- \frac{1}{8\pi} \int \mathrm{d}^3x \, \left\{ \left[\partial^0 \hat{A}_\mu(x) \right] \left[\partial^0 \hat{A}^\mu(x) \right] + \boldsymbol{\nabla} \hat{A}_\mu(x) \cdot \boldsymbol{\nabla} \hat{A}^\mu(x) \right\} \,. \tag{3.8}$$

We have fermions interacting via an electromagnetic field and with external sources and the Hamiltonian of the electromagnetic field. The renormalized Hamiltonian is

$$\hat{H}_{\text{ren}} = \hat{H} - \text{VEV} + \text{CT} . \tag{3.9}$$

It involves the trivial renormalization, the subtraction of vacuum expectation values (VEV), and the serious renormalization (see above) which can be handled by the counter-term (CT) technique [4].

3.2   Foundations

The relativistic Hohenberg-Kohn theorem was first formulated by Rajagopal and Callaway [5,6] and by McDonald and Vosko [7]. As expected for a Lorentz covariant situation it states that the ground-state energy is a unique functional of the ground-state four-current

$$E_0[j^\mu] = F[j^\mu] + \int d^3x \, j^\mu(x) v_{\text{ext}\,\mu}(x) , \tag{3.10}$$

where F is an universal functional of j^μ, and the simplest contribution, the coupling to the external sources, is (as usual) made explicit. The proof has been re-examined by Engel et al. [8], who demonstrated that field theoretical aspects (that were not considered by the previous authors) do not invalidate the conclusion. The final statement is: All ground-state observables can be expressed as unique functionals of the ground-state four-current as:

$$O[j^\mu] = \langle \Phi_0[j^\mu] | \hat{O} | \Phi_0[j^\mu] \rangle + \Delta O_{\text{CT}} - \text{VEV} . \tag{3.11}$$

Again counter-terms and subtraction of vacuum expectation values have to be taken care of.

In practical applications the question arises: What is the situation if the external potential is electrostatic $\{v^\mu_{\text{ext}}(x)\} = \{v^0_{\text{ext}}(x), 0\}$? The answer is: All ground-state variables, including the spatial components of the four-current, are then functionals of the charge density alone, e.g.,

$$\boldsymbol{j}([n], x) = \langle \Phi_0[n] | \hat{\boldsymbol{j}}(x) | \Phi_0[n] \rangle . \tag{3.12}$$

The question whether these functionals are known is a different story.

The relativistic Kohn-Sham scheme starts, in complete analogy to the non-relativistic case, with a representation of the four-current and of the non-interacting kinetic energy in terms of auxiliary spinor orbitals [8]. If one calculates the four-current of a system of fermions in an external potential (as indicated above), one obtains

$$j^\mu(\boldsymbol{x}) = j^\mu_{\text{vac}}(\boldsymbol{x}) + j^\mu_{\text{D}}(\boldsymbol{x}) . \tag{3.13}$$

The vacuum polarization current (which arises from the symmetric limit) is given by the solution of a Dirac equation by

$$j^\mu_{\text{vac}}(\boldsymbol{x}) = \frac{1}{2}\left[\sum_{\epsilon_k \leq -m} \bar{\varphi}_k(\boldsymbol{x})\gamma^\mu\varphi_k(\boldsymbol{x}) - \sum_{-m < \epsilon_k} \bar{\varphi}_k(\boldsymbol{x})\gamma^\mu\varphi_k(\boldsymbol{x})\right] . \qquad (3.14)$$

It contains negative energy solutions as well as bound states and positive energy solutions. The current due to the occupied orbitals is

$$j^\mu_{\text{D}}(\boldsymbol{x}) = \sum_{-m < \epsilon_k \leq \epsilon_F} \bar{\varphi}_k(\boldsymbol{x})\gamma^\mu\varphi_k(\boldsymbol{x}) . \qquad (3.15)$$

The non-interacting kinetic energy, including the trivial rest mass term, has a corresponding structure

$$T_s[j^\mu] = T_{s,\text{vac}}[j^\mu] + T_{s,\text{D}}[j^\mu] . \qquad (3.16)$$

The contributions are obtained from the formulae given for j^μ by the replacement

$$\gamma^\mu \longrightarrow -i\boldsymbol{\gamma}\cdot\boldsymbol{\nabla} + m . \qquad (3.17)$$

The full Kohn-Sham scheme is obtained by writing the ground-state energy as

$$E_0[j^\mu] = T_s[j^\mu] + E_{\text{ext}}[j^\mu] + E_{\text{Hartree}}[j^\mu] + E_{\text{xc}}[j^\mu] , \qquad (3.18)$$

where the xc energy is defined as the difference

$$E_{\text{xc}} = F - T_s - E_{\text{Hartree}} . \qquad (3.19)$$

The Hartree energy is the covariant version

$$E_{\text{Hartree}}[j^\mu] = \frac{1}{2}\int d^3x \int d^4y\, j^\mu(\boldsymbol{x})D^{(0)}_{\mu\nu}(x-y)j^\nu(\boldsymbol{y}) , \qquad (3.20)$$

which reduces to

$$E_{\text{Hartree}}[j^\mu] = \frac{e^2}{2}\int d^3x \int d^3y\, \frac{j^\mu(\boldsymbol{x})j_\mu(\boldsymbol{y})}{|\boldsymbol{x}-\boldsymbol{y}|} \qquad (3.21)$$

for stationary currents.

The philosophy behind this addition and subtraction both in the non-relativistic as well as in the relativistic case is to isolate the (in principle) tractable, dominant contributions. The xc-energy becomes then the key quantity concerning serious many-body effects.

Minimization of the ground-state energy with respect to the auxiliary spinor orbitals leads to the general Kohn-Sham equations:

$$\gamma^0\left\{-i\boldsymbol{\gamma}\cdot\boldsymbol{\nabla} + m + \slashed{v}_{\text{ext}}(\boldsymbol{x}) + \slashed{v}_{\text{Hartree}}(\boldsymbol{x}) + \slashed{v}_{\text{xc}}(\boldsymbol{x})\right\}\varphi_k(\boldsymbol{x}) = \epsilon_k\varphi_k(\boldsymbol{x}) . \qquad (3.22)$$

This is a Dirac equation with the effective potentials

$$\not{v}_{\text{Hartree}}(\boldsymbol{x}) = \gamma_\mu v^\mu_{\text{Hartree}}(\boldsymbol{x}) = e^2 \gamma_\mu \int d^3 y \, \frac{j^\mu(\boldsymbol{y})}{|\boldsymbol{x} - \boldsymbol{y}|} \qquad (3.23)$$

$$\not{v}_{\text{xc}}(\boldsymbol{x}) \quad = \gamma_\mu \frac{\delta E_{\text{xc}}[j^\mu]}{\delta j_\mu(\boldsymbol{x})} \, , \qquad (3.24)$$

which has to be solved self-consistently.

So far nobody has solved the indicated problem. The evaluation of the vacuum contributions (in j^μ and T_s) includes the full set of solutions of the Dirac equation and renormalization at each step of the self-consistency procedure.

In the discussion of "practical" problems some approximations are commonly applied: (i) The "no-sea" approximation, where one neglects all radiative corrections

$$j^\mu_{\text{vac}} = T_{s,\text{vac}} = E_{\text{xc,vac}} = 0 \, . \qquad (3.25)$$

If these corrections are of interest, they can be calculated perturbatively with the final self-consistent solutions. (ii) The situation encountered most often in electronic structure calculations is the one where the external potential is purely *electrostatic*. In this case, the charge density is the only variable (see above) and one has

$$\tilde{E}_{\text{Hartree}}[n] \equiv E_{\text{Hartree}}[n, \boldsymbol{j}[n]] \qquad (3.26)$$

$$\tilde{E}_{\text{xc}}[n] \quad \equiv E_{\text{xc}}[n, \boldsymbol{j}[n]] \, . \qquad (3.27)$$

As a consequence, the effective potentials are also electrostatic, e.g.,

$$\{v^\mu_{\text{Hartree}}(\boldsymbol{x})\} = \{v_{\text{Hartree}}(\boldsymbol{x}), \boldsymbol{0}\} \, , \qquad (3.28)$$

with

$$v_{\text{Hartree}}(\boldsymbol{x}) = \frac{\delta E_{\text{Hartree}}[j^\mu]}{\delta n(\boldsymbol{x})} + \sum_{k=1}^{3} \int d^3 x' \, \frac{\delta E_{\text{Hartree}}[j^\mu]}{\delta j^k(\boldsymbol{x}')} \frac{\delta j^k(\boldsymbol{x}')}{\delta n(\boldsymbol{x})} \, . \qquad (3.29)$$

The additional term arises as there is an explicit functional dependence of j^k on n.

The resulting electrostatic no-sea approximation is the standard version applied in practice. It is usually written as

$$\{-\mathrm{i}\boldsymbol{\alpha} \cdot \boldsymbol{\nabla} + m\beta + v_{\text{ext}}(\boldsymbol{x}) + v_{\text{Hartree}}(\boldsymbol{x}) + v_{\text{xc}}(\boldsymbol{x})\} \, \varphi_k(\boldsymbol{x}) = \epsilon_k \varphi_k(\boldsymbol{x}) \, , \qquad (3.30)$$

where the density is given by

$$n(\boldsymbol{x}) = \sum_{-m < \epsilon_k \leq \epsilon_F} \varphi_k^\dagger(\boldsymbol{x}) \varphi_k(\boldsymbol{x}) \, . \qquad (3.31)$$

Usually the exact current $j[n]$ (wherever it occurs) is replaced by the Kohn-Sham current

$$j(x) = \sum_{-m < \epsilon_k \leq \epsilon_F} \varphi_k^\dagger(x)\, \alpha\, \varphi_k(x)\,. \tag{3.32}$$

The differences that might occur due to this replacement have not been explored.

Further possible approximations rely on the fact that the free photon propagator, which mediates the interaction between the fermions, can be split into a longitudinal (Coulomb) and a transverse part

$$D_{\mu\nu}^{(0)}(x-y) = g_{\mu 0}\, g_{\nu 0} \frac{e^2}{|\boldsymbol{x}-\boldsymbol{y}|} \delta(x^0 - y^0) + D_{\mu\nu}^{\mathrm{T}}(x-y)\,. \tag{3.33}$$

If one neglects the transverse contribution, one arrives at what is termed the Dirac-Coulomb approximation (a standard in quantum chemistry). Inclusion of the transverse term, which describes retardation and magnetic effects, in perturbation theory (weakly relativistic limit) leads to the Dirac-Coulomb-Breit Hamiltonian.

As a conclusion of this section, I just state that the full, weakly relativistic limit of the electrostatic, no-sea approximation (obtained with techniques such as the Fouldy-Wouthuysen transformation) makes contact with non-relativistic current-density functional theory (as formulated by Rasolt and Vignale [9]).

3.3 Functionals

The standard relativistic density functional expression for the ground-state energy is

$$E_0[j^\mu] = T_s[j^\mu] + E_{\mathrm{ext}}[j^\mu] + E_{\mathrm{Hartree}}[j^\mu] + E_{\mathrm{xc}}[j^\mu]\,. \tag{3.34}$$

The functional dependence of T_s and E_{xc} on j^μ needs to be established. In Kohn-Sham applications, T_s is expressed directly in terms of spinor orbitals, so only E_{xc} has to be considered. If one is aiming at setting up relativistic extensions of extended Thomas-Fermi models, one also has to consider dependence of T_s on the four-current. I shall present a few remarks on the density functional form of T_s, but first we look at the exchange and correlation energy.

The simplest approximation is the local density approximation (LDA), which is obtained from the energy density of the relativistic homogeneous electron gas (RHEG)

$$E_{\mathrm{xc}}^{\mathrm{LDA}}[n] = \int \mathrm{d}^3 x\, e_{\mathrm{xc}}^{\mathrm{RHEG}}(n_0)\big|_{n_0 = n(\boldsymbol{x})}\,. \tag{3.35}$$

Two remarks apply: (i) For this homogeneous system the spatial current vanishes $j^{\mathrm{RHEG}} = \mathbf{0}$, so there is only a dependence on the density. (ii) Knowledge of the ground-state energy of the RHEG is much less developed than for its non-relativistic counterpart. There are, for instance, no Monte Carlo results for $e_{\mathrm{xc}}^{\mathrm{RHEG}}$. The functionals that are available in this approximation are thus obtained by painstaking evaluation of the simplest diagrammatic contributions to the ground-state energy. The details are more involved than in the non-relativistic case, partly due to questions of renormalization, partly due to the Minkowski space structure. I shall only indicate the genesis of the results in terms of the corresponding diagrams.

The x contribution has been worked out as early as 1960. One can show that after proper renormalization only the contribution

$$e_{\mathrm{x}}^{\mathrm{RHEG}} = \frac{i}{2} \quad \text{(diagram)} , \tag{3.36}$$

which is the diagram with the finite contribution of the free electron propagator, remains. The arguments leading to this result can be summarized as follows. In the RLDA the x-energy density is given by

$$e_{\mathrm{x}}^{\mathrm{RHEG}} = \frac{1}{2} \int \mathrm{d}^4 y \, D_{\mu\nu}^{(0)}(x-y) \, \mathrm{Tr} \left[S_{\mathrm{F}}^{(0)}(x-y) \gamma^\nu S_{\mathrm{F}}^{(0)}(y-x) \gamma^\mu \right] +$$
$$+ \mathrm{CT} + \mathrm{VEV} . \tag{3.37}$$

In diagrammar the loop integral looks like this

$$e_{\mathrm{x}}^{\mathrm{RHEG(1)}} = \frac{i}{2} \quad \text{(diagram)} . \tag{3.38}$$

The double line stands for the free photon propagator $D_{\mu\nu}^{(0)}$, and the wiggly line for the lowest order fermion propagator of the RHEG, $S_{\mathrm{F}}^{(0)}$. The photon propagator can be split into a vacuum contribution and a direct contribution due to the occupied electron states

$$\text{(diagram)} = \text{(diagram)}_{\mathrm{vac}} + \text{(diagram)}_D \equiv \text{(diagram)} + \text{(diagram)}_D . \tag{3.39}$$

The loop integral then consists of four contributions

$$(3.40)$$

The first contribution is divergent but is removed the subtraction of the vaccum, while the next two diagrams contain the lowest order self-energy

$= \Sigma^{(1)}_{\mathrm{vac}}$, $$(3.41)$$

which has still to be renormalized to yield $\Sigma^{(1)}_{\mathrm{vac,ren}}$. However, the terms vanish as the self-energy satisfies the on-shell condition

$$\left[(\not{p} + m) \Sigma^{(1)}_{\mathrm{vac,ren}} \right]_{p^2 = m^2} = 0 .$$

$$(3.42)$$

The factor $(\not{p} + m)$ is supplied by the remaining propagator. So, finally, only the contribution due to the occupied electron states remains.

The relativistic corrections are more readily discussed if one writes [10–12]

$$e^{\mathrm{RHEG}}_{\mathrm{x}} = e^{\mathrm{NRHEG}}_{\mathrm{x}} \Phi_{\mathrm{x}} \left(\frac{k_{\mathrm{F}}}{m} \right) ,$$

$$(3.43)$$

where the relativistic correction factor can be split (due to the structure of the free photon propagator, see above) into [13]

$$e^{\mathrm{RHEG}}_{\mathrm{x}} = e^{\mathrm{NRHEG}}_{\mathrm{x}} \left[\Phi^{\mathrm{L}}_{\mathrm{x}} \left(\frac{k_{\mathrm{F}}}{m} \right) + \Phi^{\mathrm{T}}_{\mathrm{x}} \left(\frac{k_{\mathrm{F}}}{m} \right) \right] .$$

$$(3.44)$$

One finds that the longitudinal part (L) does not differ very much from the non-relativistic limit. The transverse correction (T) is negative and it is small for low densities, but grows sufficiently in magnitude, so that the x energy density changes sign at $\beta = k_{\mathrm{F}}/m \approx 2.5$ (one should keep in mind that the maximal density in the Hg atom – in the inner shells – amounts to $\beta \approx 3$).

The transverse correction factor can be decomposed into a magnetic and a retardation contribution with opposite signs. The magnetic contribution dominates at higher densities. The expansion of $e^{\mathrm{RHEG}}_{\mathrm{x}}$ in the weakly relativistic limit gives the Breit contribution to e_{x}, that reproduces the full transverse correction factor quite closely over the relevant range of $0 \le \beta \le 3$.

Concerning correlations, the only contribution that has been worked out is the random phase (RPA) limit [10,12,14,15]. After renormalization, one can write the RPA correlation energy contribution as

$$e_c^{RPA} = i \left\{ \; \ldots \; + \; \ldots \; + \ldots \right\}. \qquad (3.45)$$

The loops of the (fermion) polarization insertion involve only the direct contribution

$$= \; \ldots \; + \; \ldots \; + \; \ldots. \qquad (3.46)$$

The interaction lines correspond to the full vacuum photon propagator that is obtained by re-summation of the series

$$= \; \ldots \; + \; \ldots$$

$$+ \; \ldots \qquad (3.47)$$

in terms of the renormalized vacuum polarization insertion.

The final evaluation (involving one numerical integration) has only been achieved within two further approximations: (i) In the no-sea approximation, the full photon propagator is replaced by the free propagator

$$\approx \qquad (3.48)$$

(ii) In the no-pair approximation (a kind of standard in quantum chemistry) one also uses the free propagator

$$\approx \qquad , \qquad (3.49)$$

and in addition evaluates the polarization insertion with the electron propagator

$$(3.50)$$

There is a conceptual difference in the sense that the no-pair approximation is gauge dependent, but in the final reckoning there is only a slight difference in the results. The result can (as for exchange) be written in the form

$$e_{c,RPA}^{RHEG} = e_{c,RPA}^{NRHEG} \, \Phi_c^{RPA}\left(\frac{k_F}{m}\right) \tag{3.51}$$

The correction factor can be represented quite accurately by

$$\Phi_c^{RPA}(\beta) = \frac{1 + a_1\beta^3 \ln \beta + a_2\beta^4 + a_3(1+\beta^2)^2\beta^4}{1 + b_1\beta^3 \ln \beta + b_2\beta^4 + b_3(A \ln \beta + B)\beta^7}, \tag{3.52}$$

which incorporates the exact large density limit as well as the non-relativistic limit. A plot shows that the relativistic corrections can become quite substantial for higher densities.

There is no systematic treatment of other contributions to the ground-state energy of the RHEG. This remark also pertains to the construction of gradient expansion approximations (GEA), which in the x-only limit involves, to lowest order, the four-point contributions

$$(3.53)$$

(You should, however, note that the proper evaluation in the non-relativistic limit took the better part of 10 years).

As the hopes placed in the GEA did not materialize (in the non-relativistic case), one turned to the construction of generalized gradient approximations (GGA). These are based on the following "philosophy": (i) Use available "exact" results for atoms (x-only or on the basis of CI calculations) and fit them to a functional of the form

$$E_{xc}^{GGA} = \int d^3x \, e_{xc}^{LDA} F_{xc}(n, \varsigma) \tag{3.54}$$

with the dimensionless gradient

$$\varsigma = \frac{(\nabla n)^2}{4n^2 \, (3\pi^2 n)^{2/3}}; \tag{3.55}$$

(ii) Incorporate (depending on the philosophy) some exact properties as, e.g., the correct weakly inhomogeneous limit, etc.; (iii) Use this functional for the discussion of more complex Coulomb systems as molecules and solids (relying on the supposed universality).

If one wishes to proceed in this fashion in the relativistic case, one has to provide "accurate" atomic data. For this purpose, OPM, the optimized potential method [16] (in the present context the relativistic extension, the ROPM) is a valuable tool. The (R)OPM relies on the fact that the functional derivative with respect to the density (or the four-current) can be evaluated with the chain rule for functional derivatives if the dependence on the density is implicit via Kohn-Sham orbitals, $(E[n] = E[\varphi_k] = E[\varphi_k[n]])$

$$v(\boldsymbol{x}) = \frac{\delta E}{\delta n(\boldsymbol{x})} = \sum_k \int \mathrm{d}^3 x' \int \mathrm{d}^3 x'' \, \frac{\delta E}{\delta \varphi_k(\boldsymbol{x}')} \frac{\delta \varphi_k(\boldsymbol{x}')}{\delta v_{\mathrm{KS}}(\boldsymbol{x}'')} \frac{\delta v_{\mathrm{KS}}(\boldsymbol{x}'')}{\delta n(\boldsymbol{x})} \,. \quad (3.56)$$

The first factor is evaluated directly from the explicit functional form, the second follows from the linear response limit of the Kohn-Sham equations as does the last one (the inverse Kohn-Sham response function).

On the basis of (3.56), an integral equation for the multiplicative potential $v(\boldsymbol{x})$ can be derived. It has the form

$$\int \mathrm{d}^3 x' \, K(\boldsymbol{x}, \boldsymbol{x}') \, v(\boldsymbol{x}') = Q(\boldsymbol{x}) \,, \quad (3.57)$$

where both the kernel, $K(\boldsymbol{x}, \boldsymbol{x}')$, and the inhomogeneous term, $Q(\boldsymbol{x})$, can be expressed in terms of Kohn-Sham orbitals.

For the x-only limit the application of the OPM is rather straightforward. One starts with the definition of the covariant x-energy

$$E_{\mathrm{x}} = \frac{1}{2} \int \mathrm{d}^3 x \int \mathrm{d}^4 y \, D_{\mu\nu}^{(0)}(x - y) \, \mathrm{Tr} \left[S_{\mathrm{F}}(x, y) \gamma^\nu S_{\mathrm{F}}(y, x) \gamma^\mu \right] \quad (3.58)$$

(see diagrams above) and evaluates the fermion propagators in the Kohn-Sham (that is, effective single particle) limit. If, in addition, one applies the electrostatic no-sea approximation, one obtains

$$E_{\mathrm{x}}^{\mathrm{KS}}[n] = -\frac{e^2}{2} \int \mathrm{d}^3 x \int \mathrm{d}^3 y$$
$$\sum_{-m < \epsilon_k, \epsilon_l \leq \epsilon_{\mathrm{F}}} \frac{\cos\left(|\epsilon_k - \epsilon_l||\boldsymbol{x} - \boldsymbol{y}|\right)}{|\boldsymbol{x} - \boldsymbol{y}|} \bar{\varphi}_k(\boldsymbol{x}) \gamma_\mu \varphi_l(\boldsymbol{x}) \bar{\varphi}_l(\boldsymbol{y}) \gamma^\mu \varphi_k(\boldsymbol{y}) \,. \quad (3.59)$$

For a corresponding correlation contribution, e.g.,

$$E_{\mathrm{c}}^{\mathrm{KS}}[n] = E_{\mathrm{xc}}[n] - E_{\mathrm{x}}^{\mathrm{KS}}[n] \quad (3.60)$$

only some variants of perturbation theory on the basis of the Kohn-Sham Hamiltonian are available like,e.g., a straightforward second order perturbation theory (in the spirit of Møller-Plesset perturbation theory) or some partially re-summed versions. I shall not present the relevant equations.

The numerical implementation of the OPM scheme is rather involved. For this reason, one often applies the Krieger-Li-Iaffrate (KLI) approximation, which turns out to be (as in the non-relativistic case) rather accurate.

From the solution of the relativistic OPM problem (or some other "exact" equivalent) for atoms, one may construct relativistic GGA-type functionals following the procedure used in non-relativistic theory. One sets, for instance in the x-only limit

$$F_x(n, \zeta) = \Phi_{x,0}(\beta) + g(\zeta)\Phi_{x,2}(\beta) . \tag{3.61}$$

One then uses non-relativistic forms of the gradient correction factor g (The different forms found in the literature do give results that vary only marginally). For the relativistic correction factor $\Phi_{x,2}$ a flexible [2,2] Padé approximant

$$\Phi_{x,2}(\beta) = \frac{a_0 + a_1\beta^2 + a_2\beta^4}{1 + b_1\beta^2 + b_2\beta^4} \tag{3.62}$$

proved to be sufficient to reproduce ROPM results to high accuracy. Both the longitudinal as well as the transverse contribution can be accommodated with this ansatz. The correct weakly relativistic limit is obtained with

$$a_0^{\mathrm{L}} = 1 , \quad a_0^{\mathrm{T}} = 0 , \quad \text{and} \quad \beta \approx 0 . \tag{3.63}$$

As a conclusion of this section I offer a few remarks on the functional $T_s[j^\mu]$. This functional is used in relativistic, extended Thomas-Fermi models, which are based on the direct variational principle

$$\frac{\delta}{\delta j^\mu(\boldsymbol{x})} \left\{ E_0[j^\nu] + \mu_{\mathrm{chem}} \int \mathrm{d}^3 y \, j^0(\boldsymbol{y}) \right\} = 0 . \tag{3.64}$$

In contrast to the Kohn-Sham scheme, no auxiliary orbitals are involved. Unfortunately, the presently available approximations to $T_s[j^\mu]$ are only adequate for general estimates (rather than for results of chemical accuracy). The functional in question is derived from the exact kinetic energy

$$T[j^\mu] = -\mathrm{i} \int \mathrm{d}^3 x \, \lim_{y \to x} \mathrm{Tr} \left[-\mathrm{i}\boldsymbol{\gamma} \cdot \boldsymbol{\nabla} + m \right] S_{\mathrm{F}}(x, y) - \mathrm{VEV} + \mathrm{CT} , \tag{3.65}$$

where the exact fermion propagator is replaced by the Kohn-Sham propagator

$$\left[\mathrm{i}\partial\!\!\!/_x - m - \psi\!\!\!/_{\mathrm{KS}}(\boldsymbol{x}) \right] S_{\mathrm{F}}^{\mathrm{KS}}(x, y) = \delta^{(4)}(x - y) . \tag{3.66}$$

As indicated, renormalization is necessary. Results are available to fourth order in the gradient of the density in the electrostatic limit

$$T_s^{\mathrm{el.st.}}[n] = T_s^{[0]}[n] + T_s^{[2]}[n] + T_s^{[4]}[n] + \ldots \tag{3.67}$$

and to second order in the gradients of the four-current for the more general case

$$T_s[n, \boldsymbol{j}] = T_s^{[0]}[n] + T_s^{[2]}[n, \boldsymbol{j}] + \ldots \tag{3.68}$$

Table 3.1. Longitudinal ground state energies $(-E_{tot}^{L})$ and highest occupied eigenvalues $(-\epsilon_{mk}^{L})$ for closed sub-shell atoms from non-relativistic OPM (NROPM [17]), relativistic OPM (ROPM [18]) and relativistic HF (RHF [19]) calculations (all energies are in hartree)

Atom	$-E_{tot}^{L}$			$-\epsilon_{mk}^{L}$		
	NROPM	ROPM	RHF	NROPM	ROPM	RHF
He (1S1/2)	2.862	2.862	2.862	0.918	0.918	0.918
Be (2S1/2)	14.572	14.575	14.576	0.309	0.309	0.309
Ne (2P3/2)	128.545	128.690	128.692	0.851	0.848	0.848
Mg (3S1/2)	199.611	199.932	199.935	0.253	0.253	0.253
Ar (3P3/2)	526.812	528.678	528.684	0.591	0.587	0.588
Ca (4S1/2)	676.751	679.704	679.710	0.196	0.196	0.196
Zn (4S1/2)	1777.828	1794.598	1794.613	0.293	0.299	0.299
Kr (4P3/2)	2752.028	2788.848	2788.861	0.523	0.515	0.514
Sr (5S1/2)	3131.514	3178.067	3178.080	0.179	0.181	0.181
Pd (4D5/2)	4937.858	5044.384	5044.400	0.335	0.319	0.320
Cd (5S1/2)	5465.056	5593.299	5593.319	0.266	0.282	0.281
Xe (5P3/2)	7232.018	7446.876	7446.895	0.456	0.439	0.440
Ba (6S1/2)	7883.404	8135.625	8135.644	0.158	0.163	0.163
Yb (6S1/2)	13391.070	14067.621	14067.669	0.182	0.196	0.197
Hg (6S1/2)	18408.313	19648.826	19648.865	0.262	0.329	0.328
Rn (6P3/2)	21865.826	23601.969	23602.005	0.427	0.382	0.384
Ra (7S1/2)	23093.258	25028.027	25028.061	0.149	0.167	0.166
No (7S1/2)	32787.471	36740.625	36740.682	0.171	0.209	0.209

3.4   Results

I shall first show some results for atoms in order to illustrate the magnitude of relativistic effects and to compare the performance of the various relativistic functionals.

The first set of tables deals with the x-only limit, where in addition a direct comparison with relativistic Hartree-Fock (RHF) results is possible. Total ground-state energies for closed sub-shell atoms (Table 3.1) calculated with the longitudinal x-contribution (the straightforward Coulomb interaction), show the following features: Comparing NROPM to ROPM results one notices the growing importance of relativistic corrections as the central charge is increased (The difference between relativistic and non-relativistic energies is nearly 4000 hartree). One also notices that there is hardly any difference between ROPM and RHF energies, although in the first case the effective potential is multiplicative while in the latter it is non-local. The second column gives the energies of the highest occupied orbitals. Even the outermost orbitals experience some effect of relativity (due to the change of the effective potential generated by the inner orbitals).

Table 3.2. Single particle energies $(-\epsilon_{nlj})$ for Hg from NROPM, ROPM and RHF calculations in comparison with DFS, RLDA and RWDA results (all energies are in hartree)

Level	NROPM	ROPM	RHF	DFS	RLDA	RWDA
1S1/2	2756.925	3047.430	3074.228	3047.517	3044.410	3051.995
2S1/2	461.647	540.056	550.251	539.713	539.250	540.530
2P1/2	444.015	518.061	526.855	518.164	517.746	519.244
2P3/2	444.015	446.682	455.157	446.671	446.399	447.469
3S1/2	108.762	128.272	133.113	128.001	127.905	128.292
3P1/2	100.430	118.350	122.639	118.228	118.148	118.592
3P3/2	100.430	102.537	106.545	102.397	102.346	102.691
3D3/2	84.914	86.201	89.437	86.085	86.060	86.364
3D5/2	84.914	82.807	86.020	82.690	82.668	82.959
4S1/2	23.522	28.427	30.648	28.067	28.046	28.200
4P1/2	19.895	24.161	26.124	23.871	23.854	24.023
4P3/2	19.895	20.363	22.189	20.039	20.030	20.167
4D3/2	13.222	13.411	14.797	13.148	13.146	13.271
4D5/2	13.222	12.700	14.053	12.434	12.432	12.553
4F5/2	4.250	3.756	4.473	3.556	3.559	3.665
4F7/2	4.250	3.602	4.312	3.402	3.404	3.509
5S1/2	3.501	4.403	5.103	4.290	4.286	4.349
5P1/2	2.344	3.012	3.538	2.898	2.896	2.955
5P3/2	2.344	2.363	2.842	2.219	2.218	2.265
5D3/2	0.538	0.505	0.650	0.363	0.363	0.397
5D5/2	0.538	0.439	0.575	0.296	0.296	0.328
6S1/2	0.262	0.329	0.328	0.222	0.222	0.254

The full situation for the orbital energies is illustrated in Table 3.2 for the Hg atom (also in the longitudinal x-only limit). Comparing the orbital energies obtained from NROPM and ROPM calculations one sees once more the effect of relativistic corrections. One also notices that the orbital energies obtained from ROPM and RHF are quite different, with the exception of the last occupied orbitals, even if the total energies agree very closely. This stresses the fact that the orbitals (and their energies) are only auxiliary quantities that should be interpreted with some care (one should, e.g., not calculate excited state energies by just promoting particles from occupied to unoccupied orbitals). The exception is the highest occupied orbital whose energy corresponds (in principle) to the first ionization potential. Included in the table are Dirac-Fock-Slater (DFS) results (using the non-relativistic Slater potential in the Dirac equation) and RLDA results (using the relativistic longitudinal x-potential). These are basically two similar versions of LDA-type Kohn-Sham calculations. One sees that the differences in the orbital energies are small for the inner orbitals and become even smaller for the

Table 3.3. Longitudinal (Coulomb) x-only energies $(-E_x^L)$ for closed sub-shell atoms from NROPM, ROPM, RHF, DFS, RLDA, and RWDA-calculations [18] (all energies are in hartree)

Atom	NROPM	ROPM	RHF	DFS	RLDA	RWDA
He	1.026	1.026	1.026	0.853	0.853	1.026
Be	2.666	2.667	2.668	2.278	2.278	2.706
Ne	12.105	12.120	12.123	10.952	10.944	12.843
Mg	15.988	16.017	16.023	14.564	14.550	17.093
Ar	30.175	30.293	30.303	27.897	27.844	32.419
Ca	35.199	35.371	35.383	32.702	32.627	37.967
Zn	69.619	70.245	70.269	66.107	65.834	75.604
Kr	93.833	95.048	95.072	89.784	89.293	102.095
Sr	101.926	103.404	103.429	97.836	97.251	111.133
Pd	139.113	141.898	141.930	134.971	133.887	152.275
Cd	148.879	152.143	152.181	144.931	143.687	163.321
Xe	179.062	184.083	184.120	175.926	174.102	197.564
Ba	189.065	194.804	194.841	186.417	184.363	209.171
Yb	276.143	288.186	288.265	278.642	274.386	310.268
Hg	345.240	365.203	365.277	354.299	347.612	392.339
Rn	387.445	414.082	414.151	402.713	394.102	444.584
Ra	401.356	430.597	430.664	419.218	409.871	462.365
No	511.906	564.309	564.415	554.242	538.040	606.216

outer ones. This indicates that one is dealing with a density range for which the relativistic corrections to the longitudinal x-energy are not too large. On the other hand there are definite differences between these LDA results and the results that treat the x-effects exactly.

Corresponding results for the Coulomb energies of the closed sub-shell atoms (Table 3.2) are also of interest. Again, ROPM and RHF results agree quite closely, but one also notices that the differences between NROPM and ROPM results are not too large (of the order of 50 hartree for No, compared to the 4000 hartree for the total ground-state energy). The major part of the relativistic correction is kinetic rather than due to the structure of the interaction functionals. The RLDA versions do not perform optimally, although they reproduce the trend of the relativistic corrections.

In Table 3.3 (still in the longitudinal x-only limit) some RGGA results are included. The corresponding functional is obtained by fitting ROPM results to the parameterization that I have discussed. i.e., $g(\zeta)$ is written in a nonlocal PW91 form [20], and $\Phi_{x,2}$ in the [2,2] Padé form.

The results show that the ROPM results (which are not that easily generated) can be reproduced with very reasonable accuracy by the RGGA parameterization.

Table 3.4. Longitudinal x-only ground-state energies: Self-consistent ROPM, RHF, RLDA and RGGA results for neutral atoms with closed sub-shells (in hartree)

Atom	$-E_{\text{tot}}^{\text{L}}$	$E_{\text{tot}}^{\text{L}} - E_{\text{tot}}^{\text{L,ROPM}}$		
	ROPM	RHF	RLDA	RPW91
He	2.862	0.000	0.138	0.006
Be	14.575	−0.001	0.350	0.018
Ne	128.690	−0.002	1.062	−0.024
Mg	199.932	−0.003	1.376	−0.001
Ar	528.678	−0.005	2.341	0.041
Ca	679.704	−0.006	2.656	0.026
Zn	1794.598	−0.014	4.140	−0.262
Kr	2788.848	−0.013	5.565	−0.021
Sr	3178.067	−0.013	5.996	−0.008
Pd	5044.384	−0.016	7.707	−0.067
Cd	5593.299	−0.020	8.213	−0.033
Xe	7446.876	−0.019	9.800	0.085
Ba	8135.625	−0.019	10.289	0.059
Yb	14067.621	−0.048	13.272	−0.893
Hg	19648.826	−0.039	17.204	−0.250
Rn	23601.969	−0.035	19.677	0.004
Ra	25028.027	−0.034	20.460	−0.006

Table 3.5 includes the transverse x-contribution. For the column labeled RHF, the additional term is evaluated with the RHF density and added to the RHF ground-state energy. Otherwise one finds a similar story: RGGA results agree well with the ROPM standard while RLDA results do not. Looking at the transverse x-energy contribution (Table 3.6) one finds that any of the corrected non-relativistic GGA functionals (here ECMV92 [21] and B88 [22]) perform equally well.

I will not show any results that indicate that ROPM x-only results can be reproduced in a satisfactory fashion with the KLI approximation.

The discussion of correlation effects is more demanding. The first statement is: The RLDA does not give very satisfactory results. The functional that was used had the form (no-sea, electrostatic)

$$E_{\text{c}}^{\text{RLDA}}[n] = E_{\text{c,rel}}^{\text{RPA}}[n] - E_{\text{c,nonrel}}^{\text{RPA}}[n] + E_{\text{c,nonrel}}^{\text{LDA}}[n] . \quad (3.69)$$

Only the relativistic correction to the RPA is used and added to a more accurate non-relativistic functional (e.g., from parameterization of Monte Carlo results). For low densities the first two terms cancel, so that the correlation energy is given by the more adequate non-relativistic contribution. For high densities the non-relativistic RPA contributions cancel, so that this functional contains the relativistic RPA contribution plus the non-relativistic non-RPA terms. The failure is illustrated in Table 3.7. where RLDA results are com-

Table 3.5. Total relativistic x-only ground state energies: Self-consistent ROPM, RLDA and (R)GGA results for neutral atoms with closed sub-shells in comparison with perturbative RHF data (in hartree)

Atom	$-E_{tot}^{L+T}$	$E_{tot}^{L+T} - E_{tot}^{L+T,ROPM}$			
	ROPM	RHF(p)	RLDA	RPW91	PW91
He	2.862	0.000	0.138	0.006	0.006
Be	14.575	−0.001	0.351	0.018	0.017
Ne	128.674	−0.002	1.080	−0.024	−0.043
Mg	199.900	−0.003	1.408	−0.001	−0.037
Ar	528.546	−0.005	2.458	0.041	−0.111
Ca	679.513	−0.006	2.818	0.026	−0.195
Zn	1793.840	−0.014	4.702	−0.263	−1.146
Kr	2787.429	−0.012	6.543	−0.022	−1.683
Sr	3176.358	−0.012	7.149	−0.010	−2.014
Pd	5041.098	−0.013	9.765	−0.069	−3.953
Cd	5589.495	−0.016	10.556	−0.035	−4.538
Xe	7441.172	−0.012	13.161	0.083	−6.706
Ba	8129.160	−0.010	14.050	0.057	−7.653
Yb	14053.748	−0.023	20.886	−0.896	−17.662
Hg	19626.702	0.005	29.159	−0.260	−27.256
Rn	23573.351	0.026	35.203	−0.012	−35.149
Ra	24996.942	0.034	37.391	−0.026	−38.271

pared to results of second order perturbation theory (relativistic). Even if one estimates a conservative error of 50% in the perturbative results, obviously the RLDA does not perform too well. Improvements can, however, be expected via Kohn-Sham perturbation theory on the basis of x-only ROPM results. Further work remains to be done.

3.5   Further Results

In this section I will just list some additional available results in relativistic density functional theory:

1. Relativistic spin-density functional theory has been explored. In this case, charge as well as magnetic densities are calculated, but one has to deal with rather tricky numerical problems. Among the quantities calculated are Kohn-Sham orbital energies, ground-state energies, and the (approximate) exchange-correlation magnetic potential.
2. Relativistic DFT calculations have been performed for the noble metal dimers Cu_2 and Au_2 as well as for the transition metal compounds Fe_2 and FeO. Separation energies, equilibrium separations, and the oscillator frequency are compared: non-relativistic versus relativistic, all electron versus pseudo-potential and LDA versus GGA results.

Table 3.6. Transverse x-only energies (E_x^T) for closed sub-shell atoms: ROPM results in comparison with the values obtained by insertion of ROPM densities into the relativistic LDA (RLDA) and two relativistic GGAs (RECMV92 and RB88) (all energies are in hartree [23])

Atom	ROPM	RLDA	RECMV92	RB88
He	0.000064	0.000159	0.000060	0.000061
Be	0.00070	0.00176	0.00071	0.00072
Ne	0.0167	0.0355	0.0166	0.0167
Mg	0.0319	0.0654	0.0319	0.0319
Ar	0.132	0.251	0.132	0.132
Ca	0.191	0.356	0.191	0.191
Zn	0.759	1.328	0.760	0.759
Kr	1.420	2.410	1.421	1.419
Sr	1.711	2.878	1.712	1.710
Pd	3.291	5.374	3.291	3.291
Cd	3.809	6.180	3.809	3.809
Xe	5.712	9.114	5.712	5.713
Ba	6.475	10.282	6.475	6.477
Yb	13.900	21.597	13.895	13.900
Hg	22.169	34.257	22.169	22.169
Rn	28.679	44.382	28.681	28.680
Ra	31.151	48.275	31.149	31.151

Table 3.7. Comparison of LDA [18], CI (estimated from non-relativistic CI-calculations for the three innermost electrons and the experimental ionization potentials of all other electrons [25]) and MBPT2 [26] correlation energies for neutral atoms: E_c^{NREL} – non-relativistic correlation energy, ΔE_c^L – relativistic contribution in the longitudinal correlation energy, E_c^T – transverse correlation energy (in the case of the MBPT2 only the dominating Breit contribution to E_c^T is given (all energies are in mhartree)

Atom	$-E_c^{NREL}$			$-\Delta E_c^L$		$-E_c^T$	
	MBPT2	CI	LDA	MBPT2	LDA	MBPT2	LDA
He	37.14	42.04	111.47	0.00	0.00	0.04	0.00
Be		94.34	224.44		0.02		0.02
Ne	383.19	390.47	743.38	0.20	0.38	1.87	0.32
Mg		438.28	891.42		0.75		0.57
Ar	697.28	722.16	1429.64	0.84	2.60	7.92	1.89
Zn	1650.61		2665.20	10.51	10.97	26.43	7.92
Kr	1835.43		3282.95	11.39	19.61	41.07	13.10
Cd	2618.11		4570.56	35.86	44.79	82.32	28.58
Xe	2921.13		5200.19	37.57	64.73	108.75	39.27
Hg	5086.24		8355.68	203.23	200.87	282.74	113.08
Rn	5392.07		9026.90	195.36	257.00	352.60	138.43

3. A number of solids with heavy constituents have been treated in relativistic DFT. For Au and Pt the linearized augmented plane-wave (LAPW) method has been implemented in order to study relativistic effects in solids (relativistic versus non-relativistic, LDA versus GGA results). Similarly, effects of spin-orbit-coupling have been investigated in bulk W, Ir and Au on the basis of relativistic LDA-LAPW approaches.

4. On the basis of quantum hadron-dynamics, a field theoretical meson exchange model, Kohn-Sham and extended Thomas-Fermi investigations of nuclei, in particular trans-uranic systems, have been carried out.

Detailed results of relativistic DFT and a full list of references can be found in [27,28].

Acknowledgments

As I am about to retire from "active service", I would like to take this opportunity to thank all the people with whom I had the pleasure to work during the last 40 years. In particular, I would like to thank my Portuguese friends Carlos Fiolhais, João da Providência and José Urbano for a continuing exchange of ideas over many years. Finally, I would like to thank the organizers of this fascinating Summer School, which attracted young people from all over Europe.

References

1. J. D. Bjorken and S. D. Drell, *Relativistic Quantum Fields* (Mc-Graw Hill, New York, 1965).
2. C. Itzykson and J.-B. Zuber, *Quantum Field Theory* (Mc-Graw Hill, New York, 1980).
3. S. Weinberg, *The Quantum Theory of Fields* (Cambridge University Press, London, 1995).
4. G. t'Hooft and M. Veltman, Nucl. Phys. B **44**, 189 (1972).
5. A. K. Rajagopal and J. Callaway, Phys. Rev. B **7**, 1912 (1973).
6. A. K. Rajagopal, J. Phys. C **11**, L943 (1978).
7. A. H. MacDonald and S. H. Vosko, J. Phys. C: Solid State Phys. **12**, 2977 (1979).
8. E. Engel, H. Müller, C. Speicher, and R. M. Dreizler, in *NATO ASI Series B*, edited by E. K. U. Gross and R. M. Dreizler (Plenum, New York, 1995), Vol. 337, p. 65.
9. G. Vignale, in *NATO ASI Series B*, edited by E. K. U. Gross and R. M. Dreizler (Plenum, New York, 1995), Vol. 337, p. 485.
10. I. A. Akhiezer and S. V. Peletminskii, Sov. Phys. JETP **11**, 1316 (1960).
11. H. S. Zapolsky, Cornell University LNS Report (unpublished).
12. B. Jancovici, Nuovo Cim. **XXV**, 428 (1962).
13. M. V. Ramana, A. K. Rajagopal, and W. R. Johnson, Phys. Rev. A **25**, 96 (1982).

14. M. V. Ramana and A. K. Rajagopal, Phys. Rev. A **24**, 1689 (1981).
15. H. Müller and B. D. Serot (unpublished).
16. See chapter by E. Engel in this volume and references therein.
17. E. Engel and S. H. Vosko, Phys. Rev. A **47**, 2800 (1993).
18. E. Engel *et al.*, Phys. Rev. A **52**, 2750 (1995).
19. K. G. Dyall *et al.*, Comp. Phys. Comm. **55**, 425 (1989).
20. J. Perdew, in *Electronic Structure of Solids*, edited by P. Ziesche and H. Eschrig (Akademie, Berlin, 1991), p. 11.
21. E. Engel, J. Chevary, L. Macdonald, and S. Vosko, Z. Phys. D **23**, 7 (1992).
22. A. D. Becke, Phys. Rev. A **38**, 3089 (1988).
23. In all our calculations the nuclei were represented by uniformly charged spheres with nuclear radii given by $R_{\text{nucl}} = 1.0793 A^{1/3} + 0.73587$ F, A being the atomic mass (weighted by isotopic abundance) taken from Table III.7 of [24] unless explicitly stated otherwise. The speed of light has been set to $c = 137.0359895$.
24. K. Hisaka *et al.* (Particle Data Group), Phys. Rev. D **45**, Part 2 (1992).
25. S. J. Chakravorty *et al.*, Phys. Rev. A **47**, 3649 (1993).
26. Y. Ishikawa and K. Koc, Phys. Rev. A **50**, 4733 (1994).
27. E. Engel and R. M. Dreizler, in *Density Functional Theory II*, Vol. 181 of *Topics in Current Chemistry*, edited by R. F. Nalewajski (Springer, Berlin, 1996), p. 1.
28. R. M. Dreizler and E. Engel, in *Density Functionals: Theory and Applications*, edited by D. P. Joubert (Springer, Berlin, 1998), p. 147.

4 Time-Dependent Density Functional Theory

Miguel A.L. Marques*
and Eberhard K.U. Gross[†]

* Donostia International Physics Center (DIPC),
P. Manuel Lardizábal 4,
20080 San Sebastián, Spain
marques@tddft.org

[†] Institut für Theoretische Physik,
Freie Universität Berlin,
Arnimallee 14, 14195 Berlin, Germany
hardy@physik.fu-berlin.de

Eberhard Gross

4.1 Introduction

Time-dependent density-functional theory (TDDFT) extends the basic ideas of ground-state density-functional theory (DFT) to the treatment of excitations and of more general time-dependent phenomena. TDDFT can be viewed as an alternative formulation of time-dependent quantum mechanics but, in contrast to the normal approach that relies on wave-functions and on the many-body Schrödinger equation, its basic variable is the one-body electron density, $n(r, t)$. The advantages are clear: The many-body wave-function, a function in a $3N$-dimensional space (where N is the number of electrons in the system), is a very complex mathematical object, while the density is a simple function that depends solely on the 3-dimensional vector r. The standard way to obtain $n(r, t)$ is with the help of a fictitious system of non-interacting electrons, the Kohn-Sham system. The final equations are simple to tackle numerically, and are routinely solved for systems with a large number of atoms. These electrons feel an effective potential, the time-dependent Kohn-Sham potential. The exact form of this potential is unknown, and has therefore to be approximated.

The scheme is perfectly general, and can be applied to essentially any time-dependent situation. Two regimes can however be observed: If the time-dependent potential is weak, it is sufficient to resort to linear-response theory to study the system. In this way it is possible to calculate e.g. optical absorption spectra. It turns out that, even with the simplest approximation to the Kohn-Sham potential, spectra calculated within this framework are in very good agreement with experimental results. However, if the time-dependent potential is strong, a full solution of the Kohn-Sham equations is required. A canonical example of this regime is the treatment of atoms or molecules in strong laser fields. In this case, TDDFT is able to describe non-linear phenomena like high-harmonic generation, or multi-photon ionization.

Our purpose in this chapter is to provide a pedagogical introduction to TDDFT[1]. With that in mind, we present, in Sect. 4.2, a quite detailed proof of the Runge-Gross theorem [5], i.e. the time-dependent generalization of the Hohenberg-Kohn theorem [6], and the corresponding Kohn-Sham construction [7]. These constitute the mathematical foundations of TDDFT. Several approximate exchange-correlation (xc) functionals are then reviewed. In Sect. 4.3 we are concerned with linear-response theory, and with its main ingredient, the xc kernel. The calculation of excitation energies is treated in the following section. After giving a brief overlook of the competing density-functional methods to calculate excitations, we present some results obtained from the full solution of the Kohn-Sham scheme, and from linear-response theory. Section 4.5 is devoted to the problem of atoms and molecules in strong laser fields. Both high-harmonic generation and ionization are discussed. Finally, the last section is reserved for some concluding remarks.

For simplicity, we will write all formulae for spin-saturated systems. Obviously, spin can be easily included in all expressions when necessary. Hartree atomic units ($e = \hbar = m = 1$) will be used throughout this chapter.

4.2 Time-Dependent DFT

4.2.1 Preliminaries

A system of N electrons with coordinates $\underline{r} = (r_1 \cdots r_N)$ is known to obey the time-dependent Schrödinger equation

$$i\frac{\partial}{\partial t}\Psi(\underline{r},t) = \hat{H}(\underline{r},t)\Psi(\underline{r},t) \ , \tag{4.1}$$

This equation expresses one of the most fundamental postulates of quantum mechanics, and is one of the most remarkable discoveries of physics during the 20th century. The absolute square of the electronic wave-function, $|\Psi(\underline{r},t)|^2$, is interpreted as the probability of finding the electrons at positions $\underline{r}$.

The Hamiltonian can be written in the form

$$\hat{T}(\underline{r}) + \hat{W}(\underline{r}) + \hat{V}_{\text{ext}}(\underline{r},t) \ . \tag{4.2}$$

The first term is the kinetic energy of the electrons

$$\hat{T}(\underline{r}) = -\frac{1}{2}\sum_{i=1}^{N}\nabla_i^2 \ , \tag{4.3}$$

while $\hat{W}$ accounts for the Coulomb repulsion between the electrons

$$\hat{W}(\underline{r}) = \frac{1}{2}\sum_{\substack{i,j=1 \\ i \neq j}}^{N}\frac{1}{|r_i - r_j|} \ . \tag{4.4}$$

---

[1] The reader interested in a more technical discussion is therefore invited to read [1–4], where also very complete and updated lists of references can be found.

Furthermore, the electrons are under the influence of a generic, time-dependent potential, $\hat{V}_{\text{ext}}(\underline{r}, t)$. The Hamiltonian (4.2) is completely general and describes a wealth of physical and chemical situations, including atoms, molecules, and solids in arbitrary time-dependent electric or magnetic fields, scattering experiments, etc. In most of the situations dealt with in this chapter we will be concerned with the interaction between a laser and matter. In that case, we can write the time-dependent potential as the sum of the nuclear potential and a laser field, $\hat{V}_{\text{TD}} = \hat{U}_{\text{en}} + \hat{V}_{\text{laser}}$. The term $\hat{U}_{\text{en}}$ accounts for the Coulomb attraction between the electrons and the nuclei,

$$\hat{U}_{\text{en}}(\underline{r}, t) = -\sum_{\nu=1}^{N_{\text{n}}} \sum_{i=1}^{N} \frac{Z_\nu}{|r_i - R_\nu(t)|} , \qquad (4.5)$$

where Z_ν and R_ν denote the charge and position of the nucleus ν, and N_{n} stands for the total number of nuclei in the system. Note that by allowing the R_ν to depend on time we can treat situations where the nuclei move along a classical path. This may be useful when studying, e.g., scattering experiments, chemical reactions, etc. The laser field, $\hat{V}_{\text{laser}}$, reads, in the length gauge,

$$\hat{V}_{\text{laser}}(\underline{r}, t) = E f(t) \sin(\omega t) \sum_{i=1}^{N} r_i \cdot \alpha , \qquad (4.6)$$

where α, ω and E are the polarization, the frequency and the amplitude of the laser, respectively. The function $f(t)$ is an envelope that shapes the laser pulse during time. Note that, in writing (4.6), we use two approximations: i) We treat the laser field classically, i.e., we do not quantize the photon field. This is a well justified procedure when the density of photons is large and the individual (quantum) nature of the photons can be disregarded. In all cases presented in this chapter this will be the case. ii) Expression (4.6) is written within the dipole approximation. The dipole approximation holds whenever (a) The wavelength of the light ($\lambda = 2\pi c/\omega$, where c is the velocity of light in vacuum) is much larger than the size of the system. This is certainly true for all atoms and most molecules we are interested in. However, one has to be careful when dealing with very large molecules (e.g. proteins) or solids. (b) The path that the particle travels in one period of the laser field is small compared to the wavelength. This implies that the average velocity of the electrons, v, fulfills $vT \ll \lambda \Rightarrow v \ll \lambda/T = c$, where T stands for the period of the laser. In these circumstances we can treat the laser field as a purely electric field and completely neglect its magnetic component. This approximation holds if the intensity of the laser is not strong enough to accelerate the electrons to relativistic velocities. (c) The total duration of the laser pulse should be short enough so that the molecule does not leave the focus of the laser during the time the interaction lasts.

Although the many-body Schrödinger equation, (see 4.1), achieves a remarkably good description of nature, it poses a tantalizing problem to scientists. Its exact (in fact, numerical) solution has been achieved so far only for a disappointingly small number of particles. In fact, even the calculation of a "simple" two electron system (the helium atom) in a laser field takes several months in a modern computer [8] (see also the work on the H_2^+ [9] molecule and the H_3^{++} molecule [10]). The effort to solve (4.1) grows exponentially with the number of particles. Therefore, rapid developments regarding the exact solution of the Schrödinger equation are not expected.

In these circumstances, the natural approach of the theorist is to transform and approximate the basic equations to a manageable level that still retains the qualitative and (hopefully) quantitative information about the system. Several techniques have been developed throughout the years in the quantum chemistry and physics world. One such technique is TDDFT. Its goal, like always in density-functional theories, is to replace the solution of the complicated many-body Schrödinger equation by the solution of the much simpler one-body Kohn-Sham equations, thereby relieving the computational burden.

The first step of any DFT is the proof of a Hohenberg-Kohn type theorem [6]. In its traditional form, this theorem demonstrates that there exists a one-to-one correspondence between the external potential and the (one-body) density. The first implication is clear: With the external potential it is always possible (in principle) to solve the many-body Schrödinger equation to obtain the many-body wave-function. From the wave-function we can trivially obtain the density. The second implication, i.e. that the knowledge of the density is sufficient to obtain the external potential, is much harder to prove. In their seminal paper, Hohenberg and Kohn used the variational principle to obtain a proof by *reductio ad absurdum*. Unfortunately, their method cannot be easily generalized to arbitrary DFTs. The Hohenberg-Kohn theorem is a very strong statement: From the density, a simple property of the quantum mechanical system, it is possible to obtain the external potential and therefore the many-body wave-function. The wave-function, in turn, determines every observable of the system. This implies that *every observable can be written as a functional of the density*.

Unfortunately, it is very hard to obtain the density of an interacting system. To circumvent this problem, Kohn and Sham introduced an auxiliary system of non-interacting particles [7]. The dynamics of these particles are governed by a potential chosen such that the density of the Kohn-Sham system equals the density of the interacting system. This potential is local (multiplicative) in real space, but it has a highly non-local functional dependence on the density. In non-mathematical terms this means that the potential at the point r can depend on the density of all other points (e.g. through gradients, or through integral operators like the Hartree potential). As we are now dealing with non-interacting particles, the Kohn-Sham equations are quite simple to solve numerically. However, the complexities of the many-body system are still present in the so-called exchange-correlation (xc) functional that needs to be approximated in any application of the theory.

4.2.2   The Runge–Gross Theorem

In this section, we will present a detailed proof of the Runge-Gross theorem [5], the time-dependent extension of the ordinary Hohenberg-Kohn theorem [6]. There are several "technical" differences between a time-dependent and a static quantum-mechanical problem that one should keep in mind while trying to prove the Runge-Gross theorem. In static quantum mechanics, the ground-state of the system can be determined through the minimization of the total energy functional

$$E[\Phi] = \langle \Phi | \hat{H} | \Phi \rangle \; . \tag{4.7}$$

In time-dependent systems, there is no variational principle on the basis of the total energy for it is not a conserved quantity. There exists, however, a quantity analogous to the energy, the quantum mechanical action

$$\mathcal{A}[\Phi] = \int_{t_0}^{t_1} \mathrm{d}t \; \langle \Phi(t) | \, \mathrm{i} \frac{\partial}{\partial t} - \hat{H}(t) \, | \Phi(t) \rangle \; , \tag{4.8}$$

where $\Phi(t)$ is a N-body function defined in some convenient space. From expression (4.8) it is easy to obtain two important properties of the action: i) Equating the functional derivative of (4.8) in terms of $\Phi^*(t)$ to zero, we arrive at the time-dependent Schrödinger equation. We can therefore solve the time-dependent problem by calculating the stationary point of the functional $\mathcal{A}[\Phi]$. The function $\Psi(t)$ that makes the functional stationary will be the solution of the time-dependent many-body Schrödinger equation. Note that there is no "minimum principle", as in the time-independent case, but only a "stationary principle". ii) The action is always zero at the solution point, i.e. $\mathcal{A}[\Psi] = 0$. These two properties make the quantum-mechanical action a much less useful quantity than its static counterpart, the total energy.

Another important point, often overlooked in the literature, is that a time-dependent problem in quantum mechanics is mathematically defined as an *initial value* problem. This stems from the fact that the time-dependent Schrödinger equation is a first-order differential equation in the time coordinate. The wave-function (or the density) thus depends on the initial state, which implies that the Runge-Gross theorem can only hold for a *fixed* initial state (and that the xc potential depends on that state). In contrast, the static Schrödinger equation is a second order differential equation in the space coordinates, and is the typical example of a *boundary value* problem.

From the above considerations the reader could conjecture that the proof of the Runge-Gross theorem is more involved than the proof of the ordinary Hohenberg-Kohn theorem. This is indeed the case. What we have to demonstrate is that if two potentials, $v(\boldsymbol{r}, t)$ and $v'(\boldsymbol{r}, t)$, differ by more than a purely time dependent function[2] $c(t)$, they cannot produce the same time-dependent

---

[2] If the two potentials differ solely by a time-dependent function, they will produce wave-functions which are equal up to a purely time-dependent phase. This phase

density, $n(r, t)$, i.e.

$$v(r, t) \neq v'(r, t) + c(t) \Rightarrow \rho(r, t) \neq \rho'(r, t) . \tag{4.9}$$

This statement immediately implies the one-to-one correspondence between the potential and the density. In the following we will utilize primes to distinguish the quantities of the systems with external potentials v and v'. Due to technical reasons that will become evident during the course of the proof, we will have to restrict ourselves to external potentials that are Taylor expandable with respect to the time coordinate around the initial time t_0

$$v(r, t) = \sum_{k=0}^{\infty} c_k(r)(t - t_0)^k , \tag{4.10}$$

with the expansion coefficients

$$c_k(r) = \frac{1}{k!} \frac{\partial^k}{\partial t^k} v(r, t) \Big|_{t=t_0} . \tag{4.11}$$

We furthermore define the function

$$u_k(r) = \frac{\partial^k}{\partial t^k} [v(r, t) - v'(r, t)] \Big|_{t=t_0} . \tag{4.12}$$

Clearly, if the two potentials are different by more than a purely time-dependent function, at least one of the expansion coefficients in their Taylor expansion around t_0 will differ by more than a constant

$$\exists_{k \geq 0} : u_k(r) \neq \text{constant} \tag{4.13}$$

In the first step of our proof we will demonstrate that if $v \neq v' + c(t)$, then the current densities, j and j', generated by v and v', are also different. The current density j can be written as the expectation value of the current density operator:

$$j(r, t) = \langle \Psi(t) | \hat{j}(r) | \Psi(t) \rangle , \tag{4.14}$$

where the operator $\hat{j}$ is

$$\hat{j}(r) = -\frac{1}{2i} \left\{ \left[\nabla \hat{\psi}^\dagger(r) \right] \hat{\psi}(r) - \hat{\psi}^\dagger(r) \left[\nabla \hat{\psi}(r) \right] \right\} . \tag{4.15}$$

We now use the quantum-mechanical equation of motion, which is valid for any operator, $\hat{O}(t)$,

$$i \frac{d}{dt} \langle \Psi(t) | \hat{O}(t) | \Psi(t) \rangle = \langle \Psi(t) | i \frac{\partial}{\partial t} \hat{O}(t) + \left[\hat{O}(t), \hat{H}(t) \right] | \Psi(t) \rangle , \tag{4.16}$$

will, of course, cancel while calculating the density (or any other observable, in fact).

to write the equation of motion for the current density in the primed and unprimed systems

$$i\frac{d}{dt}\boldsymbol{j}(\boldsymbol{r},t) = \langle\Psi(t)|\left[\hat{\boldsymbol{j}}(\boldsymbol{r}),\hat{H}(t)\right]|\Psi(t)\rangle \tag{4.17}$$

$$i\frac{d}{dt}\boldsymbol{j}'(\boldsymbol{r},t) = \langle\Psi'(t)|\left[\hat{\boldsymbol{j}}(\boldsymbol{r}),\hat{H}'(t)\right]|\Psi'(t)\rangle . \tag{4.18}$$

As we start from a fixed initial many-body state, at t_0 the wave-functions, the densities, and the current densities have to be equal in the primed and unprimed systems

$$|\Psi(t_0)\rangle = |\Psi'(t_0)\rangle \equiv |\Psi_0\rangle \tag{4.19}$$

$$n(\boldsymbol{r},t_0) = n'(\boldsymbol{r},t_0) \equiv n_0(\boldsymbol{r}) \tag{4.20}$$

$$\boldsymbol{j}(\boldsymbol{r},t_0) = \boldsymbol{j}'(\boldsymbol{r},t_0) \equiv \boldsymbol{j}_0(\boldsymbol{r}) . \tag{4.21}$$

If we now take the difference between the equations of motion (4.17) and (4.18) we obtain, at $t = t_0$,

$$\begin{aligned}
i\frac{d}{dt}\left[\boldsymbol{j}(\boldsymbol{r},t) - \boldsymbol{j}'(\boldsymbol{r},t)\right]_{t=t_0} &= \langle\Psi_0|\left[\hat{\boldsymbol{j}}(\boldsymbol{r}),\hat{H}(t_0) - \hat{H}'(t_0)\right]|\Psi_0\rangle \\
&= \langle\Psi_0|\left[\hat{\boldsymbol{j}}(\boldsymbol{r}),v(\boldsymbol{r},t_0) - v'(\boldsymbol{r},t_0)\right]|\Psi_0\rangle \\
&= i n_0(\boldsymbol{r})\nabla\left[v(\boldsymbol{r},t_0) - v'(\boldsymbol{r},t_0)\right] .
\end{aligned} \tag{4.22}$$

Let us assume that (4.13) is fulfilled already for $k = 0$, i.e. that the two potentials, v and v', differ at t_0. This immediately implies that the derivative on the left-hand side of (4.22) differs from zero. The two current densities $\boldsymbol{j}$ and $\boldsymbol{j}'$ will consequently deviate for $t > t_0$. If k is greater than zero, the equation of motion is applied $k + 1$ times, yielding

$$\frac{d^{k+1}}{dt^{k+1}}\left[\boldsymbol{j}(\boldsymbol{r},t) - \boldsymbol{j}'(\boldsymbol{r},t)\right]_{t=t_0} = n_0(\boldsymbol{r})\nabla u_k(\boldsymbol{r}) . \tag{4.23}$$

The right-hand side of (4.23) differs from zero, which again implies that $\boldsymbol{j}(\boldsymbol{r},t) \neq \boldsymbol{j}'(\boldsymbol{r},t)$ for $t > t_0$. This concludes the first step of the proof of the Runge-Gross theorem.

In a second step we prove that $\boldsymbol{j} \neq \boldsymbol{j}'$ implies $n \neq n'$. To achieve that purpose we will make use of the continuity equation

$$\frac{\partial}{\partial t}n(\boldsymbol{r},t) = -\nabla \cdot \boldsymbol{j}(\boldsymbol{r},t) . \tag{4.24}$$

If we write (4.24) for the primed and unprimed system and take the difference, we arrive at

$$\frac{\partial}{\partial t}\left[n(\boldsymbol{r},t) - n'(\boldsymbol{r},t)\right] = -\nabla \cdot \left[\boldsymbol{j}(\boldsymbol{r},t) - \boldsymbol{j}'(\boldsymbol{r},t)\right] . \tag{4.25}$$

As before, we would like an expression involving the kth time derivative of the external potential. We therefore take the $(k+1)$th time-derivative of the previous equation to obtain (at $t = t_0$)

$$\frac{\partial^{k+2}}{\partial t^{k+2}} [n(\mathbf{r}, t) - n'(\mathbf{r}, t)]_{t=t_0} = -\nabla \cdot \frac{\partial^{k+1}}{\partial t^{k+1}} [\mathbf{j}(\mathbf{r}, t) - \mathbf{j}'(\mathbf{r}, t)]_{t=t_0}$$

$$= -\nabla \cdot [n_0(\mathbf{r})\nabla u_k(\mathbf{r})] . \tag{4.26}$$

In the last step we made use of (4.23). By the hypothesis (4.13) we have $u_k(\mathbf{r}) \neq \text{const.}$ Hence it is clear that if

$$\nabla \cdot [n_0(\mathbf{r})\nabla u_k(\mathbf{r})] \neq 0 , \tag{4.27}$$

then $n \neq n'$, from which follows the Runge-Gross theorem. To show that (4.27) is indeed fulfilled, we will use the versatile technique of demonstration by *reductio ad absurdum*. Let us assume that $\nabla \cdot [n_0(\mathbf{r})\nabla u_k(\mathbf{r})] = 0$ with $u_k(\mathbf{r}) \neq$ constant, and look at the integral

$$\int d^3r \, n_0(\mathbf{r}) [\nabla u_k(\mathbf{r})]^2 = -\int d^3r \, u_k(\mathbf{r})\nabla \cdot [n_0(\mathbf{r})\nabla u_k(\mathbf{r})] \tag{4.28}$$

$$+ \int_S n_0(\mathbf{r})u_k(\mathbf{r})\nabla u_k(\mathbf{r}) \cdot d\mathbf{S} .$$

This equality was obtained with the help of Green's theorem. The first term on the right-hand side is zero by assumption, while the second term vanishes if the density and the function $u_k(\mathbf{r})$ decay in a "reasonable" manner when $r \to \infty$. This situation is always true for finite systems. We further notice that the integrand $n_0(\mathbf{r}) [\nabla u_k(\mathbf{r})]^2$ is always positive. These diverse conditions can only be satisfied if either the density n_0 or $\nabla u_k(\mathbf{r})$ vanish identically. The first possibility is obviously ruled out, while the second contradicts our initial assumption that $u_k(\mathbf{r})$ is not a constant. This concludes the proof of the Runge-Gross theorem.

4.2.3 Time-Dependent Kohn–Sham Equations

As mentioned in Sect. 4.2.1, the Runge-Gross theorem asserts that all observables can be calculated with the knowledge of the one-body density. Nothing is however stated on how to calculate that valuable quantity. To circumvent the cumbersome task of solving the interacting Schrödinger equation, Kohn and Sham had the idea of utilizing an auxiliary system of non-interacting (Kohn-Sham) electrons, subject to an external local potential, v_{KS} [7]. This potential is unique, by virtue of the Runge-Gross theorem applied to the non-interacting system, and is chosen such that the density of the Kohn-Sham electrons is the same as the density of the original interacting system. In the time-dependent case, these Kohn-Sham electrons obey the time-dependent Schrödinger equation

$$i\frac{\partial}{\partial t}\varphi_i(\mathbf{r}, t) = \left[-\frac{\nabla^2}{2} + v_{KS}(\mathbf{r}, t)\right] \varphi_i(\mathbf{r}, t) . \tag{4.29}$$

The density of the interacting system can be obtained from the time-dependent Kohn-Sham orbitals

$$n(\boldsymbol{r}, t) = \sum_{i}^{\text{occ}} |\varphi_i(\boldsymbol{r}, t)|^2 \; . \tag{4.30}$$

Equation (4.29), having the form of a one-particle equation, is fairly easy to solve numerically. We stress, however, that the Kohn-Sham equation *is not* a mean-field approximation: If we knew the exact Kohn-Sham potential, v_{KS}, we would obtain from (4.29) the exact Kohn-Sham orbitals, and from these the correct density of the system.

The Kohn-Sham potential is conventionally separated in the following way

$$v_{\text{KS}}(\boldsymbol{r}, t) = v_{\text{ext}}(\boldsymbol{r}, t) + v_{\text{Hartree}}(\boldsymbol{r}, t) + v_{\text{xc}}(\boldsymbol{r}, t) \; . \tag{4.31}$$

The first term is again the external potential. The Hartree potential accounts for the classical electrostatic interaction between the electrons

$$v_{\text{Hartree}}(\boldsymbol{r}, t) = \int \text{d}^3 r' \, \frac{n(\boldsymbol{r}, t)}{|\boldsymbol{r} - \boldsymbol{r}'|} \; . \tag{4.32}$$

The last term, the xc potential, comprises all the non-trivial many-body effects. In ordinary DFT, v_{xc} is normally written as a functional derivative of the xc energy. This follows from a variational derivation of the Kohn-Sham equations starting from the total energy. It is not straightforward to extend this formulation to the time-dependent case due to a problem related to causality [11,2]. The problem was solved by van Leeuwen in 1998, by using the Keldish formalism to define a new action functional, $\tilde{A}$ [12]. The time-dependent xc potential can then be written as the functional derivative of the xc part of $\tilde{A}$,

$$v_{\text{xc}}(\boldsymbol{r}, t) = \left. \frac{\delta \tilde{A}_{\text{xc}}}{\delta n(\boldsymbol{r}, \tau)} \right|_{n(\boldsymbol{r}, t)} , \tag{4.33}$$

where τ stands for the Keldish pseudo-time.

Inevitably, the exact expression of v_{xc} as a functional of the density is unknown. At this point we are obliged to perform an approximation. It is important to stress that this is the *only* fundamental approximation in TDDFT. In contrast to stationary-state DFT, where very good xc functionals exist, approximations to $v_{\text{xc}}(\boldsymbol{r}, t)$ are still in their infancy. The first and simplest of these is the adiabatic local density approximation (ALDA), reminiscent of the ubiquitous LDA. More recently, several other functionals were proposed, from which we mention the time-dependent exact-exchange (EXX) functional [13], and the attempt by Dobson, Bünner, and Gross [14] to construct an xc functional with memory. In the following section we will introduce the above mentioned functionals.

4.2.4 XC Functionals

Adiabatic Approximations. There is a very simple procedure that allows the use of the plethora of existing xc functionals for ground-state DFT in the time-dependent theory. Let us assume that $\tilde{v}_{xc}[n]$ is an approximation to the ground-state xc density functional. We can write an adiabatic time-dependent xc potential as

$$v_{xc}^{\text{adiabatic}}(\boldsymbol{r},t) = \tilde{v}_{xc}[n](\boldsymbol{r})\big|_{n=n(t)} , \qquad (4.34)$$

i.e. we employ the same functional form but evaluated at each time with the density $n(\boldsymbol{r},t)$. The functional thus constructed is obviously local in time. This is, of course, a quite dramatic approximation. The functional $\tilde{v}_{xc}[n]$ is a ground-state property, so we expect the adiabatic approximation to work only in cases where the temporal dependence is small, i.e., when our time-dependent system is locally close to equilibrium. Certainly this is not the case if we are studying the interaction of strong laser pulses with matter.

By inserting the LDA functional in (4.34) we obtain the so-called adiabatic local density approximation (ALDA)

$$v_{xc}^{\text{ALDA}}(\boldsymbol{r},t) = v_{xc}^{\text{HEG}}(n)\big|_{n=n(\boldsymbol{r},t)} . \qquad (4.35)$$

The ALDA assumes that the xc potential at the point $\boldsymbol{r}$, and time t is equal to the xc potential of a (static) homogeneous electron gas (HEG) of density $n(\boldsymbol{r},t)$. Naturally, the ALDA retains all problems already present in the LDA. Of these, we would like to mention the erroneous asymptotic behavior of the LDA xc potential: For neutral finite systems, the exact xc potential decays as $-1/r$, whereas the LDA xc potential falls off exponentially. Note that most of the generalized-gradient approximations (GGAs), or even the newest meta-GGAs have asymptotic behaviors similar to the LDA. This problem gains particular relevance when calculating ionization yields (the ionization potential calculated with the ALDA is always too small), or in situations where the electrons are pushed to regions far away from the nuclei (e.g., by a strong laser) and feel the incorrect tail of the potential.

Despite this problem, the ALDA yields remarkably good excitation energies (see Sects. 4.4.2 and 4.4.3) and is probably the most widely used xc functional in TDDFT.

Time-Dependent Optimized Effective Potential. Unfortunately, when trying to write v_{xc} as an explicit functional of the density, one encounters some difficulties. As an alternative, the so-called orbital-dependent xc functionals were introduced several years ago. These functionals are written explicitly in terms of the Kohn-Sham orbitals, albeit remaining *implicit* density functionals by virtue of the Runge-Gross theorem. A typical member of this family is the exact-exchange (EXX) functional. The EXX action is obtained by expanding $\mathcal{A}_{xc}$ in powers of e^2 (where e is the electronic charge), and retaining the lowest order term, the exchange term. It is given by the Fock integral

$$\mathcal{A}_{x}^{\text{EXX}} = -\frac{1}{2}\sum_{j,k}^{\text{occ}}\int_{t_0}^{t_1}\!dt\int d^3r\int d^3r'\,\frac{\varphi_j^*(r',t)\varphi_k(r',t)\varphi_j(r,t)\varphi_k^*(r,t)}{|r-r'|}\,. \tag{4.36}$$

From such an action functional, one seeks to determine the local Kohn-Sham potential through a series of chain rules for functional derivatives. The procedure is called the optimized effective potential (OEP) or the optimized potential method (OPM) for historical reasons [15,16]. The derivation of the time-dependent version of the OEP equations is very similar to the ground-state case. Due to space limitations we will not present the derivation in this chapter. The interested reader is advised to consult the original paper [13], one of the more recent publications [17,18], or the chapter by E. Engel contained in this volume. The final form of the OEP equation that determines the EXX potential is

$$\sum_j^{\text{occ}}\int_{-\infty}^{t_1}\!dt'\int d^3r'\,[v_x(r',t')-u_{x\,j}(r',t')] \tag{4.37}$$

$$\times\varphi_j(r,t)\varphi_j^*(r',t')G_R(rt,r't')+\text{c.c.}=0$$

The kernel, G_R, is defined by

$$\mathrm{i}\,G_R(rt,r't')=\sum_{k=1}^{\infty}\varphi_k^*(r,t)\varphi_k(r',t')\theta(t-t')\,, \tag{4.38}$$

and can be identified with the retarded Green's function of the system. Moreover, the expression for u_x is essentially the functional derivative of the xc action with respect to the Kohn-Sham wave-functions

$$u_{x\,j}(r,t)=\frac{1}{\varphi_j^*(r,t)}\frac{\delta\mathcal{A}_{\text{xc}}[\varphi_j]}{\delta\varphi_j(r,t)}\,. \tag{4.39}$$

Note that the xc potential is still a local potential, albeit being obtained through the solution of an extremely non-local and non-linear integral equation. In fact, the solution of (4.37) poses a very difficult numerical problem. Fortunately, by performing an approximation first proposed by Krieger, Li, and Iafrate (KLI) it is possible to simplify the whole procedure, and obtain an semi-analytic solution of (4.37) [19]. The KLI approximation turns out to be a very good approximation to the EXX potential. Note that both the EXX and the KLI potential have the correct $-1/r$ asymptotic behavior for neutral finite systems.

A Functional with Memory. There is a very common procedure for the construction of approximate xc functionals in ordinary DFT. It starts with the derivation of *exact* properties of v_{xc}, deemed important by physical arguments. Then an analytical expression for the functional is proposed, such that it satisfies those rigorous constraints. We will use this recipe to generate a time-dependent xc potential which is non-local in time, i.e. that includes the "memory" from previous times [14].

A very important condition comes from Galilean invariance. Let us look at a system from the point of view of a moving reference frame whose origin is given by $\boldsymbol{x}(t)$. The density seen from this moving frame is simply the density of the reference frame, but shifted by $\boldsymbol{x}(t)$

$$n'(\boldsymbol{r}, t) = n(\boldsymbol{r} - \boldsymbol{x}(t), t) \; . \tag{4.40}$$

Galilean invariance then implies [20]

$$v_{xc}[n'](\boldsymbol{r}, t) = v_{xc}[n](\boldsymbol{r} - \boldsymbol{x}(t), t) \; . \tag{4.41}$$

It is obvious that potentials that are both local in space and in time, like the ALDA, trivially fulfill this requirement. However, when one tries to deduce an xc potential which is non-local in time, one finds condition (4.41) quite difficult to satisfy.

Another rigorous constraint follows from Ehrenfest's theorem which relates the acceleration to the gradient of the external potential

$$\frac{d^2}{dt^2} \langle \boldsymbol{r} \rangle = - \langle \nabla v_{ext}(\boldsymbol{r}) \rangle \; . \tag{4.42}$$

For an interacting system, Ehrenfest's theorem states

$$\frac{d^2}{dt^2} \int d^3r \, \boldsymbol{r} \, n(\boldsymbol{r}, t) = - \int d^3r \, n(\boldsymbol{r}, t) \nabla v_{ext}(\boldsymbol{r}) \; . \tag{4.43}$$

In the same way we can write Ehrenfest's theorem for the Kohn-Sham system

$$\frac{d^2}{dt^2} \int d^3r \, \boldsymbol{r} \, n(\boldsymbol{r}, t) = - \int d^3r \, n(\boldsymbol{r}, t) \nabla v_{KS}(\boldsymbol{r}) \; . \tag{4.44}$$

By the very construction of the Kohn-Sham system, the interacting density is equal to the Kohn-Sham density. We can therefore equate the right-hand sides of (4.43) and (4.44), and arrive at

$$\int d^3r \, n(\boldsymbol{r}, t) \nabla v_{ext}(\boldsymbol{r}) = \int d^3r \, n(\boldsymbol{r}, t) \nabla v_{KS}(\boldsymbol{r}, t) \; . \tag{4.45}$$

If we now insert the definition of the Kohn-Sham potential, (see 4.31), and note that $\int d^3r \, n(\boldsymbol{r}, t) \nabla v_{Hartree}(\boldsymbol{r}) = 0$, we obtain the condition

$$\int d^3r \, n(\boldsymbol{r}, t) \nabla v_{xc}(\boldsymbol{r}, t) = \int d^3r \, n(\boldsymbol{r}, t) \boldsymbol{F}_{xc}(\boldsymbol{r}, t) = 0 \; , \tag{4.46}$$

i.e. the total xc force of the system, F_{xc}, is zero. This condition reflects Newton's third law: The xc effects are only due to internal forces, the Coulomb interaction among the electrons, and should not give rise to any net force on the system.

A functional that takes into account these exact constraints can be constructed [14]. The condition (4.46) is simply ensured by the expression

$$F_{xc}(r, t) = \frac{1}{n(r, t)} \nabla \int dt' \, \Pi_{xc}(n(r, t'), t - t') . \qquad (4.47)$$

The function Π_{xc} is a pressure-like scalar memory function of two variables. In practice, Π_{xc} is fully determined by requiring it to reproduce the scalar linear response of the homogeneous electron gas. Expression (4.47) is clearly non-local in the time-domain but still local in the spatial coordinates. From the previous considerations it is clear that it must violate Galilean invariance. To correct this problem we use a concept borrowed from hydrodynamics. It is assumed that, in the electron liquid, memory resides not with each fixed point r, but rather within each separate "fluid element". Thus the element which arrives at location r at time t "remembers" what happened to it at earlier times t' when it was at locations $R(t'|r, t)$, different from its present location r. The trajectory, R, can be determined by demanding that its time derivative equals the fluid velocity

$$\frac{\partial}{\partial t'} R(t'|r, t) = \frac{j(R, t')}{n(R, t')} , \qquad (4.48)$$

with the boundary condition

$$R(t|r, t) = r . \qquad (4.49)$$

We then correct the (4.47) by evaluating n at point R

$$F_{xc}(r, t) = \frac{1}{n(r, t)} \nabla \int dt' \, \Pi_{xc}(n(R, t'), t - t') . \qquad (4.50)$$

Finally, an expression for v_{xc} can be obtained by direct integration of F_{xc} (see [14] for details).

4.2.5 Numerical Considerations

As mentioned before, the solution of the time-dependent Kohn-Sham equations is an initial value problem. At $t = t_0$ the system is in some initial state described by the Kohn-Sham orbitals $\varphi_i(r, t_0)$. In most cases the initial state will be the ground state of the system (i.e., $\varphi_i(r, t_0)$ will be the solution of the ground-state Kohn-Sham equations). The main task of the computational physicist is then to propagate this initial state until some final time, t_f.

The time-dependent Kohn-Sham equations can be rewritten in the integral form

$$\varphi_i(\boldsymbol{r}, t_f) = \hat{U}(t_f, t_0)\varphi_i(\boldsymbol{r}, t_0) ,\tag{4.51}$$

where the time-evolution operator, $\hat{U}$, is defined by

$$\hat{U}(t', t) = \hat{T} \exp\left[-i \int_t^{t'} d\tau\, \hat{H}_{KS}(\tau)\right] .\tag{4.52}$$

Note that $\hat{H}_{KS}$ is explicitly time-dependent due to the Hartree and xc potentials. It is therefore important to retain the time-ordering propagator, $\hat{T}$, in the definition of the operator $\hat{U}$. The exponential in expression (4.52) is clearly too complex to be applied directly, and needs to be approximated in some suitable manner. To reduce the error in the propagation from t_0 to t_f, this large interval is usually split into smaller sub-intervals of length Δt. The wave-functions are then propagated from $t_0 \to t_0 + \Delta t$, then from $t_0 + \Delta t \to t_0 + 2\Delta t$ and so on.

The simplest approximation to (4.52) is a direct expansion of the exponential in a power series of Δt

$$\hat{U}(t + \Delta t, t) \approx \sum_{l=0}^{k} \frac{\left[-i\hat{H}(t + \Delta t/2)\Delta t\right]^l}{l!} + \mathcal{O}(\Delta t^{k+1}) .\tag{4.53}$$

Unfortunately, the expression (4.53) does not retain one of the most important properties of the Kohn-Sham time-evolution operator: unitarity. In other words, if we apply (4.53) to a normalized wave-function the result will no longer be normalized. This leads to an inherently unstable propagation.

Several different propagation methods exist in the market. We will briefly mention two of these: a modified Crank-Nicholson scheme, and the split-operator method.

A Modified Crank–Nicholson Scheme. This method is derived by imposing time-reversal symmetry to an approximate time-evolution operator. It is clear that we can obtain the state at time $t + \Delta t/2$ either by forward propagating the state at t by $\Delta t/2$, or by backward propagating the state at $t + \Delta t$

$$\varphi(t + \Delta t/2) = \hat{U}(t + \Delta t/2, t)\varphi(t)$$
$$= \hat{U}(t - \Delta t/2, t + \Delta t)\varphi(t + \Delta t) .\tag{4.54}$$

This equality leads to

$$\varphi(t + \Delta t) = \hat{U}(t + \Delta t/2, t + \Delta t)\hat{U}(t + \Delta t/2, t)\varphi(t) ,\tag{4.55}$$

where we used the fact that the inverse of the time-evolution operator $\hat{U}^{-1}(t + \Delta t, t) = \hat{U}(t - \Delta t, t)$. To propagate a state from t to $t + \Delta t$ we follow the steps: i) Obtain an estimate of the Kohn-Sham wave-functions at time $t+\Delta t$ by propagating from time t using a "low quality" formula for $\hat{U}(t+\Delta t, t)$. The expression (4.53) expanded to third or forth order is well suited for this purpose. ii) With these wave-functions construct an approximation to $\hat{H}(t+\Delta t)$ and to $\hat{U}(t+\Delta t/2, t+\Delta t)$. iii) Apply (4.55). This procedure leads to a very stable propagation.

The Split-Operator Method. In a first step we neglect the time-ordering in (4.52), and approximate the integral in the exponent by a trapezoidal rule

$$\hat{U}(t+\Delta t, t) \approx \exp\left[-i\hat{H}_{\mathrm{KS}}(t)\Delta t\right] = \exp\left[-i(\hat{T}+\hat{V}_{\mathrm{KS}})\Delta t\right] . \quad (4.56)$$

We note that the operators $\exp\left(-i\hat{V}_{\mathrm{KS}}\Delta t\right)$ and $\exp\left(-i\hat{T}\Delta t\right)$ are diagonal in real and Fourier space respectively, and therefore trivial to apply in those spaces. It is possible to decompose the exponential (4.56) into a form involving only these two operators. The two lowest order decompositions are

$$\exp\left[-i(\hat{T}+\hat{V}_{\mathrm{KS}})\Delta t\right] = \exp\left(-i\hat{T}\Delta t\right)\exp\left(-i\hat{V}_{\mathrm{KS}}\Delta t\right) + \mathcal{O}(\Delta t^2) , \quad (4.57)$$

and

$$\exp\left[-i(\hat{T}+\hat{V}_{\mathrm{KS}})\Delta t\right] = \exp\left(-i\hat{T}\frac{\Delta t}{2}\right)\exp\left(-i\hat{V}_{\mathrm{KS}}\Delta t\right)\exp\left(-i\hat{T}\frac{\Delta t}{2}\right)$$
$$+ \mathcal{O}(\Delta t^3) . \quad (4.58)$$

For example, to apply the splitting (4.58) to $\varphi(\boldsymbol{r}, t)$ we start by Fourier transforming the wave-function to Fourier space. We then apply $\exp\left(-i\hat{T}\frac{\Delta t}{2}\right)$ to $\varphi(\boldsymbol{k}, t)$ and Fourier transform back the result to real space. We proceed by applying $\exp\left(-i\hat{V}\Delta t\right)$, Fourier transforming, etc. This method can be made very efficient by the use of fast Fourier transforms.

As a better approximation to the propagator (4.52) we can use a mid-point rule to estimate the integral in the exponential

$$\hat{U}(t+\Delta t, t) \approx \exp\left[-i\hat{H}_{\mathrm{KS}}(t+\Delta t/2)\Delta t\right] . \quad (4.59)$$

It can be shown that the same procedure described above can be applied with only a slight modification: The Kohn-Sham potential has to be updated after applying the first kinetic operator [21].

4.3   Linear Response Theory

4.3.1   Basic Theory

In circumstances where the external time-dependent potential is small, it may not be necessary to solve the full time-dependent Kohn-Sham equations.

Instead perturbation theory may prove sufficient to determine the behavior of the system. We will focus on the linear change of the density, that allows us to calculate, e.g., the optical absorption spectrum.

Let us assume that for $t < t_0$ the time-dependent potential v_{TD} is zero – i.e. the system is subject only to the nuclear potential, $v^{(0)}$ – and furthermore that the system is in its ground-state with ground-state density $n^{(0)}$. At t_0 we turn on the perturbation, $v^{(1)}$, so that the total external potential now consists of $v_{ext} = v^{(0)} + v^{(1)}$. Clearly $v^{(1)}$ will induce a change in the density. If the perturbing potential is sufficiently well-behaved (like almost always in physics), we can expand the density in a perturbative series

$$n(\boldsymbol{r}, t) = n^{(0)}(\boldsymbol{r}) + n^{(1)}(\boldsymbol{r}, t) + n^{(2)}(\boldsymbol{r}, t) + \cdots, \tag{4.60}$$

where $n^{(1)}$ is the component of $n(\boldsymbol{r}, t)$ that depends linearly on $v^{(1)}$, $n^{(2)}$ depends quadratically, etc. As the perturbation is weak, we will only be concerned with the linear term, $n^{(1)}$. In frequency space it reads

$$n^{(1)}(\boldsymbol{r}, \omega) = \int d^3 r' \, \chi(\boldsymbol{r}, \boldsymbol{r}', \omega) \, v^{(1)}(\boldsymbol{r}', \omega). \tag{4.61}$$

The quantity χ is the linear density-density response function of the system. In other branches of physics it has other names, e.g., in the context of many-body perturbation theory it is called the reducible polarization function. Unfortunately, the evaluation of χ through perturbation theory is a very demanding task. We can, however, make use of TDDFT to simplify this process.

We recall that in the time-dependent Kohn-Sham framework, the density of the *interacting* system of electrons is obtained from a fictitious system of *non-interacting* electrons. Clearly, we can also calculate the linear change of density using the Kohn-Sham system

$$n^{(1)}(\boldsymbol{r}, \omega) = \int d^3 r' \, \chi_{KS}(\boldsymbol{r}, \boldsymbol{r}', \omega) \, v_{KS}^{(1)}(\boldsymbol{r}', \omega). \tag{4.62}$$

The response function that enters (4.62), χ_{KS}, is the density-density response function of a system of *non-interacting* electrons and is, consequently, much easier to calculate than the full interacting χ. In terms of the unperturbed stationary Kohn-Sham orbitals it reads

$$\chi_{KS}(\boldsymbol{r}, \boldsymbol{r}', \omega) = \lim_{\eta \to 0^+} \sum_{jk}^{\infty} (f_k - f_j) \frac{\varphi_j(\boldsymbol{r}) \varphi_j^*(\boldsymbol{r}') \varphi_k(\boldsymbol{r}') \varphi_k^*(\boldsymbol{r})}{\omega - (\epsilon_j - \epsilon_k) + i\eta}, \tag{4.63}$$

where f_m is the occupation number of the mth orbital in the Kohn-Sham ground-state. Note that the Kohn-Sham potential, v_{KS}, includes all powers of the external perturbation due to its non-linear dependence on the density. The potential that enters (4.62) is however just the *linear* change of v_{KS},

$v_{\text{KS}}^{(1)}$. This latter quantity can be calculated explicitly from the definition of the Kohn-Sham potential

$$v_{\text{KS}}^{(1)}(\boldsymbol{r}, t) = v^{(1)}(\boldsymbol{r}, t) + v_{\text{Hartree}}^{(1)}(\boldsymbol{r}, t) + v_{\text{xc}}^{(1)}(\boldsymbol{r}, t) . \qquad (4.64)$$

The variation of the external potential is simply $v^{(1)}$, while the change in the Hartree potential is

$$v_{\text{Hartree}}^{(1)}(\boldsymbol{r}, t) = \int \mathrm{d}^3 r' \frac{n^{(1)}(\boldsymbol{r}', t)}{|\boldsymbol{r} - \boldsymbol{r}'|} . \qquad (4.65)$$

Finally $v_{\text{xc}}^{(1)}(\boldsymbol{r}, t)$ is the linear part in $n^{(1)}$ of the functional $v_{\text{xc}}[n]$,

$$v_{\text{xc}}^{(1)}(\boldsymbol{r}, t) = \int \mathrm{d}t' \int \mathrm{d}^3 r' \frac{\delta v_{\text{xc}}(\boldsymbol{r}, t)}{\delta n(\boldsymbol{r}', t')} n^{(1)}(\boldsymbol{r}', t') . \qquad (4.66)$$

It is useful to introduce the exchange-correlation kernel, f_{xc}, defined by

$$f_{\text{xc}}(\boldsymbol{r}t, \boldsymbol{r}'t') = \frac{\delta v_{\text{xc}}(\boldsymbol{r}, t)}{\delta n(\boldsymbol{r}', t')} . \qquad (4.67)$$

The kernel is a well known quantity that appears in several branches of theoretical physics. For example, evaluated for the electron gas, f_{xc} is, up to a factor, the "local field correction". To emphasize the correspondence to the effective interaction of Landau's Fermi-liquid theory, to which it reduces in the appropriate limit, f_{xc} plus the bare Coulomb interaction is sometimes called the "effective interaction", while in the theory of classical liquids the same quantity is referred to as the Ornstein-Zernicke function.

Combining the previous results, and transforming to frequency space we arrive at:

$$n^{(1)}(\boldsymbol{r}, \omega) = \int \mathrm{d}^3 r' \chi_{\text{KS}}(\boldsymbol{r}, \boldsymbol{r}', \omega) v^{(1)}(\boldsymbol{r}', \omega) \qquad (4.68)$$

$$+ \int \mathrm{d}^3 x \int \mathrm{d}^3 r' \chi_{\text{KS}}(\boldsymbol{r}, \boldsymbol{x}, \omega) \left[\frac{1}{|\boldsymbol{x} - \boldsymbol{r}'|} + f_{\text{xc}}(\boldsymbol{x}, \boldsymbol{r}', \omega) \right] n^{(1)}(\boldsymbol{r}', \omega) .$$

From (4.61) and (4.68) trivially follows the relation

$$\chi(\boldsymbol{r}, \boldsymbol{r}', \omega) = \chi_{\text{KS}}(\boldsymbol{r}, \boldsymbol{r}', \omega) \qquad (4.69)$$

$$+ \int \mathrm{d}^3 x \int \mathrm{d}^3 x' \chi(\boldsymbol{r}, \boldsymbol{x}, \omega) \left[\frac{1}{|\boldsymbol{x} - \boldsymbol{x}'|} + f_{\text{xc}}(\boldsymbol{x}, \boldsymbol{x}', \omega) \right] \chi_{\text{KS}}(\boldsymbol{x}', \boldsymbol{r}', \omega) .$$

This equation is a formally exact representation of the linear density response in the sense that, if we possessed the exact Kohn-Sham potential (so that we could extract f_{xc}), a self-consistent solution of (4.69) would yield the response function, χ, of the interacting system.

4.3.2 The XC Kernel

As we have seen in the previous section, the main ingredient in linear response theory is the xc kernel. f_{xc} is, as expected, a very complex quantity that includes – or, in other words, hides – all non-trivial many-body effects. Many approximate xc kernels have been proposed in the literature over the past years. The most ancient, and certainly the simplest is the ALDA kernel

$$f_{xc}^{ALDA}(rt, r't') = \delta(r - r')\delta(t - t') f_{xc}^{HEG}(n)\big|_{n=n(r,t)} , \qquad (4.70)$$

where

$$f_{xc}^{HEG}(n) = \frac{d}{dn} v_{xc}^{HEG}(n) \qquad (4.71)$$

is just the derivative of the xc potential of the homogeneous electron gas. The ALDA kernel is local both in the space and time coordinates.

Another commonly used xc kernel was derived by Petersilka $et\ al.$ in 1996, and is nowadays referred to as the PGG kernel [22]. Its derivation starts from a simple analytic approximation to the EXX potential. This approximation, called the Slater approximation in the context of Hartree-Fock theory, only retains the leading term in the expression for EXX, which reads

$$v_x^{PGG}(r, t) = \sum_k^{occ} \frac{|\varphi_k(r, t)|^2}{n(r, t)} [u_{x\,k}(r, t) + \text{c.c.}] . \qquad (4.72)$$

Using the definition (4.67) and after some algebra, we arrive at the final form of the PGG kernel

$$f_x^{PGG}(rt, r't') = -\delta(t - t')\frac{1}{2}\frac{1}{|r - r'|}\frac{|\sum_k^{occ} \varphi_k(r)\varphi_k^*(r')|^2}{n(r)n(r')} . \qquad (4.73)$$

As in the case of the ALDA, the PGG kernel is local in time.

Noticing the crudeness of the ALDA, especially the complete neglect of any frequency dependence, one could expect it to yield very inaccurate results in most situations. Surprisingly, this is not the case as we will show in Sect. 4.4. To understand this numerical evidence, we have to take a step back and study more thoroughly the properties of the xc kernel for the homogeneous electron gas [1].

In this simple system, f_{xc}^{HEG} only depends on $r - r'$ and on $t - t'$, so it is convenient to work in Fourier space. Our knowledge of the function $f_{xc}^{HEG}(q, \omega)$ is quite limited. Several of its exact features can nevertheless be obtained through analytical manipulations. The long-wavelength limit at zero frequency is given by

$$\lim_{q \to 0} f_{xc}^{HEG}(q, \omega = 0) = \frac{d^2}{dn^2} \left[n\epsilon_{xc}^{HEG}(n) \right] \equiv f_0(n) , \qquad (4.74)$$

where ϵ_{xc}^{HEG}, the xc energy per particle of the homogeneous electron gas, is known exactly from Monte-Carlo calculations [23]. Also the infinite frequency limit can be written as a simple expression

$$\lim_{q \to 0} f_{xc}^{HEG}(q, \omega = \infty) = -\frac{4}{5} n^{\frac{2}{3}} \frac{d}{dn} \left[\frac{\epsilon_{xc}^{HEG}(n)}{n^{2/3}} \right] + 6 n^{\frac{1}{3}} \frac{d}{dn} \left[\frac{\epsilon_{xc}^{HEG}(n)}{n^{1/3}} \right] \quad (4.75)$$
$$\equiv f_\infty(n) .$$

From these two expression, one can prove that the zero frequency limit is always smaller than the infinite frequency limit, and that both these quantities are smaller than zero (according to the best approximations known for E_{xc}^{HEG}), i.e.

$$f_0(n) < f_\infty(n) < 0 . \quad (4.76)$$

From the fact that f_{xc} is a real function when written in real space and in real time, one can deduce the following symmetry relations

$$\Re f_{xc}^{HEG}(q, \omega) = \Re f_{xc}^{HEG}(q, -\omega) \quad (4.77)$$
$$\Im f_{xc}^{HEG}(q, \omega) = -\Im f_{xc}^{HEG}(q, -\omega) .$$

From causality follow the Kramers-Kronig relations:

$$\Re f_{xc}^{HEG}(q, \omega) - f_{xc}^{HEG}(q, \infty) = \mathcal{P} \int_{-\infty}^{\infty} \frac{d\omega'}{\pi} \frac{\Im f_{xc}^{HEG}(q, \omega)}{\omega - \omega'} \quad (4.78)$$

$$\Im f_{xc}^{HEG}(q, \omega) = -\mathcal{P} \int_{-\infty}^{\infty} \frac{d\omega'}{\pi} \frac{\Re f_{xc}^{HEG}(q, \omega) - f_{xc}^{HEG}(q, \infty)}{\omega - \omega'} ,$$

where $\mathcal{P}$ denotes the principal value of the integral. Note that, as the infinite frequency limit of the xc kernel is different from zero, one has to subtract $f_{xc}^{HEG}(q, \infty)$ in order to apply the Kramers-Kronig relations.

Furthermore, by performing a perturbative expansion of the irreducible polarization to second order in e^2, one finds

$$\lim_{\omega \to \infty} \Im f_{xc}^{HEG}(q = 0, \omega) = -\frac{23\pi}{15 \omega^{3/2}} . \quad (4.79)$$

The real part can be obtained with the help of the Kramers-Kronig relations

$$\lim_{\omega \to \infty} \Re f_{xc}^{HEG}(q = 0, \omega) = f_\infty(n) + \frac{23\pi}{15 \omega^{3/2}} . \quad (4.80)$$

It is possible to write an analytical form for the long-wavelength limit of the imaginary part of f_{xc} that incorporates all these exact limits [24]

$$\Im f_{xc}^{HEG}(q = 0, \omega) \approx \frac{\alpha(n)\omega}{(1 + \beta(n)\omega^2)^{\frac{5}{4}}} . \quad (4.81)$$

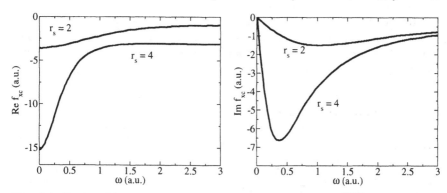

Fig. 4.1. Real and imaginary part of the parametrization for f_{xc}^{HEG}. Figure reproduced from [25]

The coefficients α and β are functions of the density, and can be determined uniquely by the zero and high frequency limits. A simple calculation yields

$$\alpha(n) = -A\left[f_\infty(n) - f_0(n)\right]^{\frac{5}{3}} \tag{4.82}$$

$$\beta(n) = B\left[f_\infty(n) - f_0(n)\right]^{\frac{4}{3}}, \tag{4.83}$$

where $A, B > 0$ and independent of n. By applying the Kramers-Kronig relations we can obtain the corresponding real part of f_{xc}^{HEG}

$$\Re f_{xc}^{HEG}(\boldsymbol{q} = 0, \omega) = f_\infty + \frac{2\sqrt{2}\alpha}{\pi\sqrt{\beta}r^2}\left[2E\left(\frac{1}{\sqrt{2}}\right)\right. \tag{4.84}$$

$$\left. - \frac{1+r}{2}\Pi\left(\frac{1-r}{2}, \frac{1}{\sqrt{2}}\right) - \frac{1-r}{2}\Pi\left(\frac{1+r}{2}, \frac{1}{\sqrt{2}}\right)\right],$$

where $r = \sqrt{1 + \beta\omega^2}$ and E and Π are the elliptic integrals of second and third kind. In Fig. 4.1 we plot the real and imaginary part of f_{xc}^{HEG} for two different densities ($r_s = 2$ and $r_s = 4$, where r_s is the Wigner-Seitz radius, $1/n = 4\pi r_s^3/3$). The ALDA corresponds to approximating these curves by their zero frequency value. For very low frequencies, the ALDA is naturally a good approximation, but at higher frequencies it completely fails to reproduce the behavior of f_{xc}^{HEG}.

To understand how the ALDA can yield such good excitation energies, albeit exhibiting such a mediocre frequency dependence, we will look at a specific example, the process of photo-absorption by an atom. At low excitation frequencies, we expect the ALDA to work. As we increase the laser frequency, we start exciting deeper levels, promoting electrons from the inner shells of the atom to unoccupied states. The atomic density increases monotonically as we approach the nucleus. The f_{xc} corresponding to that larger density (lower r_s) has a much weaker frequency dependence, and is much better approximated by the ALDA than the low density curve. In short, by noticing that high frequencies are normally related to high densities, we realize that for practical applications the ALDA is often a reasonably good approximation. One should however keep in mind that these are simple heuristic arguments that may not hold in a real physical system.

4.4 Excitation Energies

4.4.1 DFT Techniques to Calculate Excitations

In this section we will present a short overview of the several techniques to calculate excitation energies that have appeared in the context of DFT over the past years. Indeed, quite a lot of different approaches have been tried. Some are more or less *ad hoc*, others rely on a solid theoretical basis. Moreover, the degree of success varies considerably among the different techniques. The most successful of all is certainly TDDFT that has become the *de facto* standard for the calculation of excitations for finite systems. We will leave the discussion of excitation energies in TDDFT to the following sections, and concentrate for now on the "competitors". The first group of methods is based on a single determinant calculation, i.e. only one ground-state like calculation is performed, subject to the restriction that the Kohn-Sham occupation numbers are either 0 or 1.

As a first approximation to the excitation energies, one can simply take the differences between the ground-state Kohn-Sham eigenvalues. This procedure, although not entirely justifiable, is often used to get a rough idea of the excitation spectrum. We stress that the Kohn-Sham eigenvalues (as well as the Kohn-Sham wave-functions) *do not* have any physical interpretation. The exception is the eigenvalue of the highest occupied state that is equal to minus the ionization potential of the system [26].

The second scheme is based on the observation that the Hohenberg-Kohn theorem and the Kohn-Sham scheme can be formulated for the lowest state of each symmetry class [27]. In fact, the single modification to the standard proofs is to restrict the variational principle to wave-functions of a specific symmetry. The unrestricted variation will clearly yield the ground-state. The states belonging to different symmetry classes will correspond to excited states. The excitations can then be calculated by simple total-energy differences. This approach suffers from two serious drawbacks: i) Only the lowest lying excitation for each symmetry class is obtainable. ii) The xc functional that now enters the Kohn-Sham equations depends on the particular symmetry we have chosen. As specific approximations for a symmetry dependent xc functional are not available, one is relegated to use ground-state functionals. Unfortunately, the excitation energies calculated in this way are only of moderate quality.

Another promising method was recently proposed by A. Görling [28]. The so-called generalized adiabatic connection Kohn-Sham formalism is no longer based on the Hohenberg-Kohn theorem but on generalized adiabatic connections associating a Kohn-Sham state with each state of the real system. This formalism was later extended to allow for a proper treatment of the symmetry of the Kohn-Sham states [29]. The quality of the results obtained so far with this procedure varies: For alkali atoms the agreement with experimental excitation energies is quite good [28], but for the carbon atom and the CO molecule the situation is considerably worse [29]. We note however that this method is still in its infancy, so further developments can be expected in the near future.

It is also possible to calculate excitation energies from the ground-state energy functional. In fact, it was proved by Perdew and Levy [30] that "every extremum density $n_i(r)$ of the ground-state energy functional $E_v[n]$ yields the energy E_i of a stationary state of the system." The problem is that not every excited-state density, $n_i(r)$, corresponds to an extremum of $E_v[n]$, which implies that not all excitation energies can be obtained from this procedure.

The last member of the first group of methods was proposed by Ziegler, Rauk and Baerends in 1977 [31] and is based on an idea borrowed from multi-configuration Hartree-Fock. The procedure starts with the construction of many-particle states with good symmetry, Ψ_i, by taking a finite superposition of states

$$\Psi_i = \sum_\alpha c_{i\alpha}\Phi_\alpha , \qquad (4.85)$$

where Φ_α are Slater determinants of Kohn-Sham orbitals, and the coefficients $c_{i\alpha}$ are determined from group theory. Through a simple matrix inversion we can express the determinants as linear combinations of the many-body wavefunctions

$$\Phi_\beta = \sum_j a_{\beta j}\Psi_j . \qquad (4.86)$$

By taking the expectation value of the Hamiltonian in the state Φ_β we arrive at

$$\langle\Phi_\beta|\hat{H}|\Phi_\beta\rangle = \sum_j |a_{\beta j}|^2 E_j , \qquad (4.87)$$

where E_j is the energy of the many-body state Ψ_j. The "recipe" to calculate excitation energies is then: a) Build Φ_β from n Kohn-Sham orbitals (not necessarily the lowest); b) Make an ordinary Kohn-Sham calculation for each Φ_β, and associate the corresponding total energy E_β^{DFT} with $\langle\Phi_\beta|\hat{H}|\Phi_\beta\rangle$; c) Determine E_j by solving the system of linear equations (4.87).

This method works quite well in practice, and was frequently used in quantum chemistry till the advent of TDDFT. We should nevertheless indicate two of its limitations: i) The decomposition (4.85) is not unique and the system of linear equations can be under- or overdetermined. ii) The whole procedure of the "recipe" is not rigorously founded.

The next technique, known as ensemble DFT, makes use of fractional occupation numbers. Ensemble DFT, first proposed by Theophilou in 1979 [32], evolves around the concept of an ensemble. In the simplest case it consists of a "mixture" of the ground state, Ψ_1, and the first excited state, Ψ_2, described by the density matrix [33–35]

$$\hat{D} = (1 - \omega) |\Psi_1\rangle \langle \Psi_1| + \omega \, |\Psi_2\rangle \langle \Psi_2| \; , \tag{4.88}$$

where the weight, ω, is between 0 and 1/2 (in this last case the ensemble is called "equiensemble"). We can further define the ensemble energy and density

$$E(\omega) = (1 - \omega)E_1 + \omega \, E_2 \tag{4.89}$$

$$n_\omega(\boldsymbol{r}) = (1 - \omega)n_1(\boldsymbol{r}) + \omega \, n_2(\boldsymbol{r}) \; . \tag{4.90}$$

At $\omega = 0$ the ensemble energy clearly reduces to the ground-state energy. Using the ensemble density, it is possible to construct a DFT, i.e. to prove a Hohenberg-Kohn theorem and construct a Kohn-Sham scheme. The main features of the Kohn-Sham scheme are: i) The one-body orbitals have fractional occupations determined by the weight ω. ii) The xc functional depends on the weight, $E_{\text{xc}}(\omega)$. To calculate the excitation energies from ensemble DFT we can follow two paths. The first involves obtaining the ground-state energy and the ensemble energy for some fixed ω, from which the excitation energy $E_2 - E_1$ trivially follows

$$E_2 - E_1 = \frac{E(\omega) - E(0)}{\omega} \; . \tag{4.91}$$

The second path is obtained by taking the derivative of (4.89)

$$E_2 - E_1 = \frac{dE(\omega)}{d\omega} \; . \tag{4.92}$$

It is then possible to prove

$$E_2 - E_1 = \epsilon_\omega^{N+1} - \epsilon_\omega^N + \left. \frac{\partial E_{\text{xc}}(\omega)}{\partial \omega} \right|_{n=n_\omega} \; . \tag{4.93}$$

Naturally, we need approximations to the xc energy functional, $E_{\text{xc}}(\omega)$. An ensemble LDA was developed for the equiensemble by W. Kohn in 1986 [36], by treating the ensemble as a reminiscent of a thermal ensemble. He then related $E_{\text{xc}}(\omega)$ to the finite temperature xc energy of the homogeneous electron gas by equating the entropies of both systems. Unfortunately, the results obtained with this functional were not very encouraging. A promising approach, recently proposed, is the use of orbital functionals within an ensemble OEP method [37,38].

4.4.2 Full Solution of the Kohn–Sham Equations

One of the most important uses of TDDFT is the calculation of photoabsorption spectra. This problem can be solved in TDDFT either by propagating the time-dependent Kohn-Sham equations [39] or by using linear-response theory. In this section we will be concerned by the former, relegating the latter to the next section.

Let $\tilde{\varphi}_j(\boldsymbol{r})$ be the ground-state Kohn-Sham wave-functions for the system under study. We prepare the initial state for the time propagation by exciting the electrons with the electric field $v(\boldsymbol{r}, t) = -k_0 x_\nu \delta(t)$, where $x_\nu = x, y, z$. The amplitude k_0 must be small in order to keep the response of the system linear and dipolar. Through this prescription *all* frequencies of the system are excited with equal weight. At $t = 0^+$ the initial state for the time evolution reads

$$\varphi_j(\boldsymbol{r}, t = 0^+) = \hat{T} \exp\left\{-\mathrm{i} \int_0^{0^+} \mathrm{d}t \left[\hat{H}_{\mathrm{KS}} - k_0 x_\nu \delta(t)\right]\right\} \tilde{\varphi}_j(\boldsymbol{r})$$
$$= \exp\left(\mathrm{i} k_0 x_\nu\right) \tilde{\varphi}_j(\boldsymbol{r}) . \tag{4.94}$$

The Kohn-Sham orbitals are then further propagated during a finite time. The dynamical polarizability can be obtained from

$$\alpha_\nu(\omega) = -\frac{1}{k} \int \mathrm{d}^3 r \, x_\nu \, \delta n(\boldsymbol{r}, \omega) . \tag{4.95}$$

In the last expression $\delta n(\boldsymbol{r}, \omega)$ stands for the Fourier transform of $n(\boldsymbol{r}, t) - \tilde{n}(\boldsymbol{r})$, where $\tilde{n}(\boldsymbol{r})$ is the ground-state density of the system. The quantity that is usually measured in experiments, the photo-absorption cross-section, is essentially proportional to the imaginary part of the dynamical polarizability averaged over the three spatial directions

$$\sigma(\omega) = \frac{4\pi\omega}{c} \frac{1}{3} \Im \sum_\nu \alpha_\nu(\omega) , \tag{4.96}$$

where c stands for the velocity of light. Although computationally more demanding than linear-response theory, this method is very flexible, and is easily extended to incorporate temperature effects, non-linear phenomena, etc. Note also that this approach only requires an approximation to the xc potential and not to f_{xc}.

To illustrate the method, we present, in Fig. 4.2, the excitation spectrum of benzene calculated within the LDA/ALDA[3]. The agreement with experiment is quite remarkable, especially when looking at the $\pi \to \pi^*$ resonance at around $7\,\mathrm{eV}$. The spurious peaks that appear in the calculation at higher energies are artifacts caused by an insufficient treatment of the unbound states. We furthermore observe that such good results are routinely obtained when applying the LDA/ALDA to several finite systems, from small molecules to metallic clusters and biological systems.

---

[3] We will use the notation "A/B" consistently throughout the rest of this article to indicate that the ground-state xc potential used to calculate the initial state was "A", and that this state was propagated with the time-dependent xc potential "B". In the case of linear-response theory, "B" will denote the xc kernel.

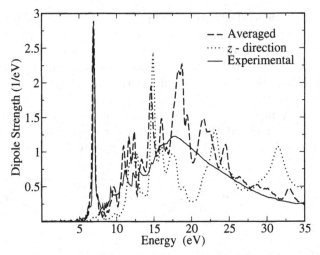

Fig. 4.2. Optical absorption of the benzene molecule. Experimental results from [40]. Figure reproduced from [41]

4.4.3   Excitations from Linear-Response Theory

The first self-consistent solution of the linear response (see 4.69) was performed by Zangwill and Soven in 1980 using the LDA/ALDA [42]. Their results for the photo-absorption spectrum of xenon for energies just above the ionization threshold are shown in Fig. 4.3. Once more the theoretical curve compares very well to experiments.

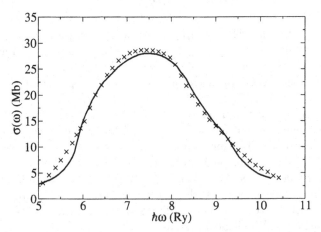

Fig. 4.3. Total photo-absorption cross-section of xenon versus photon energy in the vicinity of the 4d threshold. The solid line represents TDDFT calculations and the crosses are the experimental results of [43]. Figure adapted from [42]

Unfortunately, a full solution of (4.69) is still quite difficult numerically. Besides the large effort required to solve the integral equation, we need the non-interacting response function as an input. To obtain this quantity it is usually necessary to perform a summation over all states, both occupied and unoccupied [cf. (4.63)]. Such summations are sometimes slowly convergent and require the inclusion of many unoccupied states. There are however approximate frameworks that circumvent the solution of (4.69). The one we will present in the following was proposed by Petersilka *et al.* [22].

The density response function can be written in the Lehmann representation

$$\chi(\boldsymbol{r},\boldsymbol{r}',\omega)= \lim_{\eta\to0^+} \sum_m \left[\frac{\langle 0|\,\hat{n}(\boldsymbol{r})\,|m\rangle\,\langle m|\,\hat{n}(\boldsymbol{r}')\,|0\rangle}{\omega-(E_m-E_0)+i\eta} - \frac{\langle 0|\,\hat{n}(\boldsymbol{r}')\,|m\rangle\,\langle m|\,\hat{n}(\boldsymbol{r})\,|0\rangle}{\omega+(E_m-E_0)+i\eta} \right] ,$$

$$(4.97)$$

where $|m\rangle$ is a complete set of many-body states with energies E_m. From this expansion it is clear that the full response function has poles at frequencies that correspond to the excitation energies of the interacting system

$$\Omega = E_m - E_0 . \qquad (4.98)$$

As the external potential does not have any special pole structure as a function of ω, (4.61) implies that also $n^{(1)}(\boldsymbol{r},\omega)$ has poles at the excitation energies, Ω. On the other hand, χ_{KS} has poles at the excitation energies of the non-interacting system, i.e. at the Kohn-Sham orbital energy differences $\epsilon_j - \epsilon_k$ [cf. (4.63)].

By rearranging the terms in (4.68) we obtain the fairly suggestive equation

$$\int \mathrm{d}^3 r' \, [\delta(\boldsymbol{r}-\boldsymbol{r}') - \Xi(\boldsymbol{r},\boldsymbol{r}',\omega)]\, n^{(1)}(\boldsymbol{r}',\omega) = \int \mathrm{d}^3 r' \, \chi_{\mathrm{KS}}(\boldsymbol{r},\boldsymbol{r}',\omega)\, v^{(1)}(\boldsymbol{r}',\omega) ,$$

$$(4.99)$$

where the function Ξ is defined by

$$\Xi(\boldsymbol{r},\boldsymbol{r}',\omega) = \int \mathrm{d}^3 r'' \, \chi_{\mathrm{KS}}(\boldsymbol{r},\boldsymbol{r}'',\omega) \left[\frac{1}{|\boldsymbol{r}''-\boldsymbol{r}'|} + f_{\mathrm{xc}}(\boldsymbol{r}'',\boldsymbol{r}',\omega) \right] . \qquad (4.100)$$

As noted previously, in the limit $\omega \to \Omega$ the linear density $n^{(1)}$ has a pole, while the right-hand side of (4.99) remains finite. For the equality (4.99) to hold, it is therefore required that the operator multiplying $n^{(1)}$ on the left-hand of (4.99) side has zero eigenvalues at the excitation energies Ω. This implies $\lambda(\omega) \to 1$ when $\omega \to \Omega$, where $\lambda(\omega)$ is the solution of the eigenvalue equation

$$\int \mathrm{d}^3 r' \, \Xi(\boldsymbol{r},\boldsymbol{r}',\omega)\xi(\boldsymbol{r}',\omega) = \lambda(\omega)\xi(\boldsymbol{r},\omega) . \qquad (4.101)$$

This is a rigorous statement, that allows the determination of the excitation energies of the systems from the knowledge of χ_{KS} and f_{xc}. It is possible to transform this equation into another eigenvalue equation having the true excitation energies of the system, Ω, as eigenvalues [44]. We start by defining the quantity

$$\zeta_{jk}(\omega) = \int d^3r' \int d^3r'' \, \varphi_j^*(r'')\varphi_k(r'') \left[\frac{1}{|r'' - r'|} + f_{xc}(r'', r', \omega)\right] \xi(r', \omega) .$$
(4.102)

With the help of ζ_{jk}, (4.101) can be rewritten in the form

$$\sum_{jk} \frac{(f_k - f_j)\,\varphi_j(r)\varphi_k^*(r)}{\omega - (\epsilon_j - \epsilon_k) + i\eta}\zeta_{jk}(\omega) = \lambda(\omega)\xi(r, \omega) .$$
(4.103)

By solving this equation for $\xi(r, \omega)$ and inserting the result into (4.102), we arrive at

$$\sum_{j'k'} \frac{M_{jk,j'k'}}{\omega - (\epsilon_{j'} - \epsilon_{k'}) + i\eta}\zeta_{j'k'}(\omega) = \lambda(\omega)\zeta_{jk}(\omega) ,$$
(4.104)

where we have defined the matrix element

$$M_{jk,j'k'}(\omega) = (f_{k'} - f_{j'}) \int d^3r \int d^3r' \, \varphi_j^*(r)\varphi_k(r)\varphi_j(r')\varphi_k^*(r') \times$$

$$\left[\frac{1}{|r - r'|} + f_{xc}(r, r', \omega)\right] .$$
(4.105)

Introducing the new eigenvector

$$\beta_{jk} = \frac{\zeta_{jk}(\Omega)}{\Omega - (\epsilon_{j'} - \epsilon_{k'})} ,$$
(4.106)

taking the $\eta \to 0$ limit, and by using the condition $\lambda(\Omega) = 1$, it is straightforward to recast (4.104) into the eigenvalue equation

$$\sum_{j'k'} [\delta_{jj'}\delta_{kk'}(\epsilon_{j'} - \epsilon_{k'}) + M_{jk,j'k'}(\Omega)] \beta_{j'k'} = \Omega\beta_{jk} .$$
(4.107)

It is also possible to derive an operator whose eigenvalues are the *square* of the true excitation energies, thereby reducing the dimension of the matrix equation (4.107) [45]. The oscillator strengths can then be obtained from the eigenfunctions of the operator.

The eigenvalue equation (4.107) can be solved in several different ways. For example, it is possible to expand all quantities in a suitable basis and solve numerically the resulting matrix-eigenvalue equation. As an alternative, we can perform a Laurent expansion of the response function around the excitation energy

$$\chi_{KS}(r, r', \omega) = \lim_{\eta \to 0^+} \frac{\varphi_{j_0}(r)\varphi_{j_0}^*(r')\varphi_{k_0}(r')\varphi_{k_0}^*(r)}{\omega - (\epsilon_{j_0} - \epsilon_{k_0}) + i\eta} + \text{higher orders} . \quad (4.108)$$

By neglecting the higher-order terms, a simple manipulation of (4.101) yields the so-called single-pole approximation (SPA) to the excitation energies

$$\Omega = \Delta\epsilon + K(\Delta\epsilon) \,, \tag{4.109}$$

where $\Delta\epsilon$ is the difference between the Kohn-Sham eigenvalue of the unoccupied orbital j_0 and the occupied orbital k_0,

$$\Delta\epsilon = \epsilon_{j_0} - \epsilon_{k_0} \,, \tag{4.110}$$

and K is a correction given by

$$K(\Delta\epsilon) = 2\Re \int d^3r \int d^3r' \, \varphi_{j_0}(r)\varphi_{j_0}^*(r')\varphi_{k_0}(r')\varphi_{k_0}^*(r) \times \tag{4.111}$$
$$\left[\frac{1}{|r - r'|} + f_{xc}(r, r', \Delta\epsilon) \right] .$$

Although not as precise as the direct solution of the eigenvalue equation, (see 4.107), this formula provides us with a simple and fast way to calculate the excitation energies.

To assert how well this approach works in practice we list, in Table 4.1, the $^1S \rightarrow ^1P$ excitation energies for several atoms [22]. Surprisingly perhaps, the eigenvalue differences, $\Delta\epsilon$, are already of the proper order of magnitude. For other systems they can be even much closer (cf. Table 4.3). Adding the correction K then brings the numbers indeed very close to experiments for both xc functionals tried. We furthermore notice that the EXX/PGG functional gives clearly superior results than the LDA/ALDA. This is related to the different quality of the unoccupied states generated with the two ground-state xc functionals. The unoccupied states typically probe the farthest regions from the system, where the LDA potential exhibits severe deficiencies (as previously mentioned in Sect. 4.2.4). As the EXX potential does not suffer from this problem, it yields better unoccupied orbitals and consequently better excitation energies.

Table 4.1. $^1S \rightarrow ^1P$ excitation energies for selected atoms. Ω_{exp} denotes the experimental results from [46]. All energies are in hartrees. Table adapted from [22]

Atom	$\Delta\epsilon_{LDA}$	$\Omega_{LDA/ALDA}$	$\Delta\epsilon_{EXX}$	$\Omega_{EXX/PGG}$	Ω_{exp}
Be	0.129	0.200	0.130	0.196	0.194
Mg	0.125	0.176	0.117	0.164	0.160
Ca	0.088	0.132	0.079	0.117	0.108
Zn	0.176	0.239	0.157	0.211	0.213
Sr	0.082	0.121	0.071	0.105	0.099
Cd	0.152	0.214	0.135	0.188	0.199

In Table 4.2 we show the excitation energies of the CO molecule. This case is slightly more complicated than the previous example due to the existence of degeneracies in the eigenspectrum of the CO molecule. Although the Kohn-Sham eigenvalue differences are equal for all transitions involving degenerate states, the true excitation energies depend on the symmetry of the initial and final many-body states. As is clearly seen from the table, this splitting of the excitations is correctly described by the correction factor K.

Table 4.2. Excitation energies for the CO molecule. $\Omega_{\text{LDA/ALDA}}^{\text{SPA}}$ are the LDA/ALDA excitation energies obtained from (4.109), and $\Omega_{\text{LDA/ALDA}}^{\text{full}}$ are obtained from the solution of (4.107) neglecting continuum states. Ω_{exp} are the experimental results from [47]. All energies are in hartrees. Table reproduced from [48]

State		$\Delta\epsilon_{\text{LDA}}$	$\Omega_{\text{LDA/ALDA}}^{\text{SPA}}$	$\Omega_{\text{LDA/ALDA}}^{\text{full}}$	Ω_{exp}
A $^1\Pi$	$5\sigma \to 2\pi$	0.2523	0.3268	0.3102	0.3127
a $^3\Pi$			0.2238	0.2214	0.2323
B $^1\Sigma^+$	$5\sigma \to 6\sigma$	0.3332	0.3389	0.3380	0.3962
b $^3\Sigma^+$			0.3315	0.3316	0.3822
I $^1\Sigma^-$			0.3626	0.3626	0.3631
e $^3\Sigma^-$			0.3626	0.3626	0.3631
a' $^3\Sigma^+$	$1\pi \to 2\pi$	0.3626	0.3181	0.3149	0.3127
D $^1\Delta$			0.3812	0.3807	0.3759
d $^3\Delta$			0.3404	0.3396	0.3440
c $^3\Pi$	$4\sigma \to 2\pi$	0.4388	0.4204	0.4202	0.4245
E $^1\Pi$	$1\pi \to 6\sigma$	0.4436	0.4435	0.4435	0.4237

We remember that several approximations have been made to produce the previous results. First, a static Kohn-Sham calculation was performed with an approximate v_{xc}. Then the resulting eigenfunctions and eigenvalues were used in (4.109) to obtain the excitation energies. In the last step, we used an approximate form for the xc kernel, f_{xc}, and we neglected the higher order terms in the Laurent expansion of the response functions. To assert which of these approximations is more important, we can look at the lowest excitation energies of the He atom. For this simple system the *exact* stationary Kohn-Sham potential is known [49], so we can eliminate the first source of error. We can then test different approximations for f_{xc}, both by performing the single-pole approximation or not. The results are summarized in Table 4.3. We first note that the quality of the results is almost insensitive to the xc kernel used. Both using the ALDA or the PGG yield the same mean error. This statement seems to hold not only for atoms but also for molecular systems [50]. From the table it is also clear that the SPA is an excellent approximation and that the calculated excitation energies are in very close agreement to the exact

Table 4.3. Comparison of the excitation energies of neutral helium, calculated from the exact xc potential [49] by using approximate xc kernels. SPA stands for "single pole approximations", while "full" means the solution of (4.107) neglecting continuum states. The exact values are from a non-relativistic variational calculation [53]. The mean absolute deviation and mean percentage errors also include the transitions from the 1s until the 9s and 9p states. All energies are in hartrees. Table adapted from [17]

State $k_0 \rightarrow j_0$	$\Delta\epsilon_{KS}$	exact/ALDA (xc)		exact/PGG		exact
		SPA	full	SPA	full	
2^3S $1s \rightarrow 2s$	0.7460	0.7357	0.7351	0.7232	0.7207	0.7285
2^1S		0.7718	0.7678	0.7687	0.7659	0.7578
3^3S $1s \rightarrow 3s$	0.8392	0.8366	0.8368	0.8337	0.8343	0.8350
3^1S		0.8458	0.8461	0.8448	0.8450	0.8425
4^3S $1s \rightarrow 4s$	0.8688	0.8678	0.8679	0.8667	0.8671	0.8672
4^1S		0.8714	0.8719	0.8710	0.8713	0.8701
2^3P $1s \rightarrow 2p$	0.7772	0.7702	0.7698	0.7693	0.7688	0.7706
2^1P		0.7764	0.7764	0.7850	0.7844	0.7799
3^3P $1s \rightarrow 3s$	0.8476	0.8456	0.8457	0.8453	0.8453	0.8456
3^1P		0.8483	0.8483	0.8500	0.8501	0.8486
4^3P $1s \rightarrow 4s$	0.8722	0.8714	0.8715	0.8712	0.8713	0.8714
4^1P		0.8726	0.8726	0.8732	0.8733	0.8727
Mean abs. dev.		0.0011	0.0010	0.0010	0.0010	
Mean % error		0.15%	0.13%	0.13%	0.13%	

values. Why, and under which circumstances this is the case is discussed in detail in [51,52]. This leads us to conclude that the crucial approximation to obtain excitation energies in TDDFT is the choice of the static xc potential used to calculate the Kohn-Sham eigenfunctions and eigenvalues.

4.4.4 When Does It Not Work?

In the previous sections we showed the results of several TDDFT calculations, most of them agreeing quite well with experiment. Clearly no physical theory works for all systems and situations, and TDDFT is not an exception. It is the purpose of this section to show some examples where the theory does not work. However, before proceeding with our task, we should specify what we mean by "failures of TDDFT". TDDFT is an *exact* reformulation of the time-dependent many-body Schrödinger equation – it can only fail in situations where quantum-mechanics also fails. The key approximation made in practical applications is the approximation for the xc potential. Errors in the calculations should therefore be imputed to the functional used. As a large majority of TDDFT calculations use the ALDA or the adiabatic GGA, we will be mainly interested in the errors caused by these approximate functionals. Furthermore, and as we already mentioned in the previous section, there are usually two sources of errors in the calculation: i) the functional used to obtain the Kohn-Sham ground-state; ii) the approximate time-dependent xc potential/kernel. In any discussion on the errors of TDDFT the effects of these two sources have to be clearly separated. With these arguments in mind let us then proceed.

Our first example is the calculation of optical properties of long conjugated molecular chains [54]. For these systems, the local or gradient-corrected approximations can give overestimations of several orders of magnitude. The problem is related to a non-local dependence of the xc potential: In a system with an applied electric field, the exact xc potential develops a linear part that counteracts the applied field [54,55]. This term is completely absent in both the LDA and the GGA, but is present in more non-local functionals like the EXX.

A related problem occurs in solids [56]. In fact, the ALDA does not work properly for the calculation of excitations of non-metallic solids, especially in systems like wide-band gap semiconductors. For infinite systems, the Coulomb potential is (in momentum space) $4\pi/q^2$. It is then clear from the response (see 4.69) that if f_{xc} is to correct the non-interacting response for $q \to 0$ it will have to contain a term that behaves asymptotically as $1/q^2$ when $q \to 0$. This is not the case for the local or gradient-corrected approximations. Several attempts have been made to correct this problem from which we mention [57–60].

Another problematic system for the ALDA is the streched H_2 molecule [61,62]. From a comparison with exact results it was found that the ALDA fails to reproduce even qualitatively the shape of the potential curves for the $^3\Sigma_u^+$ and $^1\Sigma_u^+$ states. A detailed analysis of the problem shows that the failure is related to the breakdown of the simple local approximation to the kernel.

Furthermore, the ALDA yields a large error in the calculation of singlet-triplet separation energies [63], underestimates the onset of absorption for some clusters [50], etc.

However, and despite these limitations, we would like to emphasize that the ALDA does work very well for the calculation of excitations in a large class of systems.

4.5   Atoms and Molecules in Strong Laser Fields

4.5.1   What Is a "Strong" Laser?

Before discussing the behavior of atoms and molecules in strong laser fields, we have to specify what the adjective "strong" means in this context. The electric field that an electron feels in a hydrogen atom, at the distance of one Bohr from the nucleus, is

$$E = \frac{1}{4\pi\epsilon_0} \frac{e}{a_0^2} = 5.1 \times 10^9 \, \text{V/m} \,. \tag{4.112}$$

The laser intensity that corresponds to this field is given by

$$I = \frac{1}{2}\epsilon_0 c E^2 = 3.51 \times 10^{16}\,\text{W}/\text{cm}^2 \; . \tag{4.113}$$

We can clearly consider a laser to be "strong" when its intensity becomes comparable to (4.113). In this regime, perturbation theory is no longer applicable, and the theorist has to resort to non-perturbative methods. When approaching these high intensities, a wealth of non-linear phenomena appear, like multi-photon ionization, above threshold ionization (ATI), high harmonic generation, etc.

The fact that allowed systematic investigation of these high-intensity phenomena was the remarkable evolution in laser technology during the past four decades. Through a series of technological breakthroughs, scientists were able to boost the peak intensity of pulsed lasers from $10^9\,\text{W}/\text{cm}^2$ in the 1960s, to more than $10^{21}\,\text{W}/\text{cm}^2$ of the current systems – 12 orders of magnitude! Besides this increase in laser intensity, very short pulses – sometimes of the order of hundreds of attoseconds ($1\,\text{as} = 10^{-18}\,\text{s}$) – became available at ultraviolet or soft X-ray frequencies [64,65]. In the present context we are concerned mainly with intensities in the range $10^{13} - 10^{16}\,\text{W}/\text{cm}^2$. For higher intensities many-body effects associated with the electron-electron interaction – which are the main interest of DFT – become less and less important due to the strongly dominant external field.

TDDFT is a tool particularly suited for the study of systems under the influence of strong lasers. We recall that the time-dependent Kohn-Sham equations yield the *exact* density of the system, including *all* non-linear effects. To simulate laser induced phenomena it is customary to start from the ground-state of the system, which is then propagated under the influence of the potential

$$v_{\text{TD}}(\boldsymbol{r}, t) = E f(t) z \sin(\omega t) \; . \tag{4.114}$$

v_{TD} describes a laser of frequency ω and amplitude[4] E. The function $f(t)$, typically a Gaussian or the square of a sinus, defines the temporal shape of the laser pulse. From the time-dependent density it is then possible to calculate the photon spectrum using the relation

$$\sigma(\omega) \propto |d(\omega)|^2 \; , \tag{4.115}$$

where $d(\omega)$ is the Fourier transform of the time-dependent dipole of the system

$$d(t) = \int \text{d}^3 r \, z \, n(\boldsymbol{r}, t) \; . \tag{4.116}$$

Other observables, such as the total ionization yield or the ATI spectrum, are much harder to calculate within TDDFT. Even though these observables (as all others) are functionals of the density by virtue of the Runge-Gross theorem, the explicit functional dependence is unknown and has to be approximated.

---

[4] The amplitude is related to the laser intensity by the relation $I = \frac{1}{2}\epsilon_0 c E^2$.

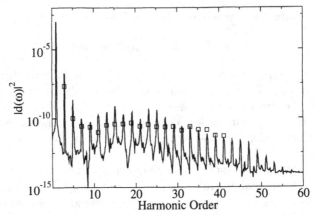

Fig. 4.4. Harmonic spectrum for He at $\lambda = 616\,\mathrm{nm}$ and $I = 3.5 \times 10^{14}\,\mathrm{W/cm^2}$. The squares represent experimental data taken from [66] normalized to the value of the 33rd harmonic of the calculated spectrum. Figure reproduced from [67]

4.5.2   High-Harmonic Generation

If we shine a high-intensity laser onto an atom (or a molecule, or even a surface), an electron may absorb several photons and then return to its ground-state by emitting a single photon. The photon will have a frequency which is an integer multiple of the external laser frequency. This process, known as high-harmonic generation, has received a great deal of attention from both theorists and experimentalists. As the outgoing high-energy photons maintain a fairly high coherence, they can be used as a source for X-ray lasers.

A typical high-harmonic spectrum is shown in Fig. 4.4 for the helium atom. The squares represent experimental data taken from [66], and the solid line was obtained from a calculation using the EXX/EXX functional [67]. The spectrum consists of a series of peaks, first decreasing in amplitude and then reaching a plateau that extends to very high frequency. The peaks are placed at the odd multiples of the external laser frequency (the even multiples are dipole forbidden by symmetry). We note that any approach based on perturbation theory would yield a harmonic spectrum that decays exponentially, i.e. such a theory could never reproduce the measured peak intensities. TDDFT, on the other hand, gives a quite satisfactory agreement with experiment.

As mentioned above, high-harmonics can be used as a source of soft X-ray lasers. For such purpose, one tries to optimize the laser parameters, the frequency, intensity, etc., in order to increase the intensity of the emitted harmonics, and to extend the plateau the farthest possible. By performing "virtual experiments", TDDFT can be once more used to tackle such an important problem. As an illustration, we show in Fig. 4.5 the result of irradiating a hydrogen atom with lasers of the same frequency but with different intensities. For clarity, we only show the position of the peaks, and the points were connected by straight lines. As we increase the intensity of the laser,

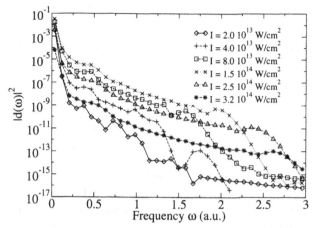

Fig. 4.5. Harmonic spectra of hydrogen at a laser wavelength of $\lambda = 1064\,\text{nm}$ for various laser intensities. Figure reproduced from [68]

the amplitude of the harmonics also increases, until reaching a maximum at $I = 1.5 \times 10^{14}\,\text{W/cm}^2$. A further increase of the intensity will, however, decrease the produced harmonics. This reflects the two competing processes that happen upon multiple absorption of photons: The electron can either ionize, or fall back into the ground-state emitting a highly energetic photon. Beyond a certain threshold intensity the ionization channel begins to predominate, thereby reducing the production of harmonics. Other laser parameters, like the intensity or the spectral composition of the laser, are also found to influence the generation of high-harmonics in atoms [67,68].

4.5.3   Multi-photon Ionization

To better understand the process of ionization of an atom in strong laser fields, it is convenient to resort to a simple quasi-static picture. In Fig. 4.6

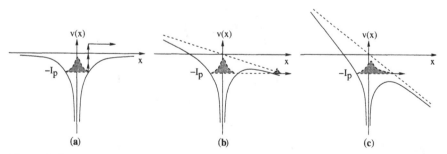

Fig. 4.6. Ionization in strong laser fields: (a) Multi-photon ionization; (b) Tunneling; (c) Over the barrier

we have depicted a one-electron atom at a time t after the beginning of the laser pulse. The dashed line represents the laser potential felt by the electron and the solid line the total (i.e. the nuclear plus the laser) potential. Three different regimes of ionizations are governed by the Keldish parameter,

$$\gamma = \frac{\omega}{E} \, . \tag{4.117}$$

At low intensities ($I < 10^{14}\,\mathrm{W/cm^2}$, $\gamma \gg 1$) the electron has to absorb several photons before leaving the atom. This is the so-called multi-photon ionization regime. At higher intensities ($I \le 10^{15}\,\mathrm{W/cm^2}$, $\gamma \approx 1$) we enter the tunneling regime. If we further increase the strength of the laser field ($I > 10^{16}\,\mathrm{W/cm^2}$, $\gamma \ll 1$), then the electron can simply pass over the barrier.

The measured energy spectrum of the outgoing photo-electrons is called the above threshold ionization (ATI) spectrum [69]. As the electron can absorb more photons than necessary for escaping the atom, an ATI spectrum will consist of a sequence of equally spaced peaks at energies

$$E = (n + s)\omega - I_\mathrm{p} \, , \tag{4.118}$$

where n is a natural integer, s is the minimum integer such that $s\omega - I_\mathrm{p} > 0$, and I_p denotes the ionization potential of the system.

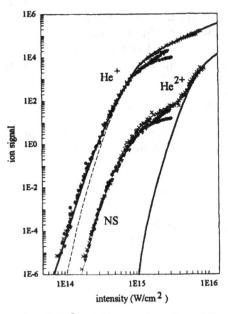

Fig. 4.7. Measured $\mathrm{He^+}$ and $\mathrm{He^{2+}}$ yields as a function of the laser intensity. The solid curve on the right is the $\mathrm{He^{2+}}$ yield, calculated under the assumption of a sequential mechanism. Figure reproduced from [70]

Another interesting observable is the number of outgoing charged atoms as a function of the laser intensity. The two sets of points in Fig. 4.7 represent the yield of singly ionized and doubly ionized helium. The solid curve on the right is the result of a calculation assuming a sequential mechanism for the double ionization of helium, i.e., the He^{2+} is generated by first removing one electron from He, and then a second from He^+. Strikingly, this naïve sequential mechanism is wrong by six orders of magnitude for some intensities.

Similar experimental results were found for a variety of molecules. Furthermore, in these more complex systems, the coupling of the nuclear and the electronic degrees of freedom gives rise to new physical phenomena. As illustrative examples of such phenomena, we refer to the so-called ionization induced Coulomb explosion [71], and the production of even harmonics as a consequence of beyond-Born-Oppenheimer dynamics [72].

4.5.4   Ionization Yields from TDDFT

It is apparent from Fig. 4.7 that a simple sequential mechanism is insufficient to describe the double ionization of helium. In this section we will show how one can try to go beyond this simple picture with the use of TDDFT [73].

To calculate the helium yields we invoke a geometrical picture of ionization. We divide the three-dimensional space, $\mathbb{R}^3$, into a (large) box, A, containing the helium atom, and its complement, $B = \mathbb{R}^3 \backslash A$. Normalization of the (two-body) wave function of the helium atom, $\Psi(\boldsymbol{r}_1, \boldsymbol{r}_2, t)$, then implies

$$1 = \int_A\!\!\int_A d^3r_1 d^3r_2 \, |\Psi(\boldsymbol{r}_1, \boldsymbol{r}_2, t)|^2 + 2 \int_A\!\!\int_B d^3r_1 d^3r_2 \, |\Psi(\boldsymbol{r}_1, \boldsymbol{r}_2, t)|^2 \tag{4.119}$$
$$+ \int_B\!\!\int_B d^3r_1 d^3r_2 \, |\Psi(\boldsymbol{r}_1, \boldsymbol{r}_2, t)|^2 \ ,$$

where the subscript "X" has the meaning that the space integral is only over region X. A long time after the end of the laser excitation, we expect that all ionized electrons are in region B. This implies that the first term in the right-hand side of (4.119) measures the probability that an electron remains close to the nucleus; Similarly, the second term is equal to the probability of finding an electron in region A and simultaneously another electron far from the nucleus, in region B. This is interpreted as single ionization; Likewise, the final term is interpreted as the probability for double ionization. Accordingly, we will refer to these terms as $p^{(0)}(t)$, $p^{(+1)}(t)$, and $p^{(+2)}(t)$.

To this point of the derivation we have utilized the many-body wavefunction to define the ionization probabilities. Our goal is however to construct a density functional. For that purpose, we introduce the pair-correlation function

$$g[n](\boldsymbol{r}_1, \boldsymbol{r}_2, t) = \frac{2\,|\Psi(\boldsymbol{r}_1, \boldsymbol{r}_2, t)|^2}{n(\boldsymbol{r}_1, t)n(\boldsymbol{r}_2, t)} \ , \tag{4.120}$$

and rewrite

$$p^{(0)}(t) = \frac{1}{2} \int_A \int_A d^3 r_1 d^3 r_2\, n(\mathbf{r}_1, t) n(\mathbf{r}_2, t) g[n](\mathbf{r}_1, \mathbf{r}_2, t)$$
$$p^{(+1)}(t) = \int_A d^3 r\, n(\mathbf{r}, t) - 2 p^{(0)}(t) \qquad (4.121)$$
$$p^{(+2)}(t) = 1 - p^{(0)}(t) - p^{(+1)}(t)\,.$$

We recall that by virtue of the Runge-Gross theorem g is a functional of the time-dependent density. Separating g into an exchange part (which is simply $1/2$ for a two electron system) and a correlation part,

$$g[n](\mathbf{r}_1, \mathbf{r}_2, t) = \frac{1}{2} + g_c[n](\mathbf{r}_1, \mathbf{r}_2, t) \qquad (4.122)$$

we can cast (4.121) into the form

$$p^{(0)}(t) = [N_{1s}(t)]^2 + K(t)$$
$$p^{(+1)}(t) = 2 N_{1s}(t)\, [1 - N_{1s}(t)] - 2 K(t) \qquad (4.123)$$
$$p^{(+2)}(t) = [1 - N_{1s}(t)]^2 + K(t)\,,$$

with the definitions

$$N_{1s}(t) = \frac{1}{2} \int_A d^3 r\, n(\mathbf{r}, t) = \int_A d^3 r\, |\varphi_{1s}(\mathbf{r}, t)|^2 \qquad (4.124)$$

$$K(t) = \frac{1}{2} \int_A \int_A d^3 r_1 d^3 r_2\, n(\mathbf{r}_1, t) n(\mathbf{r}_2, t) g_c[n](\mathbf{r}_1, \mathbf{r}_2, t)\,. \qquad (4.125)$$

In Fig. 4.8 we depict the probability for double ionization of helium calculated from (4.123) by neglecting the correlation part of g. It is clear that all functionals tested yield a significant improvement over the simple sequential model. Due to the incorrect asymptotic behavior of the ALDA potential, the ALDA overestimates ionization: The outermost electron of helium is not sufficiently bound and ionizes too easily.

To compare the TDDFT results with experiment it is preferable to look at the ratio of double- to single-ionization yields. This simple procedure eliminates the experimental error in determining the absolute yields. Clearly all TDDFT results presented in Fig. 4.9 are of very low quality, sometimes wrong by two orders of magnitude. We note that *two* approximations are involved in the calculation: The time-dependent xc potential used to propagate the Kohn-Sham equations, and the neglect of the correlation part of the pair-correlation function. By using a one-dimensional helium model, Lappas and van Leeuwen were able to prove that even the simplest approximation for g was able to reproduce the knee structure [74]. As neither of the TDDFT calculations depicted in Fig. 4.9 show the knee structure, the approximation used for the time-dependent xc potential appears to be more important in obtaining proper ionization yields.

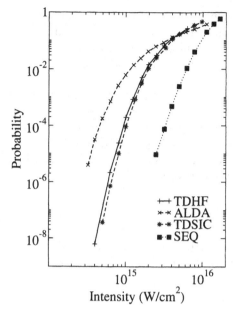

Fig. 4.8. Calculated double-ionization probabilities from the ground-state of helium irradiated by a 16 fs, 780 nm laser pulse for different choices of the time-dependent xc potentials. Figure reproduced from [73]

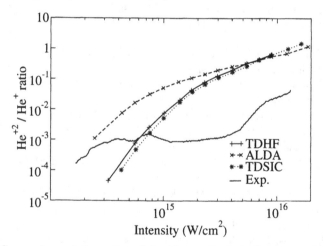

Fig. 4.9. Comparison of the ratios of double- to single-ionization probability calculated for different choices of the time-dependent xc potential. Figure reproduced from [73]

4.6 Conclusion

In this chapter we tried to give a brief, yet pedagogical, overview of TDDFT, from its mathematical foundations – the Runge-Gross theorem and the time-dependent Kohn-Sham scheme – to some of its applications, both in the linear and in the non-linear regimes. In the linear regime, TDDFT has become the standard tool to calculate excitation energies within DFT, and is by now incorporated in all of the major quantum-chemistry codes. In the non-linear regime, TDDFT is able to describe extremely non-linear effects, like high-harmonic generation, or multi-photon ionization. Unfortunately, some problems, like the knee structure in the yield of doubly ionized helium, are still beyond the reach of modern time-dependent xc potentials. In our opinion, we should not dismiss these problems as failures of TDDFT, but as a challenge to the next generation of "density-functionalists", in their quest for better approximations to the elusive xc potential.

Acknowledgments

We would like to thank L. Wirtz, A. Rubio and N. Helbig for their useful suggestions and comments, and A. Castro for his invaluable help in producing the figures.

References

1. E. K. U. Gross and W. Kohn, Adv. Quantum Chem. **21**, 255 (1990).
2. E. K. U. Gross, J. F. Dobson, and M. Petersilka, in *Topics in Current Chemistry*, edited by R. F. Nalewajski (Springer Verlag, Heidelberg, 1996), Vol. 181, p. 81.
3. G. Onida, L. Reinig, and A. Rubio, Rev. Mod. Phys. **74**, 601 (2002).
4. N. T. Maitra *et al.*, in *Reviews in Modern Quantum Chemistry: A Celebration of the Contributions of R. G. Parr*, edited by K. D. Sen (World Scientific, Singapore, 2002).
5. E. Runge and E. K. U. Gross, Phys. Rev. Lett. **52**, 997 (1984).
6. P. Hohenberg and W. Kohn, Phys. Rev. **136**, B864 (1964).
7. W. Kohn and L. J. Sham, Phys. Rev. **140**, A1133 (1965).
8. D. Dundas, K. T. Taylor, J. S. Parker, and E. S. Smyth, J. Phys. B **32**, L231 (1999).
9. S. Chelkowsky, T. Zuo, O. Atabek, and A. D. Brandrauk, Phys. Rev. A **52**, 2977 (1995).
10. T. Zuo and A. D. Bandrauk, Phys. Rev. A **54**, 3254 (1996).
11. E. K. U. Gross, C. A. Ullrich, and U. J. Gossman, in *Density Functional Theory*, Vol. 337 of *NATO ASI, Ser. B*, edited by E. K. U. Gross and R. Dreizler (Plenum Press, New York, 1995),.
12. R. van Leeuwen, Phys. Rev. Lett. **80**, 1280 (1998).
13. C. A. Ullrich, U. Gossmann, and E. K. U. Gross, Phys. Rev. Lett. **74**, 872 (1995).
14. J. F. Dobson, M. J. Bünner, and E. K. U. Gross, Phys. Rev. Lett. **79**, 1905 (1997).
15. R. T. Sharp and G. K. Horton, Phys. Rev. **90**, 317 (1953).

16. J. D. Talman and W. F. Shadwick, Phys. Rev. A **14**, 36 (1976).

17. M. Petersilka, U. J. Gossmann, and E. K. U. Gross, in *Electronic Density Functional Theory: Recent Progress and New Directions*, edited by J. F. Dobson, G. Vignale, and M. P. Das (Plenum Press, New York, 1998).

18. T. Grabo, T. Kreibich, S. Kurth, and E. K. U. Gross, in *Strong Coulomb Correlations in Electronic Structure Calculations: Beyond the Local Density Approximation*, edited by V. I. Anisimov (Gordon and Breach, Amsterdam, 2000).

19. J. B. Krieger, Y. Li, and G. J. Iafrate, Phys. Rev. A **45**, 101 (1992).

20. G. Vignale, Phys. Rev. Lett. **74**, 3233 (1995).

21. H. Appel and E. Gross, in *Quantum Simulations of Complex Many-Body Systems: From Theory to Algorithms*, Vol. 10 of *NIC Series*, edited by J. Grotendorst, D. Marx, and A. Muramatsu (John von Neumann Institute for Computing, FZ Jülich, 2002), p. 255.

22. M. Petersilka, U. J. Gossmann, and E. K. U. Gross, Phys. Rev. Lett. **76**, 1212 (1996).

23. D. M. Ceperley and B. J. Alder, Phys. Rev. Lett. **45**, 566 (1980).

24. E. K. U. Gross and W. Kohn, Phys. Rev. Lett. **55**, 2850 (1985).

25. N. Iwamoto and E. K. U. Gross, Phys. Rev. B **35**, 3003 (1987).

26. C.-O. Almbladh and U. von Barth, Phys. Rev. B **31**, 3231 (1985).

27. O. Gunnarsson and B. I. Lundqvist, Phys. Rev. B **13**, 4274 (1976).

28. A. Görling, Phys. Rev. A **47**, 3359 (1999).

29. A. Görling, Phys. Rev. Lett. **85**, 4229 (2000).

30. J. P. Perdew and M. Levy, Phys. Rev. B **31**, 6264 (1985).

31. T. Ziegler, A. Rauk, and E. J. Baerends, Theoret. Chim. Acta **43**, 261 (1977).

32. A. Theophilou, J. Phys. C **12**, 5419 (1979).

33. E. K. U. Gross, L. N. Oliveira, and W. Kohn, Phys. Rev. **A34**, 2805 (1988).

34. E. K. U. Gross, L. N. Oliveira, and W. Kohn, Phys. Rev. **A34**, 2809 (1988).

35. L. N. Oliveira, E. K. U. Gross, and W. Kohn, Phys. Rev. **A34**, 2821 (1988).

36. W. Kohn, Phys. Rev. A **34**, 737 (1986).

37. Á. Nagy, Int. J. Quantum Chem. **69**, 247 (1998).

38. N. I. Gidopoulos, P. G. Papaconstantinou, and E. K. U. Gross, Phys. Rev. Lett. **88**, 033003 (2002).

39. K. Yabana and G. F. Bertsch, Phys. Rev. B **54**, 4484 (1996).

40. E. E. Koch and A. Otto, Chem. Phys. Lett. **12**, 476 (1972).

41. M. A. L. Marques, A. Castro, G. F. Bertsch, and A. Rubio, Comput. Phys. Commun. **151**, 60 (2003).

42. A. Zangwill and P. Soven, Phys. Rev. A **21**, 1561 (1980).

43. R. Haensel, G. Keitel, P. Schreiber, and C. Kunz, Phys. Rev. **188**, 1375 (1969).

44. T. Grabo, M. Petersilka, and E. K. U. Gross, Journal of Molecular Structure (Theochem) **501**, 353 (2000).

45. M. Casida, in *Recent developments and applications in density functional theory*, edited by J. M. Seminario (Elsevier, Amsterdam, 1996), p. 391.

46. C. E. Moore, *Nat. Stand. Ref. Data Ser.* **35** (United States Government Printing Office, Washington, 1971), Vol. I-III.

47. E. S. Nielsen, P. Jørgensen, and J. Oddershede, J. Chem. Phys. **73**, 6238 (1980), erratum: ibid **75**, 499 (1981).

48. E. K. U. Gross, T. Kreibich, M. Lein, and M. Petersilka, in *Electron Correlations and Materials Properties*, edited by A. Gonis, N. Kioussis, and M. Ciftan (Plenum Press, New York, 1999).

49. C. J. Umrigar and X. Gonze, Phys. Rev. A **50**, 3827 (1994).

50. M. A. L. Marques, A. Castro, and A. Rubio, J. Chem. Phys. **115**, 3006 (2001).

51. X. Gonze and M. Scheffler, Phys. Rev. Lett. **82**, 4416 (1999).

52. H. Appel, E. K. U. Gross, and K. Burke, submitted to Phys. Rev. Lett., cond-mat/0203027 (2002).

53. A. Kono and S. Hattori, Phys. Rev. A **29**, 2981 (1984).

54. S. J. A. van Gisbergen *et al.*, Phys. Rev. Lett. **83**, 694 (1999).

55. O. V. Gritsenko and E. J. Baerends, Phys. Rev. A **64**, 042506 (2001).

56. X. Gonze, P. Ghosez, and R. W. Godby, Phys. Rev. Lett. **74**, 4035 (1995).

57. Y.-H. Kim and A. Görling, Phys. Rev. Lett. **89**, 096402 (2002).

58. L. Reining, V. Olevano, A. Rubio, and G. Onida, Phys. Rev. Lett. **88**, 066404 (2002).

59. P. L. de Boeij, F. Kootstra, J. A. Berger, R. van Leeuwen, and J. G. Snijders, J. Chem. Phys. **115**, 1995 (2001).

60. G. F. Bertsch, J.-I. Iwata, A. Rubio, and K. Yabana, Phys. Rev. B **62**, 7998 (2000).

61. F. Aryasetiawan, O. Gunnarson, and A. Rubio, Europhys. Lett. **57**, 683 (2002).

62. O. V. Gritsenko, S. J. A. van Gisbergen, A. Görling, and E. J. Baerends, J. Chem. Phys. **113**, 8478 (2000).

63. M. Petersilka, E. K. U. Gross, and K. Burke, Int. J. Quantum Chem. **80**, 534 (2000).

64. Z. Chang, A. Rundquist, H. Wang, M. M. Murnane, and H. C. Kapteyn, Phys. Rev. Lett. **79**, 2967 (1997).

65. Ch. Spielmann, N. H. Burnett, S. Sartania, R. Koppitsch, M. Schnürer, and C. Kan, Science **278**, 661 (1997).

66. K. Miyazaki and H. Sakai, J. Phys. B **25**, L83 (1992).

67. C. A. Ullrich, S. Erhard, and E. K. U. Gross, in *Super Intense Laser Atom Physics (SILAP IV)*, edited by H. G. Muller and M. V. Fedorov (Kluwer Publishing Company, Amsterdam, 1996).

68. S. Erhard and E. K. U. Gross, in *Multiphoton Processes*, edited by P. Lambropoulos and H. Walther (IOP Publishing, Bristol, 1996).

69. P. Agostini, F. Fabre, G. Mainfray, G. Petite, and N. K. Rahman, Phys. Rev. Lett. **42**, 1127 (1979).

70. B. Walker, B. Sheehy, L. F. DiMauro, P. Agostini, K. J. Schafer, and K. C. Kulander, Phys. Rev. Lett. **73**, 1227 (1994).

71. S. Chelkowski and A. D. Bandrauk, J. Phys. B **28**, L723 (1995).

72. T. Kreibich, M. Lein, V. Engel, and E. K. U. Gross, Phys. Rev. Lett. **87**, 103901 (2001).

73. M. Petersilka and E. K. U. Gross, Laser Physics **9**, 105 (1999).

74. D. G. Lappas and R. van Leeuwen, J. Phys. B **31**, L249 (1998).

5 Density Functional Theories and Self-energy Approaches

Rex W. Godby* and Pablo García-González[†]

* Department of Physics, University of York,
Heslington, York YO10 5DD,
United Kingdom
rwg3@york.ac.uk

† Departamento de Física Fundamental,
Universidad Nacional de Educación
a Distancia, Apto. 60141, 28080 Madrid, Spain
pgarcia@fisfun.uned.es

Rex Godby

5.1 Introduction

One of the fundamental problems in condensed-matter physics and quantum chemistry is the theoretical study of electronic properties. This is essential to understand the behaviour of systems ranging from atoms, molecules, and nanostructures to complex materials. Since electrons are governed by the laws of quantum mechanics, the many-electron problem is, in principle, fully described by a Schrödinger equation (supposing the nuclei to be fixed). However, the electrostatic repulsion between the electrons makes its numerical resolution an impossible task in practice, even for a relatively small number of particles.

Fortunately, we seldom need the full solution of the Schrödinger equation. When one is interested in *structural* properties, the ground-state total energy of the system is sufficient. In other cases, we want to study how the system responds to some external probe, and then knowledge of a few excited-state properties must be added. For instance, in a direct photoemission experiment a photon impinges on the system and an electron is removed. In an inverse photoemission process, an electron is absorbed and a photon is ejected. In both cases we have to deal with the gain or loss of energy of the N electron system when a single electron is added or removed, i.e. with the one-particle spectra. If the electron is not removed after the absorption of the photon, the system had evolved from its ground-state to an excited state, and the process is described by a set of *electron-hole* excitation energies. These few examples reflect the fact that practical applications of quantum theory are actually based on more elaborated and specialised techniques than simply trying to solve directly the Schrödinger equation. As we may see in other chapters of this book, the ground-state energy can be obtained – in principle exactly – using density functional theory (DFT) [1,2]. Regarding excited states, the information about single particle spectra is contained in the so called one-electron Green's function, whereas the electron-hole properties are described by the two-electron Green's function. Many-body perturbation theory (MBPT) [3–7], which focuses on these Green's functions directly, is a natural tool for the study of these phenomena.

Interestingly, the one-electron Green's function can also be used to calculate the ground-state energy as well as the expectation value of any one-particle observable (like the density or the kinetic energy) which is that DFT most naturally addresses[1]. This opens an appealing possibility: the use of MBPT instead of DFT in those cases in which the latter – because of the lack of knowledge of the *exact* exchange-correlation (xc) energy functional $E_{xc}[n]$ – does not provide accurate results. For example, systems in which van der Waals bonds are important are completely outside the scope of the familiar local-density (LDA) or generalised gradient (GGA) approximations. However, we shall see that these van der Waals forces can be studied through MBPT within Hedin's GW approximation [8,4] which is the most widely used many-body method in solid-state physics.

In this chapter, after a brief introduction to MBPT and Hedin's GW approximation, we will summarise some peculiar aspects of the Kohn-Sham xc energy functional, showing that some of them can be illuminated using MBPT. Then, we will discuss how to obtain ground-state total energies from GW. Finally, we will present a way to combine techniques from many-body and density functional theories within a generalised version of Kohn-Sham (KS) DFT.

5.2   Many-Body Perturbation Theory

Our discussion focuses on the concepts from MBPT that will be useful in this chapter. We will also present a short overview of some current problems in *ab-initio* calculations of quasiparticle properties. We refer the reader to [3–7] and the review articles [9–13] for further information on theoretical foundations and applications to solid-state physics, respectively.

5.2.1   Green's Function and Self-energy Operator

Green's functions are the key ingredients in many-body theory from which relevant physical information can be extracted. Given a non-relativistic N electron system under an external potential $v_{ion}(x)$, the one-particle Green's function (for simplicity we henceforth omit the prefix "one-particle") is defined as

$$G(x, x'; t - t') = -i \left\langle \Psi_N^{(0)} \left| \mathcal{T} \left[\widehat{\psi}(x, t) \, \widehat{\psi}^\dagger(x', t') \right] \right| \Psi_N^{(0)} \right\rangle ; \qquad (5.1)$$

---

[1] Similarly, two-particle ground-state quantities, like the pair correlation function, can be obtained from the two-electron Green's function.

where $x \equiv (r, \xi)$ symbols the space and spin coordinates, $\left| \Psi_N^{(0)} \right\rangle$ is the ground-state of the system, $\widehat{\psi}(x, t)$ is the annihilation operator in the Heisenberg picture, and $\mathcal{T}$ is Wick's time-ordering operator[2]. We may see that for $t > t'$, the Green's function is the probability amplitude to find an electron with spin ξ at point r and time t if the electron was added to the system with spin ξ' at point r' and time t'. When $t' > t$, the Green's function describes the propagation of a hole created at t.

As commented, the Green's function contains the information about one-particle excitations (we will see in Sect. 5.4 how to obtain ground-state properties). We start from the Lehmann representation of the Green's function:

$$G(x, x'; \omega) = \sum_n \frac{f_n(x) f_n^*(x')}{\omega - \mathcal{E}_n - i\eta \operatorname{sgn}(\mu - \mathcal{E}_n)} . \tag{5.2}$$

Here, $G(\omega)$ is the Fourier transform with respect to $\tau = t - t'$, η is a positive infinitesimal, μ is the Fermi energy of the system, and

$$\begin{aligned} f_n(x) &= \left\langle \Psi_N \left| \widehat{\psi}(x) \right| \Psi_{N+1}^{(n)} \right\rangle , \quad \mathcal{E}_n = E_{N+1}^{(n)} - E_N^{(0)} \text{ if } \mathcal{E}_n > \mu \\ f_n(x) &= \left\langle \Psi_{N-1}^{(n)} \left| \widehat{\psi}(x) \right| \Psi_N \right\rangle , \quad \mathcal{E}_n = E_N^{(0)} - E_{N-1}^{(n)} \text{ if } \mathcal{E}_n < \mu \end{aligned} , \tag{5.3}$$

with $E_N^{(0)}$ the ground-state energy and $\left| \Psi_{N\pm 1}^{(n)} \right\rangle$ the n-th eigenstate with energy $E_{N\pm 1}^{(n)}$ of the $N\pm 1$ electron system. By taking the imaginary part of (5.2) we have the so-called spectral function:

$$A(x, x'; \omega) = \frac{1}{\pi} |\Im G(x, x'; \omega)| = \sum_n f_n(x) f_n^*(x') \delta(\omega - \mathcal{E}_n) . \tag{5.4}$$

We may see that $A(x, x'; \omega)$ is just the superposition of delta functions with weights given by the amplitudes $f_n(x)$ centred at each of the one-particle excitation energies $\mathcal{E}_n$. That is, as anticipated above, the Green's function reflects the one-particle excitation spectra. Moreover, such weights – see (5.3), depend on the density of available eigenstates after the addition/removal of one electron. Further details about the role of $A(\omega)$ in the interpretation of photoemission experiments can be found in [14].

The spectral function – actually selected diagonal matrix elements $A_{nn}(\omega)$ in a suitable one-electron basis representation – may exhibit well-defined structures reflecting the existence of highly probable one-electron excitations. Due to the Coulomb interaction, we cannot assign each excitation to an independent particle (electron or hole) added to the system with the excitation energy. Nonetheless, some of these structures can be explained *approximately* in terms of a particle-like behaviour, so having a *quasiparticle* (QP) peak. Where a second peak is required we may have what is called a *satellite*.

---

[2] Note that G depends on $t - t'$ due to translational time invariance.

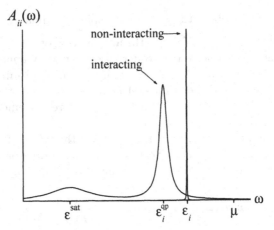

Fig. 5.1. Comparison between the non-interacting spectral function for a hole and the interacting one. Note how the interaction shifts down and broadens the QP peak and the appearance of a satellite at $\omega = \varepsilon^{\text{sat}}$

Of course this distinction is somewhat arbitrary, but a way of doing it is the following. Let us suppose that we switch off the interaction, so having a system of independent particles whose eigenstates can be described using one-electron orbitals $\phi_j(\boldsymbol{r})$ with eigenenergies ε_j. In this case, the matrix elements of the spectral function in the orbital basis set are

$$A_{ij}(\omega) = \langle \phi_i | A(\boldsymbol{x}, \boldsymbol{x}'; \omega) | \phi_j \rangle = \delta_{ij}\delta(\omega - \varepsilon_i)$$

That is, for the non-interacting system $A_{ii}(\omega)$ is just a delta function centred at $\omega = \varepsilon_i$ and the orbital energies are the one-electron excitation energies. If now we turn on the interaction, we may see that the delta function changes its position, broadens, and loses spectral weight which is transferred into the spectral background of the interacting $A_{ii}(\omega)$ – see Fig. 5.1. At the end of the process, the delta function has become a QP peak – in the sense that it originates from an independent single-particle state – and further structures that might have appeared would be the satellites. Note that the width of the QP peak reflects the finite lifetime of the added-particle state since it is not longer a real eigenstate of the system, whereas the satellites often reflect its resonant coupling with other elementary excitations like plasmons.

This one-electron picture can be formally introduced with the aid of the so-called self-energy operator Σ, which is defined through the Dyson equation

$$G^{-1}(\boldsymbol{x}, \boldsymbol{x}'; \omega) = G_{\text{H}}^{-1}(\boldsymbol{x}, \boldsymbol{x}'; \omega) - \Sigma(\boldsymbol{x}, \boldsymbol{x}'; \omega) . \qquad (5.5)$$

Here, we have used the Hartree Green's function

$$G_{\text{H}}^{-1}(\boldsymbol{x}, \boldsymbol{x}'; \omega) = \delta(\boldsymbol{x} - \boldsymbol{x}')[\omega - h_0(\boldsymbol{x})] , \qquad (5.6)$$

that corresponds to the non-interacting system in which $h_0(x)$ is the one-electron Hamiltonian under the external potential $v_{ion}(x)$ plus the classical Hartree potential $v_H(r)$. Then, it is evident that the self-energy contains the many-body effects due to Pauli exchange and Coulomb correlation, and that sharp structures in $G(\omega)$ are related to small expectation values of the frequency-dependent operator $\omega - \hat{h}_0 - \widehat{\Sigma}(\omega)$. Moreover, if we extend the ω-dependence of the self-energy to complex frequencies, such structures can be attributed to zeros of the operator $\omega - \hat{h}_0 - \widehat{\Sigma}(\omega)$, that is, to solutions of the *non-Hermitian* eigenvalue problem

$$h_0(x)\,\phi_n^{qp}(x) + \int dx'\, \Sigma(x, x', E_n^{qp})\,\phi_n^{qp}(x') = E_n^{qp}\phi_n^{qp}(x) \ , \qquad (5.7)$$

with complex energies E_n^{qp}. This is the *quasiparticle* equation, where Σ plays the role of an effective frequency-dependent and non-local potential. We may see that the self-energy has a certain resemblance with the DFT xc potential $v_{xc}(x)$ but, of course, the two objects are not equivalent. We have to bear in mind that the local and static $v_{xc}(x)$ is part of the potential of the fictitious KS non-interacting system, whereas the self-energy may be thought of as the potential felt by an added/removed electron to/from the interacting system.

Now, it is easy to see the correspondence between the QP peaks in the spectral function and the quasiparticle states ϕ_n^{qp}. If we expand $\Sigma_n(\omega) = \langle \phi_n^{qp} | \Sigma(\omega) | \phi_n^{qp} \rangle$ around $\omega = E_n^{qp}$ we have that

$$G_n(\omega) = \langle \phi_n^{qp} | G(\omega) | \phi_n^{qp} \rangle \simeq \frac{Z_n}{\omega - (\varepsilon_n^{qp} + i\Gamma_n)} \ , \qquad (5.8)$$

with $\varepsilon_n^{qp} = \Re E_n^{qp}$, $\Gamma_n = \Im E_n^{qp}$, and Z_n the complex QP renormalisation factor given by

$$Z_n^{-1} = 1 - \left.\frac{\partial \Sigma_n(\omega)}{\partial \omega}\right|_{\omega = E_n^{qp}} . \qquad (5.9)$$

As a consequence, if Γ_n is small, the spectral function $\Im G_n(\omega)$ is expected to have a well defined peak centred at ε_n^{qp} of width Γ_n and weight $|\Re Z_n|$. Therefore, the real part ε_n^{qp} is the QP energy itself, and it provides the band-structure of the system. The inverse of the imaginary part Γ_n^{-1} gives the corresponding QP lifetime.

5.2.2 Many-Body Perturbation Theory and the *GW* Approximation

In practical applications, we have to obtain (under certain unavoidable approximations) the self-energy operator. From this we calculate the QP spectrum using (5.7) and, if required, the full Green's function given by (5.5). MBPT provides a tool for such a task but, as in any other perturbation theory, we have to define the unperturbed system and the perturbation itself. In the above discussion, the unperturbed system seemed to be the non-interacting system of electrons under the potential $v_{\mathrm{ion}}(\boldsymbol{x}) + v_{\mathrm{H}}(\boldsymbol{r})$. However, due to the obvious problems that arise when trying to converge a perturbation series, it is much better to start from a different non-interacting scenario, like the LDA or GGA KS system, which already includes an attempt to describe exchange and correlation in the actual system. Considering the perturbation, the bare Coulomb potential w is very strong and, besides, we know that in a many-electron system the Coulomb interaction between two electrons is readily screened by a dynamic rearrangement of the other electrons [15], reducing its strength. Therefore, it is much more natural to describe the Coulomb interaction in terms of a *screened* Coulomb potential W and then write down the self-energy as a perturbation series in terms of W. If we just keep the first term of such an expansion, we will have the GW approximation.

The self-energy can be obtained from a self-consistent set of Dyson-like equations known as Hedin's equations:

$$P(1\,2) = -\mathrm{i}\int \mathrm{d}(3\,4)\, G(1\,3)\, G\left(4\,1^+\right) \Gamma(3\,4,2) \tag{5.10a}$$

$$W(1\,2) = w(1\,2) + \int \mathrm{d}(3\,4)\, W(1\,3)\, P(3\,4)\, w(4\,2) \tag{5.10b}$$

$$\Sigma(1\,2) = \mathrm{i}\int \mathrm{d}(3\,4)\, G\left(1\,4^+\right) W(1\,3)\, \Gamma(4\,2,3) \tag{5.10c}$$

$$G(1\,2) = G_{\mathrm{KS}}(1\,2) \tag{5.10d}$$
$$+ \int \mathrm{d}(3\,4)\, G_{\mathrm{KS}}(1\,3)\left[\Sigma(3\,4) - \delta(3\,4)\, v_{\mathrm{xc}}(4)\right] G(4\,2)$$

$$\Gamma(1\,2,3) = \delta(1\,2)\,\delta(1\,3) \tag{5.10e}$$
$$+ \int \mathrm{d}(4\,5\,6\,7)\, \frac{\delta\Sigma(1\,2)}{\delta G(4\,5)} G(4\,6)\, G(7\,5)\, \Gamma(6\,7,3)\ ,$$

where we have used the simplified notation $1 \equiv (\boldsymbol{x}_1, t_1)$ etc. Above, P is the irreducible polarisation, Γ is the so-called vertex function, and

$$G_{\mathrm{KS}}(\boldsymbol{x}, \boldsymbol{x}'; \omega) = \sum_n \frac{\phi_n(\boldsymbol{x})\, \phi_n^*(\boldsymbol{x}')}{\omega - \varepsilon_n^{\mathrm{KS}} - \mathrm{i}\eta\, \mathrm{sgn}\left(\mu - \varepsilon_n^{\mathrm{KS}}\right)}\ , \tag{5.11}$$

with G_{KS} the Green's function of the KS system and ϕ_n the corresponding KS wavefunctions with eigenenergies $\varepsilon_n^{\mathrm{KS}}$. We arrive at the GW approximation by eliminating the second term in the vertex function (5.10e) (i.e. neglecting "vertex corrections") in such a way that (5.10a) and (5.10b) reduces to

$$P(1\,2) = -\mathrm{i}G(1\,2)\, G\left(2\,1^+\right) \tag{5.12a}$$
$$\Sigma(1\,2) = \mathrm{i}G\left(1\,2^+\right) W(1\,2)\ . \tag{5.12b}$$

That is, in GW the screened Coulomb potential is calculated at the RPA level and Σ is just the direct product of G and W (hence the name). Also note

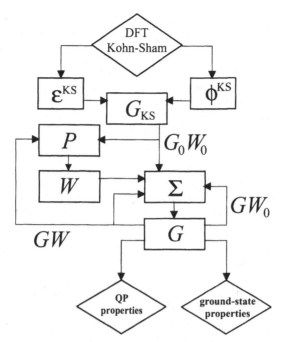

Fig. 5.2. Flow diagram sketching the practical implementation of the GW method. The partially self-consistent GW_0 updates the self-energy operator Σ, whereas the fully self-consistent GW also updates the screened Coulomb potential W

that in the Hartree-Fock approximation the Fock operator Σ_x is obtained as in (5.12b) but with W replaced by the static bare Coulomb potential w. Based on this, GW may be understood as a physically motivated generalisation of the Hartree-Fock method in which the Coulomb interaction is dynamically screened. A flow diagram sketching the practical implementation of the GW method is shown in Fig. 5.2.

In most GW applications, self-consistency is set aside, and P and Σ are obtained by setting $G = G_{KS}$ in (5.12a) and (5.12b). The interacting Green's function is then obtained by solving (5.10e) once. Furthermore, in many cases there is an almost complete overlap between the QP and the KS wavefunctions, and the full resolution of the QP equation (5.7) may be circumvented. Thus, E_n^{qp} is given as a first-order perturbation of the KS energy ε_n^{KS}:

$$E_n^{qp} \simeq \varepsilon_n^{KS} + \left\langle \phi_n \left| \Sigma \left(\varepsilon_n^{KS} \right) - v_{xc} - \Delta\mu \right| \phi_n \right\rangle , \qquad (5.13)$$

where a constant $\Delta\mu$ has been added to align the chemical potential before (KS level) and after the inclusion of the GW correction. As long as we are just interested on band-structures, further approximations, generally through a *plasmon-pole* ansatz [16], may be used to evaluate W in real materials. However, these models prevent us from calculating the whole Green's function so losing important spectral features like QP lifetimes and they can hardly be justified in systems others than *sp* metals. An efficient procedure to find out the entire spectral representation of the self-energy is the so-called *space-time* method [17], in which dynamical dependencies are represented in terms of imaginary times and frequencies, and each of Hedin's GW equations is solved in the most favourable spatial representation. As a final step, the self-energy for real frequencies can be obtained using analytical continuation from its values at imaginary frequencies after a fitting to a suitable analytical function. This method shows a favourable scaling with system size and avoids fine ω-grids that were needed to represent sharp spectral features in $G_{\mathrm{KS}}(\omega)$ and $W(\omega)$ [17,18].

Since the first *ab-initio* calculations performed by Hybertsen and Louie in 1985 [19], non-self-consistent GW has been applied to calculate QP properties (band-structures and lifetimes) of a wide variety of systems. The most striking success of this "G_0W_0" approximation is the fairly good reproduction (to within 0.1 eV of experiments) of experimental band gaps for many semiconductors and insulators, so circumventing the well-known failure of LDA when calculating excitation gaps. It is also worth emphasising that G_0W_0 gives much better ionisation energies than LDA in localised systems [20–22], and its success when studying lifetimes of hot electrons in metals and image states at surfaces (see [11,12] and references therein).

In spite of its overall success, G_0W_0 has some limitations. For instance, agreement with experiment for energy gaps and transitions may mask an overall additive error in the value of the self-energy; satellite structures are not well described in detail; and agreement with experiment worsens away from the Fermi energy (notably the bandwidth of alkali metals). Further approximations not related to MBPT, like the use of pseudopotentials in practical *ab-initio*calculations and those simplifications made when interpreting experimental results have been also considered. The main conclusions can be summarised as follows:

- Inclusion of vertex corrections improves the description of the absolute position of QP energies in semiconductors [23] and the homogeneous electron gas (HEG) [24], although the amount of such corrections depends very sensitively on the model used for the vertex [25]. Vertex corrections constructed using the so-called cumulant expansion [26], reproduce the multiple-plasmon satellite structure in alkali metals [27] (the GW spectral function only shows an isolated satellite).

- On the other hand, the absence of vertex corrections does not seem to be the full explanation of the differences (0.3–0.4 eV) between the measured valence bandwidth for alkali metals [28,29] and the G_0W_0 values [30,31]. The inclusion of vertex effects slightly changes the occupied bandwidth of the HEG, but this correction is not enough to fit the experimental results [24,32,33]. Of course these results are not conclusive because any effect due to the crystal structure is neglected. Nonetheless, the fact that

G_0W_0 plus vertex barely changes the valence bandwidth of Si [23], gives further indirect support to the existence of other mechanisms explaining this discrepancy. It seems plausible that specific details of the photoemission process could be the ultimate reason of the discrepancies between theory and experiment [9,33–35].

- The "bandwidth problem" mentioned above was the primary motivation of the first complete study of the role of self-consistency in GW performed by von Barth and Holm for the HEG [36,37]. Partially self-consistent "GW_0" calculations – in which W is calculated only once using the RPA, so that (5.12a) is not included in the iterative process – slightly *increase* the G_0W_0 occupied bandwidth. Results are even worse at full self-consistency in which, besides, there is not any well-defined plasmon structure in W and, as a consequence, the plasmon satellite in the spectral function practically vanishes. These results were confirmed by Schöne and Eguiluz [31] for bulk K where they obtained that the GW bandwidth is 0.6 eV broader than that of G_0W_0. These authors found another important result: self-consistency overestimates by 0.7 eV the experimental fundamental gap of Si, which is an error (but of the opposite sign) comparable with the one given by LDA. As a consequence, it does not seem a good idea to perform self-consistent GW calculations to obtain QP properties. The effects resulting from an unphysical screened Coulomb potential must be necessarily balanced by the proper inclusion of vertex corrections along the self-consistent procedure. However, as we will see in Sect. 5.4, such a self-consistency is essential to evaluate absolute ground-state energies.

- Very recently, a fully self-consistent calculation including vertex corrections has been reported for the HEG by Takada [35]. Compared with a G_0W_0 calculation (see Fig. 5.3), both methods give practically the same bandwidth, although the QP peak is much broader than in G_0W_0. The latter reflects a more effective damping of the QP due to the multiple electron-hole excitations that are included in diagrams beyond GW. A similar broadening can be observed in the first plasmon satellite peak, which it is fairly well located at the expected position (ω_p below the QP peak, ω_p being the bulk plasmon frequency). Interestingly, there is no significant change in the width of the valence band, but excellent agreement is obtained by including the self-energy corrections for the final state of the photoemitted electron. However, application of this self-consistent procedure to inhomogeneous systems appears to be very challenging.

- Finally, core electrons, that are absent from routine pseudopotential (PP) calculations, could be important in the final determination of spectral properties. Nonetheless, the inclusion of core-electrons in *ab-initio* MBPT schemes should be done with caution. For instance, an all-electron G_0W_0 calculation reduces the corresponding PP value of the fundamental gap of bulk Si at least 0.3 eV [38–40]. This effect has been interpreted as a result of exchange coupling between core and valence electrons [40] which,

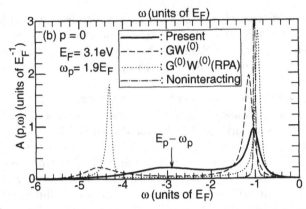

Fig. 5.3. Spectral function at the bottom of the valence band of the HEG ($r_\mathrm{s} = 4$) given by a non-interacting picture, G_0W_0, GW_0, and a full self-consistent procedure with the inclusion of vertex corrections. After Takada [35]

of course, is described in a PP calculation only at the level of the underlying atomic LDA calculation. However, the occupied bandwidth only suffers a marginal change of 0.1 eV after an all-electron calculation (note that the experimental value is 12.5 eV). This might suggest that vertex corrections, that are almost irrelevant when determining the band gap of *sp* semiconductors under the PP approximation [23,41], could be more important in those situations in which valence states coexist with more localised core states. Furthermore, the performance of G_0W_0 in transition metals, with the corresponding appearance of more localised *d* states, has not been fully assessed yet [9]. For these reasons, the striking coincidence between the experimental Si band gap and the all-electron self-consistent *GW* result reported by Ku and Eguiluz [40] might be fortuitous.

In summary, G_0W_0 is an excellent approximation for the evaluation of QP properties of simple systems and, very likely, able to provide the main trends in more complex systems. Theories beyond G_0W_0 are required to study other spectral features.

5.3   Pathologies of the Kohn–Sham xc Functional

The Kohn-Sham formalism [2] relies on the link between an actual N electron system and a fictitious non-interacting counterpart through the xc potential $v_{\mathrm{xc}}\left(\boldsymbol{r}\right) = \delta E_{\mathrm{xc}}\left[n\right]/\delta n\left(\boldsymbol{r}\right)^3$. Hence, $v_{\mathrm{xc}}\left(\boldsymbol{r}\right)$ contains essential information about many-body correlations which, as we have seen in the previous section, MBPT describes in terms of non-local dynamical functions. Then, we

---

[3] For simplicity, in v_{xc} we omit the explicit functional dependence on the density.

can easily realise that the mapping between ground-state densities and KS potentials,

$$n\left(\boldsymbol{r}\right) \to v_{\mathrm{KS}}\left[n\right]\left(\boldsymbol{r}\right) = v_{\mathrm{ion}}\left(\boldsymbol{r}\right) + \int \mathrm{d}\boldsymbol{r}' \, \frac{n\left(\boldsymbol{r}'\right)}{\left|\boldsymbol{r} - \boldsymbol{r}'\right|} + v_{\mathrm{xc}}\left(\boldsymbol{r}\right) \ , \qquad (5.14)$$

must depend on $n\left(\boldsymbol{r}\right)$ in a very peculiar and sensitive way. In fact, the actual functional relation between $n\left(\boldsymbol{r}\right)$ and $v_{\mathrm{xc}}\left(\boldsymbol{r}\right)$ (or $E_{\mathrm{xc}}\left[n\right]$) is:

- highly non-analytical: small or even infinitesimal changes in the density may induce substantial variations of the xc potential;
- highly non-local[4]: changes in the density at a given point $\boldsymbol{r}$ may induce substantial variations of the xc potential at a very distant point $\boldsymbol{r}'$.

These conditions are the origin of some special features that we will review in this section, and show how difficult is to construct a fully reliable approximation to the exact xc energy or potential that is an explicit functional of the density. Note that the LDA does not fulfil either requirement, and GGAs are just analytical semi-local approaches. The novel meta-GGAs (see the chapter by John P. Perdew in this book) are interesting in the sense that they include further non-analytical and non-local behaviour through the explicit appearance of the KS wavefunctions. Their performance remains to be explored, but it is likely that some non-analyticities and non-locality of the exact functional remains beyond their grasp. In fact, the virtue of these models is their ability to provide accurate results in many situations being, at the same time, very easy to apply. Other alternatives, like averaged and weighted density approximations [42–44], are truly non-local prescriptions but, despite its complexity, are once more limited by their explicit dependence on the density. Finally, we would like to mention the existence – as discussed by E. Engel in this book – of a very promising *third generation* of xc energy density functionals. In these models, the exchange energy – which is already non-local and non-analytical – is treated exactly [45,46], and then only Coulomb correlations remains to be approximated. The only drawback is that they do not benefit any more from the well-known cancellation between exchange and correlation effects in extended systems which, somehow, is exploited by other approximations. Therefore, the correlation part should be, in principle, more sophisticated than an LDA or GGA. To what extent these new functionals incorporate the following peculiarities of the xc functional remains to be investigated.

---

[4] The non-local *density*-dependence of the xc potential should not be confused with whether the $v_{\mathrm{xc}}\left(\boldsymbol{r}\right)$ is a local or non-local potential in its dependence on its spatial argument $\boldsymbol{r}$; in Kohn-Sham theory the xc potential is always a *local* potential in the latter sense.

5.3.1   The Band Gap Problem

We have already mentioned the inaccuracy of LDA-KS when determining the band gap of semiconductors and insulators. This failure is intimately related to a pathological non-analytical behaviour of the xc energy functional, as shown by J. P. Perdew and M. Levy and by L. J. Sham and M. Schlüter [47,48]. Namely, the xc potential may be increased by a finite constant of the order of $1\,\mathrm{eV}$ as a result of the addition of an extra electron to an extended system, that is, after an infinitesimal change of the electron density.

As is well known [49,50], the band gap E_{gap} of an N electron system is defined as the difference between the electron affinity $A = E_N^{(0)} - E_{N+1}^{(0)} \equiv -\mathcal{E}_{\mathrm{LUMO}}$ and the ionisation potential $I = E_{N-1}^{(0)} - E_N^{(0)} \equiv -\mathcal{E}_{\mathrm{HOMO}}$:

$$E_{\mathrm{gap}} = I - A = \mathcal{E}_{\mathrm{LUMO}} - \mathcal{E}_{\mathrm{HOMO}} \,, \tag{5.15}$$

where HOMO and LUMO stand for highest occupied and lowest unoccupied molecular orbital respectively. We may see that the band gap (or the HOMO-LUMO gap in a finite system) is just the difference between two single-electron removal/addition energies, so it is immediately addressed by MBPT. We can also calculate E_{gap} using KS-DFT through the expression

$$E_{\mathrm{gap}} = \varepsilon_{N+1}^{\mathrm{KS}}(N+1) - \varepsilon_N^{\mathrm{KS}} \,, \tag{5.16}$$

where $\varepsilon_{N+1}^{\mathrm{KS}}(N+1)$ is the energy of the highest occupied KS orbital of the $N+1$ electron system, and $\varepsilon_N^{\mathrm{KS}}$ is the HOMO level of the KS N-particle system – note that we keep the notation introduced in the previous section, in which $\varepsilon_i^{\mathrm{KS}}$ is the i-th KS orbital energy of the N electron system. It is easy to arrive at (5.16) just bearing in mind that the affinity of an N electron system is the opposite of the ionisation potential of the $N+1$ electrons, and that the Kohn-Sham HOMO level equals the actual one[5] [51].

For a non-interacting system, the gap can be readily written in terms of its orbital energies. Therefore, for the fictitious N electron KS system we have

$$E_{\mathrm{gap}}^{\mathrm{KS}} = \varepsilon_{N+1}^{\mathrm{KS}} - \varepsilon_N^{\mathrm{KS}} \,. \tag{5.17}$$

From (5.16) and (5.17), we immediately get that the actual and KS gaps are related through

$$E_{\mathrm{gap}} = \left(\varepsilon_{N+1}^{\mathrm{KS}} - \varepsilon_N^{\mathrm{KS}}\right) + \left(\varepsilon_{N+1}^{\mathrm{KS}}(N+1) - \varepsilon_{N+1}^{\mathrm{KS}}\right) \equiv E_{\mathrm{gap}}^{\mathrm{KS}} + \Delta_{\mathrm{xc}} \,, \tag{5.18}$$

an expression which is illustrated in Fig. 5.4. We may see that Δ_{xc} is just the difference between the energies of the $(N+1)$-th orbitals of the KS systems that correspond to the neutral and ionised electron systems. In a solid, in which $N \gg 0$, the addition of an extra electron only induces an infinitesimal

---

[5] That is, for an N electron system $\varepsilon_N^{\mathrm{KS}} = -I$. Remember that this is the only KS orbital energy with an explicit physical meaning.

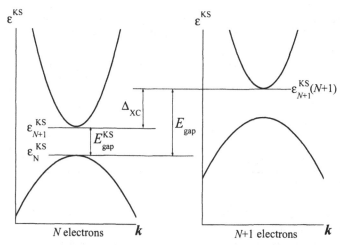

Fig. 5.4. Sketch of the Kohn-Sham band structure of a semiconductor (left panel). After the addition of an electron which occupies the empty conduction band, (right panel) the xc potential and the whole band-structure shift up a quantity Δ_{xc}

change of the density. Therefore, the two corresponding KS potentials must be practically the same inside the solid up to a constant shift and, consequently, the KS wavefunctions do not change. The energy difference Δ_{xc} is then the aforementioned rigid shift which, in addition, is entirely contained in v_{xc} because the Hartree potential depends explicitly on the density. As a conclusion, Δ_{xc} is the measure of a well-defined non-analytical behaviour of the xc energy functional. Namely, it is a finite variation of $v_{xc}(r)$ extended everywhere in the solid due to an infinitesimal variation of $n(r)$

$$\Delta_{xc} = \left(\left. \frac{\delta E_{xc}[n]}{\delta n(r)} \right|_{N+1} - \left. \frac{\delta E_{xc}[n]}{\delta n(r)} \right|_N \right) + \mathcal{O}\left(\frac{1}{N}\right) . \tag{5.19}$$

Now it is easy to see the relation between a non-analytical v_{xc} and the band gap problem. If v_{xc} were actually discontinuous, the actual band gap would not be given in terms of the KS energies of the N electron system[6]. On the contrary, if Δ_{xc} were zero (or very close to zero), the difference between the actual gap and the LDA-KS one E_{gap}^{LDA} would be just an inherent limitation of the local-density approximation. In the latter case, the formulation of more sophisticated approaches to the xc energy would allows us to calculate the gap of a real material directly from its corresponding KS band-structure. Nonetheless, the LDA is already a good approximation when calculating total energies and densities of bulk semiconductors and, moreover, improvements upon the LDA, such as the GGA or the WDA, change the KS gap very little.

---

[6] In a similar context this is what happens in a metal. Although the KS Fermi energy is equal to the actual one, the corresponding Fermi surfaces may differ.

Table 5.1. The xc discontinuity Δ_{xc}, and calculated and experimental fundamental band gaps, for four semiconductor and insulators. All energies are in eV. From Godby et al. [53]

	Si	GaAs	AlAs	Diamond
Δ_{xc}	0.58	0.67	0.65	1.12
Band gaps:				
KS-LDA	0.52	0.67	1.37	3.90
G_0W_0	1.24	1.58	2.18	5.33
Experiment	1.17	1.63	2.32	5.48

The existence of a discontinuity in v_{xc} is, then, more plausible than an error in the LDA band-structure.

The first evidence of a non-zero Δ_{xc} in real matter was given by Godby et al. [52,53] who used the so-called Sham-Schlüter equation [48,54],

$$\int d\boldsymbol{r}' \, v_{xc}\left(\boldsymbol{r}'\right) \int d\omega \, G_{KS}\left(\boldsymbol{r}, \boldsymbol{r}'; \omega\right) G\left(\boldsymbol{r}', \boldsymbol{r}; \omega\right)$$

$$= \int d\boldsymbol{r}' \, d\boldsymbol{r}'' \int d\omega \, G_{KS}\left(\boldsymbol{r}, \boldsymbol{r}''; \omega\right) \Sigma\left(\boldsymbol{r}', \boldsymbol{r}''; \omega\right) G\left(\boldsymbol{r}'', \boldsymbol{r}; \omega\right) , \quad (5.20)$$

to calculate the xc potential from the many-body self-energy operator, which was obtained under the GW approximation. This MBPT-based potential was found to be very similar to the LDA one and the corresponding band structures turned out to be practically the same. As a consequence, the xc discontinuity Δ_{xc} is the main cause of the difference between the experimental gaps and those given by the LDA. In fact, Δ_{xc} accounts for about 80% of the LDA band gap error for typical semiconductors and insulators (see Table 5.1). This result was confirmed by Knorr and Godby for a family of model semiconductors [55]. In this case, the exact potential $v_{KS}\left(\boldsymbol{r}\right)$ (and hence, the exact E_{gap}^{KS}) was calculated by imposing the reproduction of quantum Monte-Carlo densities. Again $\Delta_{xc} = E_{gap}^{KS} - E_{gap} \neq 0$, accounting for 80% of the LDA error $E_{gap} - E_{gap}^{LDA}$. Interestingly, an opposite trend (i.e. $\Delta_{xc} \simeq 0$) was found by Gunnarsson and Schönhammer in a very different scenario: a simple Hubbard-like one-dimensional semiconductor in which v_{xc} and the gap can be obtained exactly [56].

Recently, Städele and co-workers [57] calculated the fundamental band gaps for a number of standard semiconductors and insulators using the exact exchange functional together with the local approximation to the correlation energy (which we denote EXX(c)). In several of the materials studied the KS gaps within this approximation were found to be notably closer to experiment than the LDA gaps. However, the same paper also evaluated the exchange contribution to Δ_{xc} (defined as the difference between the Hartree-Fock and exact-exchange KS gaps [59]), which was several electron volts, much larger

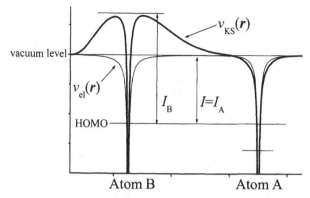

Fig. 5.5. The Kohn-Sham effective potential $v_{KS}(r)$ for two widely separated open shell atoms. Whereas the classical contributions to v_{KS} do not show any pathological behaviour, the exchange-correlation potential takes a positive value $I_B - I_A$ around the atom B

than any estimate of the total Δ_{xc}. This serves to emphasise the large degree of cancellation between exchange and correlation effects, familiar from other aspects of the electronic structure of solids, which suggests that caution must be exercised in interpreting a calculation in which exchange and correlation are treated on quite different footings. In a further paper [58], $G_0 W_0$ band gaps calculated from EXX(c) wavefunctions were found to be little different from those calculated from LDA wavefunctions, supporting the notion that a variety of reasonable descriptions of exchange and correlation provide adequate zeroth-order starting points for a MBPT calculation.

5.3.2 Widely Separated Open Shell Atoms

It is known that the xc potential is, in many cases, long ranged. For instance, in a neutral atom $v_{xc}(r)$ decays asymptotically as $-1/r$, whereas for a metal surface it exhibits an image-like behaviour $-1/4z$ [51]. What it is not so known is that, as shown by Almbladh and von Barth [60], under some special circumstances, $v_{xc}(r)$ can be macroscopically long-ranged, thus reflecting a pathological *ultra-high* non-locality.

Let us consider two atoms A and B, each of them with an unpaired electron, whose ionisation potentials are I_A and I_B with $I_A < I_B$. If the atoms are separated by a very large arbitrary distance d, the ionisation potential of the whole system I is then given by the smallest (I_A) of the two ionisation potentials. Taken into account that in a finite system the ground-state density decays as $n(r) \propto \exp\left(-2r\sqrt{2I}\right)$ [51,61], the asymptotic behaviour of the ground-state density of this "molecule" is governed by $I = I_A$ except in a region surrounding atom B, where the exponential fall-off of the density is given in terms of I_B.

If the ground-state of the N-particle system is a spin singlet[7], the highest occupied state ϕ_N of the KS fictitious system with energy $\varepsilon_N^{KS} = -I = -I_A$ is doubly occupied – remember that all the lower KS-states must be completely full. All the asymptotic behaviour of the density is determined by the HOMO, thus there is a region around the atom B in which $v_{KS} - \varepsilon_N^{KS} \simeq I_B$ whereas $v_{KS} - \varepsilon_N^{KS}$ tends to I_A in the rest of the system – see Fig. 5.5. Both electrostatic and ionic contributions to v_{KS} decay to zero everywhere. Therefore, although the xc potential decays to zero around the atom A and, in general, at sufficiently large distances, v_{xc} tends to $I_B - I_A > 0$ in the neighbourhood of the atom B. That is, v_{xc} shifts up a finite amount around B due to the presence of another electron at an arbitrary large distance. Moreover, v_{xc} must have a spatial variation in a region between A and B where the electron density is practically zero. Both features clearly illustrate that the xc potential exhibits an unphysical infinite range in this model system. Note that this behaviour cannot emerge by any means from typical non-local prescriptions which assume a finite range around a point r that depends on the density $n(r)$.

5.3.3   The Exchange-Correlation Electric Field

An insulating solid is, of course, composed of individual unit cells, each of which contains polarisable electrons which may become slightly displaced in response to an applied uniform electric field, so that each unit cell acquires an electric dipole moment. According to the well-known theory of dielectric polarisation, this *macroscopic polarisation* produces a "depolarising" electrostatic field which reduces the net electric field by a factor of ε, the macroscopic dielectric constant. In Kohn-Sham DFT, however, there is a further possible contribution to the potential felt by the Kohn-Sham electrons: the exchange-correlation potential $v_{xc}(r)$ may also acquire a long-range variation, which was termed the exchange-correlation "electric field" by Godby and Sham, and Gonze, Ghosez and Godby in a series of papers [62–67]. (Of course, it is not truly an *electric* field in the sense that it is produced by real electric charge, but its effect on the Kohn-Sham potential is the same as that of an electric field.)

Figure 5.6 shows the basic concept. The two polarised insulators shown in the central and lower parts of the Figure have identical electron densities, but different Kohn-Sham potentials: the two systems differ in their macroscopic polarisation. In order to reproduce the correct macroscopic polarisation, the exact Kohn-Sham xc potential must acquire a part which varies linearly in space: the xc field.

For our purpose, the point is that the exchange-correlation electric field is another pathological aspect of the exact Kohn-Sham xc functional: the

---

[7] If the ground-state were a triplet we should use the extension of KS-DFT to spin polarised systems, but in this case the pathology we are describing will not appear.

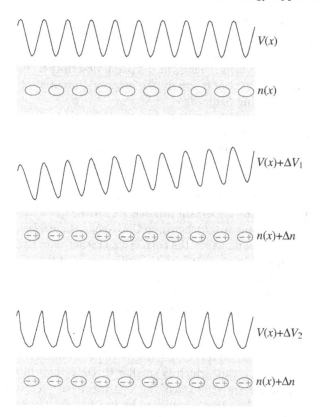

Fig. 5.6. A schematic illustration of the origin of the exchange-correlation "electric field". Top: an unpolarised insulator; the blobs represent the regions of high electron density within each unit cell. Centre: The same insulator, polarised by the addition of an external electric field, which (together with the depolarising internal electric field and any exchange-correlation "electric field") results in the total Kohn-Sham potential shown. Bottom: The same polarised electron density in the bulk crystal may be generated by a Kohn-Sham potential with zero net long-range field (as shown here), or indeed by a family of potentials, each with a different net field. Each member of the family corresponds to a different macroscopic polarisation, i.e. a different surface charge. A particular non-zero value of the Kohn-Sham exchange-correlation "electric field" is required to reproduce the correct macroscopic polarisation

electron density in the polarised insulator is the same from one unit cell to the next, while $v_{xc}(\mathbf{r})$ rises by a constant amount over the same distance. Therefore, the xc potential cannot be regarded as a functional of the electron density within its own unit cell, or indeed the electron density in any finite region. The xc-field part of the potential depends on the polarisation; that is, on the electron density at the surface of the crystal. For this reason, the xc field represents an *ultra-non-local* dependence of the xc potential on the electron density. In contrast, in a MBPT description, the self-energy operator, which is written as a perturbation series in terms of fairly local[8] quantities, is believed to have no such long-range variation from one cell to the next, and hence no "electric field". Thus, in a MBPT description, the long-range part of the effective potential is simply the external potential plus the actual electrostatic depolarising field.

A simple argument [63] indicates why the xc-field must be non-zero, and allows an estimate of its magnitude. Consider an unpolarised insulator, in which the band gap discontinuity is Δ_{xc}. Let us, for a moment, make the quasiparticle approximation in which the spectral weight in MBPT is assumed to be dominated by the quasiparticle peaks, i.e. the properties of the system emerge from the quasiparticle wavefunctions and energies in a similar way to KSDFT, with the important difference that the quasiparticles feel the non-local self-energy operator rather than the exchange-correlation potential, and the quasiparticle band gap is the correct gap rather than the Kohn-Sham gap. In the presence of an electric field, the polarisation of the electron density is described by the density response function. The same change of electron density is described by the Kohn-Sham electrons, responding to the change in their Kohn-Sham potential, as by the quasiparticles, responding to the change in the actual electrostatic potential (external plus Hartree, since the self-energy operator has no long-range part in conventional MBPT). However, in one-electron perturbation theory, the degree of polarisation is inversely proportional to the energy gap, which is smaller in DFT than in quasiparticle theory. To compensate, the strength of the long-range part of the Kohn-Sham potential must be weaker; this is achieved by the xc field. Godby and Sham, making the further approximation that the quasiparticle wavefunctions were similar to the Kohn-Sham wavefunctions, deduced that

$$\frac{\Delta V_{xc}}{\Delta V} \approx -\frac{\Delta_{xc}}{E_{gap}} \, , \tag{5.21}$$

where ΔV_{xc} is the strength of the xc field, ΔV is the strength of the actual electrostatic field, and E_{gap} is the quasiparticle band gap. This fraction is significant: about -0.5 in silicon, for instance.

In reciprocal space, Ghosez et al. [67] showed that the xc field corresponds to a $1/q^2$ divergence in the exchange-correlation kernel for small wavevectors q. This ultra-non-local density dependence is certainly missing from all density based approximations to the exact xc functional, potential or kernel. One possibility for approximating it within Kohn-Sham DFT has been explored recently by de Boeij et al. [68] by using a functional of the current rather than the density, in the low-frequency limit of time-dependent DFT.

---

[8] That is, mathematically non-local but with the range of the non-locality restricted to a few ångstroms.

5.4   Total Energies from Many-Body Theory

To apply the KS method to real problems with confidence in its predictive accuracy, we need reasonable approximations to the exchange-correlation energy functional. However, we have seen in the previous section that $E_{xc}[n]$ is a very peculiar object which is described far from properly by the common local-density or generalised gradient approximations. Thus, although the basic reason for the success of the LDA was understood many years ago [42,69], there are a number of well identified cases in which LDAs and GGAs fail dramatically. For instance, they give qualitatively wrong structural results when studying not only some strongly correlated materials [70], but also in some systems dominated by sp bonds [71,72]; or they systematically overestimate cohesive energies and underestimate the activation barrier of chemical reactions [73]. This is not a surprise because, in essence, LDA/GGAs are limited by their intrinsic semi-local nature and by the absence of self-interaction corrections.

To some extent, all the acknowledged improvements upon LDA/GGAs start from model systems (usually the homogeneous electron gas). It would be desirable for a total energy method not to rely on the similarity of a system to a particular reference, thus having a truly *ab-initio* technique. Configuration interaction (CI) and quantum Monte Carlo (QMC) [74] are examples of such methods. Both procedures are in principle exact, but the scaling of CI with system size implies an almost prohibitive computational effort even in medium-size problems. QMC calculations are less demanding, but they are still much more expensive than standard DFT.

MBPT-based schemes can be meant as an alternative for those situations in which known DFT models are inaccurate, but whose complexity makes the implementation of QMC difficult. In this section, after a brief summary of the theoretical foundations, we will review some of the recent applications which, so far, have been restricted to model systems but in which LDA/GGAs clearly show their limitations. Finally, we will present a simplified many-body theory amenable for its implementation in a DFT-fashion.

5.4.1   Theoretical Background

Although many-body theory gives *per se* enough information to obtain the ground-state energy $E^{(0)}$ of an electron system, it is useful to keep a link between MBPT- and DFT-based expressions. First, it is computationally more convenient to evaluate the *difference* between MBPT and KS results than the full energy given by MBPT. Second, a fully self-consistent calculation can be achieved in the framework of MBPT, but a first estimation of the results beyond LDA can be obtained just by evaluating the many-body corrections over the LDA-KS system, using the KS system as a zeroth-order approximation – as it is done, for instance, in the G_0W_0 method. It is convenient, on occasion, to write down many-body-based expressions for the xc energy, defined precisely as in exact KS-DFT.

MBPT provides several ways to obtain each of the different contributions to the ground-state energy $E^{(0)}$. Perhaps the best known, owing to its role in the construction of xc energy functionals, is the expression based on the adiabatic-connection-fluctuation-dissipation (ACFD) theorem [75,76]

$$E_{xc}[n_0] = \frac{1}{2}\int_0^1 d\lambda \int d\mathbf{r}\, d\mathbf{r}' \frac{1}{|\mathbf{r} - \mathbf{r}'|} \tag{5.22}$$
$$\times \left[n_0(\mathbf{r})\delta(\mathbf{r} - \mathbf{r}') - \int_0^{+\infty}\frac{d\omega}{\pi}\chi_\lambda(\mathbf{r}, \mathbf{r}'; i\omega) \right].$$

Here, $\chi_\lambda(i\omega)$ is the causal density response function at imaginary frequencies of a system in which the electrons interact through a modified Coulomb potential $w_\lambda(r) = \lambda/r$, and whose ground state density is equal to the actual one. $\chi_\lambda(i\omega)$ is related to the polarisation function $P_\lambda(i\omega)$ through the equality[9]

$$\widehat{\chi}_\lambda(i\omega) = \widehat{P}_\lambda(i\omega)\left[1 - \widehat{w}_\lambda\widehat{P}_\lambda(i\omega)\right]^{-1}, \tag{5.23}$$

where usual matrix multiplications are implied. For practical purposes, we subtract from (5.22) the exact exchange energy functional

$$E_x[n_0] = -\sum_\sigma \int d\mathbf{r}\, d\mathbf{r}' \frac{\left|\sum_j^{occ}\phi_j^*(\mathbf{r}, \sigma)\phi_j^*(\mathbf{r}', \sigma)\right|^2}{2|\mathbf{r} - \mathbf{r}'|} = \tag{5.24}$$
$$= \int d\mathbf{r}\, d\mathbf{r}' \frac{1}{2|\mathbf{r} - \mathbf{r}'|}\left[n_0(\mathbf{r})\delta(\mathbf{r} - \mathbf{r}') - \int_0^{+\infty}\frac{d\omega}{\pi}\chi_0(\mathbf{r}, \mathbf{r}'; i\omega) \right],$$

where $\chi_0(i\omega) \equiv P_{KS}(i\omega)$ is the density response of the fictitious KS system

$$\widehat{\chi}_0(\mathbf{r}, \mathbf{r}'; i\omega) = \sum_\sigma \sum_{i,j} \frac{(f_i - f_j)\phi_i^*(\mathbf{r}, \sigma)\phi_i(\mathbf{r}', \sigma)\phi_j(\mathbf{r}, \sigma)\phi_j^*(\mathbf{r}', \sigma)}{\varepsilon_i^{KS} - \varepsilon_j^{KS} + i\omega}, \tag{5.25}$$

f_j being the Fermi occupation (0 or 1) of the j-th KS orbital. As a consequence, the correlation energy can be evaluated as

$$E_c[n_0] = \int_0^1 d\lambda\, \mathrm{tr}\left\{ \widehat{w}\int_0^{+\infty}\frac{d\omega}{2\pi}[\widehat{\chi}_0(i\omega) - \widehat{\chi}_\lambda(i\omega)] \right\}, \tag{5.26}$$

where "tr" stands for the spatial trace. Note that if we set $P_\lambda \simeq P_{KS}$ in (5.23) we have the random phase approximation (RPA) – strictly speaking, an LDA-based RPA since the local density is used to obtain the one-electron orbitals and energies.

---

[9] We can establish this relation because at imaginary frequencies the causal and the time-ordered response functions coincide.

The same information can be also extracted from the self-energy operator and the Green's function. Namely, using the adiabatic-connection that led to (5.22)[10]

$$E_{\text{xc}}[n_0] = -\frac{i}{2} \int_0^1 \frac{d\lambda}{\lambda} \int_{-\infty}^{+\infty} \frac{d\omega}{2\pi} \int d\boldsymbol{x}\, d\boldsymbol{x}'\, \Sigma_\lambda(\boldsymbol{x}, \boldsymbol{x}'; \omega)\, G_\lambda(\boldsymbol{x}', \boldsymbol{x}; \omega) \ , \quad (5.27)$$

where, again, Σ_λ and G_λ refer to a fictitious system with the scaled Coulomb potential w_λ, and a convergence factor $\exp(i\eta\omega)$ is to be understood in the ω integral. Nonetheless, the one-electron density matrix $\gamma(\boldsymbol{x}, \boldsymbol{x}')$ can be obtained directly from G:

$$\gamma(\boldsymbol{x}, \boldsymbol{x}') = -i \int \frac{d\omega}{2\pi} G(\boldsymbol{x}, \boldsymbol{x}'; \omega) \ , \quad (5.28)$$

and the Green's function provides the expectation value of any one-particle operator[11]. Thus, it is more convenient to calculate explicitly the kinetic energy contribution to E_{xc} rather than making the adiabatic connection:

$$\begin{aligned} E_{\text{xc}}[n_0] = &-\frac{i}{2} \int_{-\infty}^{+\infty} \frac{d\omega}{2\pi} \int d\boldsymbol{x}\, d\boldsymbol{x}'\, \Sigma(\boldsymbol{x}, \boldsymbol{x}'; \omega)\, G(\boldsymbol{x}', \boldsymbol{x}; \omega) \\ &-i \int_{-\infty}^{+\infty} \frac{d\omega}{2\pi} \int d\boldsymbol{x}\, \lim_{\boldsymbol{x}' \to \boldsymbol{x}} \left[-\frac{\nabla^2}{2} \delta G(\boldsymbol{x}, \boldsymbol{x}'; \omega) \right] \ , \quad (5.29) \end{aligned}$$

where $\delta G = G - G_{\text{KS}}$ is the difference between the Green's function of the real system and the KS one. Finally, if we separate the exchange and correlation contributions to (5.29) using

$$E_{\text{x}}[n_0] = -\frac{i}{2} \int_{-\infty}^{+\infty} \int d\boldsymbol{x}\, d\boldsymbol{x}'\, \frac{d\omega}{2\pi} \Sigma_{\text{x}}(\boldsymbol{x}, \boldsymbol{x}')\, G_{\text{KS}}(\boldsymbol{x}', \boldsymbol{x}; \omega) \ , \quad (5.30)$$

---

[10] As shown in [75,76], the xc energy of an electron system can be written as:

$$E_{\text{xc}}[n_0] = \int_0^1 \frac{d\lambda}{\lambda} \left(\left\langle \widehat{W} \right\rangle_\lambda - E[\lambda, n_0] \right) = \int_0^1 \frac{d\lambda}{\lambda} W_{\text{xc}}[\lambda, n_0]$$

Here, $\left\langle \widehat{W} \right\rangle_\lambda$ is the expectation value of the electron-electron interaction energy of the fictitious system whose ground-state density is n_0 but interacting through the potential λ/r, and $E[\lambda, n_0] = \lambda E[n_0]$ is the corresponding Hartree classical contribution. If we evaluate W_{xc} in terms of the density response function of the fictitious system we arrive at (5.22). By using the self-energy and the Green's function instead, we get the expression (5.27).

[11] For instance, the electron density is simply given by

$$n(\boldsymbol{r}) = -i \sum_\sigma \int \frac{d\omega}{2\pi} G(\boldsymbol{x}, \boldsymbol{x}; \omega)$$

As a consequence, we might also calculate the MBPT corrections to the LDA/GGA density for those systems in which they are expected to be inaccurate and, hence, to the classical Hartree electrostatic energy.

with Σ_x the Fock operator of the KS system

$$\Sigma_x\left(\boldsymbol{x}, \boldsymbol{x}\right) = -\sum_i^{\text{occ}} \frac{\phi_i\left(\boldsymbol{x}\right)\phi_i^*\left(\boldsymbol{x}'\right)}{\left|\boldsymbol{r} - \boldsymbol{r}'\right|} ,\tag{5.31}$$

we arrive at the definite expression

$$\begin{aligned}E_c\left[n_0\right] &= -\mathrm{i}\int_{-\infty}^{+\infty}\frac{\mathrm{d}\omega}{2\pi}\,\mathrm{Tr}\left[\frac{1}{2}\widehat{\Sigma}_c\left(\omega\right)\widehat{G}\left(\omega\right) + \left(\frac{1}{2}\widehat{\Sigma}_x + \widehat{t}\right)\delta\widehat{G}\left(\omega\right)\right]\\&= \int_0^{+\infty}\frac{\mathrm{d}\omega}{\pi}\,\mathrm{Tr}\left[\frac{1}{2}\widehat{\Sigma}_c\left(\mu + \mathrm{i}\omega\right)\widehat{G}\left(\mu + \mathrm{i}\omega\right) + \left(\frac{1}{2}\widehat{\Sigma}_x + \widehat{t}\right)\delta\widehat{G}\left(\mu + \mathrm{i}\omega\right)\right] ,\end{aligned}\tag{5.32}$$

where we have deformed the contour to the imaginary axis. In (5.32), "Tr" is the total trace – including the spin, in contrast to "tr" in (5.26), $\Sigma_c = \Sigma - \Sigma_x$ is the correlation part of the self-energy, and $\widehat{t}$ is the one-particle kinetic energy operator. It is also worth noting that the whole ground-state energy can be written just in terms of the Green's function using the so-called Galitskii-Migdal formula [77]

$$E^{(0)} = \frac{1}{2}\int_{-\infty}^{\mu}\mathrm{d}\omega\,\mathrm{Tr}\left[\left(\omega + \widehat{h}_0\right)\widehat{A}\left(\omega\right)\right] ,\tag{5.33}$$

with $\widehat{A}\left(\omega\right)$ the spectral function and $\widehat{h}_0 = \widehat{t} + \widehat{v}_{\text{ion}}$, which, after the inclusion of the remaining contributions to the energy, is equivalent to (5.32). Finally, although we do not discuss them in detail, we have to mention that the ground-state energy can be also obtained from the many-body Luttinger-Ward variational functional [78] and extensions thereof like the Almbladh-von Barth-van Leeuwen theory [79], which are closely related to the Green's function-based formulation we have described here.

It is evident that if the exact theory were used, all the quoted methods would give the same result. Nonetheless, in practical implementations we have to resort to further approximations. The ACFD expression (5.26) requires the knowledge of the interacting response function χ_λ, which is a quantity that can be obtained in the framework of time-dependent DFT [80,81]. Galitskii-Migdal and Luttinger-Ward-like methods need the interacting many-body Green's function[12]. We shall focus mainly on Green's-function-based evaluations – e.g. equations (5.32) and (5.33) – of the ground-state properties, but we note that several ACFD approaches have been used to study the HEG [82], the van der Waals interaction between two thin metal slabs [83], the jellium surface energy [84], and molecular properties like atomisation energies, bond lengths, and dissociation curves [85,86]. Many-body variational functionals have not been so widely tested, and applications to electron systems have been restricted to the HEG [87], closed-shell atoms [88], and the hydrogen molecule [89].

---

[12] Because of the close relation between P and χ, the response function may be also obtained from many-body approaches.

Table 5.2. Correlation energy per particle (in Ha) for the spin-unpolarised phase of the 3D electron gas obtained using several GW schemes, QMC, and RPA. For reference, the exchange energy per particle is included in the last row

r_s	1	2	4	5	10
QMC$^{(a)}$	−0.060	−0.045	−0.032	−0.028	−0.019
QMC$^{(b)}$	−0.055	−0.042		−0.028	
$GW^{(c)}$	−0.058	−0.044	−0.031	−0.027	−0.017
$GW^{(d)}$		−0.045	−0.032		
$GW_0^{(c)}$	−0.061	−0.043	−0.028	−0.024	−0.015
$G_0W_0^{(c)}$	−0.070	−0.053	−0.038	−0.033	−0.021
RPA	−0.079	−0.062	−0.047	−0.042	−0.031
ε_x	−0.458	−0.229	−0.115	−0.092	−0.046

[a] Reference [94]
[b] Reference [95]
[c] Reference [92]
[d] Reference [91]

5.4.2   Applications

The first application of Green's function theory to the calculation of ground-state properties of the three-dimensional (3D) homogeneous electron gas (HEG) at metallic densities appeared in a seminal work by Lundqvist and Samathiyakanit in the late 1960s [90]. However, systematic studies on the performance of Hedin's GW method for the same model were published only a few years ago by von Barth and Holm [37,91], and extended by García-González and Godby [92] to the spin-polarised 3D HEG, and the 2D HEG (a system where GW might be expected to perform less well because correlation is stronger).

As we may see in Table 5.2, the non-self-consistent G_0W_0 underestimates the total energy of the spin-unpolarised 3D HEG at metallic densities around 10 mHa per electron, which is half the error in the ACFD-RPA [93]. The same trend – that is, a 50% reduction of the RPA error – also appears in the spin-polarised 3D electron gas. A better performance is achieved by using the partially self-consistent GW_0 and, strikingly, at full self-consistency there is an almost perfect agreement with the exact QMC data [94,95]. Moreover, at lower densities and at the 2D HEG – where, as commented, the diagrams not included in GW are more relevant – the GW greatly improves the RPA energies. Thus, we may conclude that *the greater the self-consistency the more accurate the total-energy results*, in marked contrast with the tendency described in Sect. 5.2 for the QP energy dispersion relation.

The accuracy of GW may be traced back to the fulfilment of all conservation rules in the framework of the theory developed by Baym and

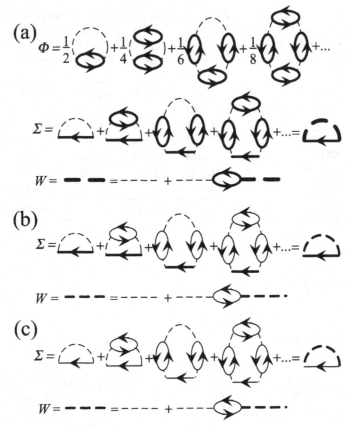

Fig. 5.7. Diagrammatic representation of the self-energy and the screened Coulomb potential in (**a**) the fully self-consistent GW approximation, (**b**) the partially self-consistent GW_0 approximation, and (**c**) the G_0W_0 approximation. The generating functional Φ is also represented in (**a**)

Kadanoff [96]. As showed by Baym [97], the self-energy operator of a conserving approach can be represented as the derivative of a generating functional Φ:

$$\widehat{\Sigma} = \frac{\delta}{\delta\widehat{G}}\Phi\left[\widehat{G}\right]\ , \qquad (5.34)$$

which has to be evaluated self-consistently at the Green's function that is the solution of the Dyson equation (5.10e). The self-consistent GW approximation does derive from a functional Φ_{GW} (see Fig. 5.7). Therefore, its implementation guarantees among other things, the conservation of the number of particles of the system[13]; the coincidence of the Fermi levels obtained by

---

[13] Since we can include the interaction between the electrons in a perturbative fashion, conservation means that GW gives the *correct* number of particles after integration of the corresponding Green's function.

solving the QP equation and by subtracting the ground-state energies of the N- and $(N-1)$- particle systems; and the equivalence of (5.22) and (5.27) when calculating xc energies.

Nonetheless, the success of GW when obtaining the ground-state energy of homogeneous systems has to be taken with caution. First, GW is not so accurate in highly correlated systems described by simple Hubbard Hamiltonians [98–100,89], and the GW correlation function of the HEG does not improve significantly on the RPA [101]. The GW polarisation function – and, hence, the density response – shows certain unphysical features. It it also worth noting that, even using the space-time procedure [17][14], a fully self-consistent resolution of Hedin's GW equations is very demanding for any inhomogeneous system. As a consequence, efforts to evaluate structural properties from MBPT should be directed towards non- or partially self-consistent schemes with further inclusion of short-ranged correlations that are absent in the GW diagrams. However, there is no guarantee that approximations other than self-consistent Φ-derivable schemes are conserving theories, and the fulfillment of exact sum rules by these models should be assessed carefully if they are intended to be used as practical tools to evaluate ground-state properties.

The most fundamental sum rule is the conservation of the number of particles which is satisfied by the partially self-consistent GW_0 method [102], even though it is not Φ-derivable. However, as demonstrated by Schindlmayr in a Hubbard model system [103], G_0W_0 does *not* yield the correct number of particles. This failure was confirmed by Rieger and Godby for bulk Si [104], where G_0W_0 slightly underestimates the total number of valence electrons. A study of particle-number violation in diagrammatic self-energy models has been recently presented by Schindlmayr *et al.* [105]. These authors provided a general criterion that allows, by simple inspection, to verify whether an approximation satisfies the particle-number sum rule. They also showed that the G_0W_0 particle-number violation is not, in practice, significant within the range of densities of physical interest (see Fig. 5.8). The same conclusion applies to models built by insertion of local vertex corrections into a G_0W_0 scheme [25].

The first evidence of the usefulness of these non-self-consistent diagrammatic schemes to evaluate structural properties has been the application of G_0W_0 to calculate the ground-state energy of confined quasi-2D systems and the interaction energy between two thin metal slabs [106]. For the quasi-2D gas, the high inhomogeneity of the density profile along the confining direction is clearly beyond the scope of local and semi-local KS-DFT approaches which, in fact, diverge when approaching the 2D limit [107–109]. RPA-ACFD does not show such a divergence but clearly underestimates the energy of quasi-2D systems. G_0W_0, whose superiority to the RPA in the 2D and 3D HEG

---

[14] Note that we just need the self-energy and the Green's function at imaginary frequencies to obtain ground-state properties from MBPT.

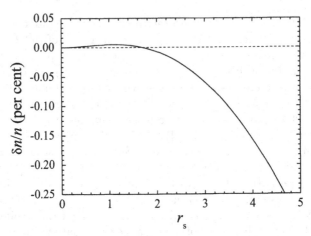

Fig. 5.8. Violation of the particle-number sum rule for the homogeneous electron gas in the G_0W_0 approximation. The relative error in the density is always negative and of the order of 0.1% in the range of metallic densities. After Schindlmayr *et al.* [105]

has been already noted, retains this superiority for these quasi-2D systems. Of more direct significance is the study of the interaction between two un-confined jellium slabs. At small distance separation d the density profiles of each subsystem overlap, so having a *covalent bond*. If $d \gg 0$, there is no such overlap and the only source of bonding is the appearance of correlation van der Waals forces which cannot be described at all by KS-LDA/GGA [110]. The xc energy per particle ε_{xc} as a function of d is depicted in the upper panel of Fig. 5.9 using the LDA, the RPA, and the G_0W_0, for two slabs of width $L = 12a_0$ and a background density corresponding to $r_s = 3.93$ – the mean density of sodium. In the lower panel, we present the correlation binding energy per particle, defined as $e_c(d) = \varepsilon_c(d) - \varepsilon_c(\infty)$, for the same system. We may see that the local density is unable to reproduce the charac-teristic asymptotic d^{-2} van der Waals behaviour[15] which, on the contrary, is present in the RPA and G_0W_0 calculations. The results from the two latter approaches are very similar at large separations, which is not a surprise be-cause van der Waals forces are already contained at the RPA level [83]. For intermediate and small separations there are slight differences between RPA and G_0W_0, but much less important than those appearing when comparing the total correlation energies.

It is worth pointing out that the remaining error in the absolute G_0W_0 correlation energy is amenable to an LDA-like correction

$$\Delta E_c[n] = \int d\mathbf{r}\, n(\mathbf{r})\, \Delta \varepsilon_c^{GW}(n(\mathbf{r}))\,, \tag{5.35}$$

---

[15] Non-local xc functionals such as the ADA or WDA also fail to describe van der Waals forces.

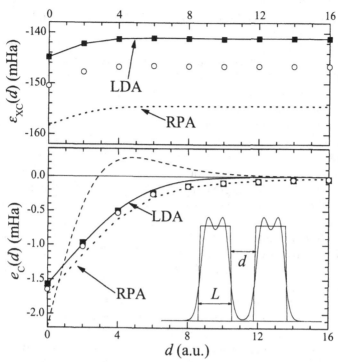

Fig. 5.9. Upper panel: xc energy per particle for two jellium slabs as a function of the distance d (see inset). Lines: LDA and RPA; empty circles: G_0W_0; squares: $G_0W_0+\Delta$. Lower panel: Correlation binding energy per particle. The exchange-only binding energy (dashed line) has been also included in this panel

with[16]

$$\Delta\varepsilon_{\mathrm{c}}^{GW}\left(r_{\mathrm{s}}\right) = \frac{0.04054}{1 + 2.086\sqrt{r_{\mathrm{s}}} + 0.1209 r_{\mathrm{s}}^2}\ \mathrm{hartree}\,. \qquad (5.36)$$

Thus, we have a *hybrid* approximation in the spirit of that proposed by Kurth and Perdew [111] for the RPA-ACFD with the further advantage that G_0W_0 and RPA require similar computational effort but $\Delta\varepsilon_{\mathrm{c}}^{GW}\left(r_{\mathrm{s}}\right) < \Delta\varepsilon_{\mathrm{c}}^{\mathrm{RPA}}\left(r_{\mathrm{s}}\right)$. As we can see in Fig. 5.9, the absolute xc energy obtained in this way (which we label as $G_0W_0+\Delta$) is in broad correspondence with the LDA energy, but the binding energy is slightly altered, and, of course, the van der Waals bonding is present. Although, as commented above, these corrections should be described through the implementation of vertex diagrams, this is a first step towards the inclusion of short-ranged correlations.

---

[16] This parameterisation has been obtained by comparing the G_0W_0 and QMC correlation energies in the range $r_{\mathrm{s}} \in [1, 20]$.

5.4.3   Generalised KS Schemes and Self-energy Models

We have seen that many-body-based methods provide an *ab-initio* way to treat the Coulomb correlation in an N electron system without the expensive cost of QMC calculations. However, they are computationally more demanding than routine LDA-KS calculations and, hence, the feasibility of their application to complex systems is unclear, especially in the context of *ab-initio* molecular dynamics calculations, where many total-energy evaluations are required. As described in Sect. 5.3, the main problem when constructing approximations to $E_{xc}[n]$ is related to its inherent non-analytical character which is due to the specific way in which the KS mapping between the real and the fictitious systems is done. However, this *is not* the only possible realisation of DFT and recently, new DFT methods have been proposed [112,113]. In these *generalised* Kohn-Sham schemes (GKS) the actual electron system is mapped onto a fictitious one in which particles move in an effective *non-local* potential. As a result of this, it is possible to describe structural properties at the same (or better) level than LDA/GGA but improving on its description of quasiparticle properties.

Specifically, as shown in [63–65], pathologies of the exact KS functional such as the band-gap discontinuity and the xc field may be understood as arising when one transforms a MBPT description, with a non-pathological but non-local self-energy operator, into the KS system with its local potential. In this sense, a non-local xc potential should be more amenable to accurate approximation as an explicit functional of the density.

The GKS approximation proposed by Sánchez-Friera and Godby [114] relies on the use of a jellium-like self-energy to describe the xc effects of inhomogeneous systems:

$$\Sigma^0\left(\boldsymbol{r},\boldsymbol{r}';\omega\right) = \frac{v_{xc}^{LDA}\left(\boldsymbol{r}\right) + v_{xc}^{LDA}\left(\boldsymbol{r}'\right)}{2} g\left(\left|\boldsymbol{r}-\boldsymbol{r}'\right|;n_0\right), \qquad (5.37)$$

where $g\left(r,n\right)$ is a parameterised spreading function and n_0 is the mean density of the system. This approximation is suggested by the fact that the frequency-dependence of the self-energy is weak for occupied states, and that for several semiconductors Σ has been shown to be almost spherical and to have the same range than the self-energy of a jellium system with the same mean density [115]. Since (5.37) is real and static, it defines a fictitious system that in this GKS scheme replaces the standard KS non-interacting one, and whose mass operator (Hartree potential plus self-energy) is

$$M_S\left(\boldsymbol{r},\boldsymbol{r}'\right) = \int d\boldsymbol{r}'' \frac{n\left(\boldsymbol{r}''\right)}{\left|\boldsymbol{r}-\boldsymbol{r}''\right|}\delta\left(\boldsymbol{r}-\boldsymbol{r}'\right) + \Sigma^0\left(\boldsymbol{r},\boldsymbol{r}';\omega\right). \qquad (5.38)$$

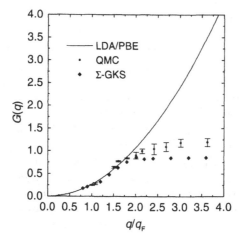

Fig. 5.10. Local field factor $G(q)$ for the linear response of jellium at $r_s = 2$ in the Σ-GKS scheme compared to the QMC results and the LDA/GGA(PBE). After Sánchez-Friera and Godby [114]

The total energy of the actual system is then approximated by

$$E^{(0)} = T_S + \frac{1}{2} \sum_n^{\text{occ}} \int d\mathbf{r}\, d\mathbf{r}'\, \phi_n^*(\mathbf{r})\, M_S(\mathbf{r}, \mathbf{r}')\, \phi_n(\mathbf{r}) \qquad (5.39)$$

$$+ E_{\text{ss}}[n] + \int d\mathbf{r}\, n(\mathbf{r})\, v_{\text{ion}}(\mathbf{r})\ ,$$

where $E_{\text{ss}}[n]$ is a local functional that is added so that the model is exact in the homogeneous limit. By minimising (5.39) with respect to variations of the one-particle wavefunctions $\phi_n(\mathbf{r})$ a set of self-consistent KS-like equations are obtained. The simple form of the non-locality ensures computational efficiency.

This approximation, labelled as Σ-GKS, shows performance similar to the LDA in the calculation of structural properties of silicon. The most striking feature of this new scheme is the significant improvement when calculating the local field factor of the HEG $G(q)$ with respect to local and semi-local approaches. As it is depicted in Fig. 5.10, Σ-GKS fits very well the QMC data by Moroni *et al.* [116] also at large values of the wavevector, where the LDA and the GGA by Perdew, Burke, and Ernzerhof (PBE) [117] fail badly. These results, as well as the efficiency of this new GKS scheme, opens the prospect of a new class of methods that yield accurate total energies and realistic QP spectra through avoiding the pathological aspects of the Kohn-Sham xc energy functional, while retaining computational efficiency comparable to Kohn-Sham DFT.

5.5   Concluding Remarks

In this chapter we have contrasted two approaches to the many-body problem. In Kohn-Sham DFT, fictitious non-interacting electrons move in an effective potential, part of which – the exchange-correlation potential – arises from a functional that in its exact form exhibits complex and sometimes pathological dependence on the electron density, but that in practice is generally approximated by an explicit functional of the density which fails to describe these pathologies. In many-body perturbation theory, electrons move in a spatially non-local, energy-dependent potential which arises from a perturbation expansion which may be evaluated to a chosen order. The calculations are more expensive because of the non-locality and energy-dependence of the self-energy operator, and the need to evaluate a complex expression to obtain it, but the pathologies of the Kohn-Sham functional have no counterparts in MBPT.

We have shown how each theory may be used to illuminate and develop the other. MBPT may be used to exhibit and explore the pathologies of Kohn-Sham DFT with the aim of appreciating the physical effects that are incorrectly described by a given approximate density based functional, and identifying prospects for addressing them in other ways (such as with the explicit wavefunction-dependence of exact-exchange KS-DFT, or current-based functionals). On the other hand, the technology of *ab-initio* DFT calculations has been adapted for MBPT, both for the calculation of quasiparticle and other spectral properties, and, more recently, for ground-state total energy calculations. Also, we have described the possibility of methods intermediate between KS-DFT and MBPT, generalised Kohn-Sham density functional theories, in which the computational efficiency of a density based functional is combined with the physically important non-locality of the self-energy operator.

References

1. P. Hohenberg and W. Kohn, Phys. Rev. **136**, B864 (1964).
2. W. Kohn and L. Sham, Phys. Rev. **140**, A1133 (1965).
3. P. Nozieres, *Theory of Interacting Fermi Systems* (Benjamin, New York, 1964).
4. L. Hedin and S. Lundqvist, in *Solid State Physics*, edited by H. Ehrenreich, F. Seitz, and D. Turnbull (Academic Press, New York, 1969), Vol. 23, p. 1.
5. A. L. Fetter and J. D. Walecka, *Quantum Theory of Many-Particle Systems*, (McGraw-Hill, New York, 1971).
6. G. D. Mahan, *Many-Particle Physics*(Plenum, New York, 1990)
7. E. K. U. Gross, E. Runge, and O. Heinonen, *Many-Particle Theory* (Adam Hilger, Bristol, 1991).
8. L. Hedin, Phys. Rev. **139**, A796 (1965).
9. F. Aryasetiawan and O. Gunnarsson, Rep. Prog. Phys. **61**, 3 (1998).

10. W. G. Aulbur, L. Jönsson, and J. W. Wilkins, in *Solid State Physics*, edited by H. Ehrenreich (Academic Press, Orlando, 2000), Vol. 54, p 1.
11. P. M. Echenique, J. M. Pitarke, E. Chulkov, and A. Rubio, Chem. Phys. **251**, 1 (2000).
12. P. García-González and R. W. Godby, Comp. Phys. Comm. **137**, 108 (2001).
13. G. Onida, L. Reining, and A. Rubio, Rev. Mod. Phys. **74**, 601 (2002).
14. C.-O. Almbladh and L. Hedin, in *Handbook on Synchrotron Radiation*, edited by E. E. Koch (North-Holland, 1983), Vol 1.
15. D. Pines, *Elementary excitations in solids*(Benjamin, New York, 1964).
16. B. I. Lundqvist, Phys. Kondens. Mater. **6**, 193 (1967). W. von der Linden and P. Horsch, Phys. Rev. B **37**, 8351 (1988). G. E. Engel and B. Farid, Phys. Rev. B **47**, 15931 (1993).
17. H. N. Rojas, R. W. Godby, and R. Needs, Phys. Rev. Lett. **74**, 1827 (1995).
18. M. M. Rieger, L. Steinbeck, I. D. White, N. H. Rojas, and R. W. Godby, Comp. Phys. Comm. **117**, 211 (1999); L. Steinbeck, A. Rubio, L. Reining, M. Torrent, I. D. White, and R. W. Godby, *ibid* **125**, 105 (2000).
19. M. S. Hybertsen and S. G. Louie, Phys. Rev. Lett. **55**, 1418 (1985).
20. S. Saito, S. B. Zhang, S. G. Louie, and M. L. Cohen, Phys. Rev. B **40**, 3643 (1989).
21. G. Onida, L. Reining, R. W. Godby, R. Del Sole, and W. Andreoni, Phys. Rev. Lett. **75**, 818 (1995).
22. P. Rinke, P. García-González, and R. W. Godby (unpublished).
23. R. Del Sole, L. Reining, and R. W. Godby, Phys. Rev. B **49**, 8024 (1994).
24. M. Hindgren and C.-O. Almbladh, Phys. Rev. B **56**, 12832 (1997).
25. P. García-González and R. W. Godby (unpublished).
26. D. Langreth, Phys. Rev. B **1**, 471 (1970).
27. F. Aryasetiawan, L. Hedin, and K. Karlsson, Phys. Rev. Lett. **77**, 2268 (1996).
28. E. Jensen and E. W. Plummer, Phys. Rev. Lett. **55**, 1912 (1985).
29. B. S. Itchkawitz, I.-W. Lyo, and E. W. Plummer, Phys. Rev. B **41**, 8075 (1990).
30. J. E. Northrup, M. S. Hybertsen, and S. G. Louie, Phys. Rev. Lett. **59**, 819 (1989).
31. W.-D. Schöne and A. G. Eguiluz, Phys. Rev. Lett. **81**, 1662 (1998).
32. G. D. Mahan and B. E. Sernelius, Phys. Rev. Lett. **62** 2718 (1989).
33. H. Yasuhara, S. Yoshinaga, and M. Higuchi, Phys. Rev. Lett. **83**, 3250 (1999).
34. K. W.-K. Shung and G. D. Mahan, Phys. Rev. B **38**, 3856 (1988).
35. Y. Takada, Phys. Rev. Lett. **87**, 226402 (2001).
36. U. von Barth and B. Holm, Phys. Rev. B **54**, 8411 (1996).
37. B. Holm and U. von Barth, Phys. Rev. B **57**, 2108 (1998).
38. N. Hamada, M. Hwang, and A. J. Freeman, Phys. Rev. B **41**, 3620 (1990).
39. B. Arnaud and M. Alouani, Phys. Rev. B 562, 4464 (2000).
40. W. Ku and A. G. Eguiluz, Phys. Rev. Lett. **89**, 126401 (2002).
41. P. A. Bobbert and W. van Haeringen, Phys. Rev. B **49**, 10326 (1994).
42. O. Gunnarsson, M. Jonson, and B. I. Lundqvist, Phys. Rev. B **20**, 3136 (1979).
43. E. Chacón and P. Tarazona, Phys. Rev. B **37**, 4013 (1988).
44. J. A. Alonso and N. A. Cordero, in *Recent Developments and Applications of Modern Density Functional Theory*, edited by J. M. Seminario (Elsevier, Amsterdam, 1996) and references therein.
45. J. D. Talman and W. F. Shadwich, Phys. Rev. A **14**, 36 (1976).

46. V. Sahni, J. Gruenebaum, and J. P. Perdew, Phys. Rev. B **26**, 4371 (1982).
47. J. P. Perdew and M. Levy, Phys. Rev. Lett. **51**,1881 (1983).
48. L. J. Sham and M. Schlüter, Phys. Rev. Lett. **51**, 1888 (1983).
49. R. G. Parr and W. Yang, *Density Functional Theory of Atoms and Molecules* (Oxford University Press, New York, 1989).
50. R. M. Dreizler and E. K. U. Gross, *Density Functional Theory: An Approach to the Quantum Many-Body Problem* (Springer-Verlag, Berlin, 1990).
51. C.-O. Almbladh and U. von Barth, Phys. Rev. B **31**, 3231 (1985).
52. R. W. Godby, L. J. Sham, and M. Schlüter, Phys. Rev. Lett. **56**, 2415 (1986).
53. R. W. Godby, L. J. Sham, and M. Schlüter, Phys. Rev. B **36**, 6497 (1987).
54. L. J. Sham and M. Schlüter, Phys. Rev. B **32**, 3883 (1985).
55. W. Knorr and R. W. Godby, Phys. Rev. Lett. **65**, 639 (1992); Phys. Rev. B **50**, 1779 (1994).
56. O. Gunnarsson and K. Schönhammer, Phys. Rev. Lett. **56**, 1968 (1986).
57. M. Städele, M. Mouraka, J. A. Majewski, P. Vogl, and A. Görling, Phys. Rev. B **59**, 10031 (1999).
58. W. G. Aulbur, M. Städele, and A. Görling, Phys. Rev. B **62**, 7121 (2000).
59. A. Görling and M. Levy, Phys. Rev. A **52**, 4493 (1995); A. Görling, *ibid.* **54**, 3912 (1996).
60. C.-O. Almbladh and U. von Barth, in *Density Functional Methods in Physics*, edited by R. M. Dreizler and J. da Providência (Plenum Press, New York, 1985), p. 209.
61. J. Katriel and E. R. Davidson, Proc. Nat. Acad. Sci. USA **77**, 4403 (1980).
62. R.W. Godby, L.J. Sham and M. Schlüter, Phys. Rev. Lett. **65**, 2083 (1990).
63. R.W. Godby and L.J. Sham, Phys. Rev. B **49**, 1849 (1994).
64. X. Gonze, P. Ghosez and R. W. Godby, Phys. Rev. Lett. **74**, 4035 (1995).
65. X. Gonze, P. Ghosez and R. W. Godby, Phys. Rev. Lett. **78**, 294 (1997).
66. X. Gonze, P. Ghosez and R. W. Godby, Phys. Rev. Lett. **78**, 2029 (1997).
67. P. Ghosez, X. Gonze and R. W. Godby, Phys. Rev. B **56**, 12811 (1997).
68. P. L. de Boeij, F. Kootstra, J. A. Berger, R. van Leeuwen, and J. G. Snijders, J. Chem. Phys. 15, 1995 (2001).
69. R. O. Jones and O. Gunnarsson, Rev. Mod. Phys **61**, 689 (1989).
70. H. Sawada and K. Terakura, Phys. Rev. B **58**, 6831 (1998).
71. J. C. Grossman, L. Mitas, and K. Raghavachari, Phys. Rev. Lett. **75**, 3870 (1995).
72. P. R. C. Kent, M. D. Towler, R. J. Needs, and G. Rajagopal, Phys. Rev. B **62**, 15394 (2000).
73. J. C. Grossman and L. Mitas, Phys. Rev. Lett. **79**, 4353 (1997).
74. W. M. C. Foulkes, L. Mitas, R. J. Needs, and G. Rajagopal, Rev. Mod. Phys. **73**, 17 (2000).
75. D. C. Langreth and J. P. Perdew, Solid State Commun. **17**, 1425 (1975); Phys. Rev. B **15**, 2884 (1977).
76. W. Kohn and P. Vashishta, in *Theory of the Inhomogeneous Electron Gas*, edited by S. Lundqvist and N. H. March (Plenum, New York, 1983).
77. V. M. Galitskii and A. B. Migdal, Zh. Éksp. Teor. Fiz. **34**, 139 (1958) [Sov. Phys. JETP **7**, 96 (1958)].
78. J. M. Luttinger and J. C. Ward, Phys. Rev. **118**, 1417 (1960).
79. C.-O. Almbladh, U. von Barth, and R. van Leeuwen, Int. J. Mod. Phys. B **13**, 535 (1999).

80. E. Runge and E. K. U. Gross, Phys. Rev. Lett. **52**, 997 (1984).
81. E. K. U. Gross, C. A. Ullrich, and U. J. Grossman, in *Density Functional Theory*, edited by E. K. U. Gross and R. M. Dreizler (Plenum, New York, 1995).
82. M. Lein, E. K. U. Gross, and J. P. Perdew, Phys. Rev. B **61**, 13431 (2000) and references therein.
83. J. F. Dobson and J. Wang, Phys. Rev. Lett. **82**, 2123 (1999); Phys. Rev. B **62**, 10038 (2000).
84. J. M. Pitarke and A. Eguiluz, Phys. Rev. B **57**, 6329 (1998); **63**, 045116 (2001).
85. F. Furche, Phys. Rev. B **64**, 195120 (2001).
86. M. Fuchs and X. Gonze, Phys. Rev. B **65**, 235109 (2002).
87. M. Hindren, PhD. thesis, Lund University, 1997.
88. N. E. Dahlen and U. von Barth (unpublished).
89. F. Aryasetiawan, T. Mikaye, and K. Terakura, Phys. Rev. Lett. **88**, 166401 (2002).
90. B. I. Lundqvist and V. Samathiyakanit, Phys. Kondens. Mater. **9**, 231 (1969).
91. B. Holm, Phys. Rev. Lett. **83**, 788 (1999).
92. P. García-González and R. W. Godby, Phys. Rev. B **63**, 075112 (2001).
93. R. F. Bishop and K. H. Lührmann, Phys. Rev. B **26**, 5523 (1982).
94. D. M. Ceperley and B. J. Adler, Phys. Rev. Lett. **45**, 566 (1980).
95. G. Ortiz, M. Harris, and P. Ballone, Phys. Rev. Lett. **82**, 5317 (1999); G. Ortiz and P. Ballone, Phys. Rev. B **50**, 1391 (1994).
96. G. Baym and L. P. Kadanoff, Phys. Rev. **124**, 287 (1961).
97. G. Baym, Phys. Rev. **127**, 1391 (1962).
98. C. Verdozzi, R. W. Godby and S. Holloway, Phys. Rev. Lett. **74**, 2327 (1995).
99. A. Schindlmayr, T. J. Pollehn and R. W. Godby, Phys. Rev. B **58** 12684-90 (1998).
100. T. J. Pollehn, A. Schindlmayr and R. W. Godby, J. Phys.: Condens. Matter **10** 1273-1283 (1998).
101. B. Holm and U. von Barth (unpublished).
102. B. Holm, PhD thesis, Lund University (1997).
103. A. Schindlmayr, Phys. Rev. B **56**, 3528 (1997).
104. M. M. Rieger and R. W. Godby, Phys. Rev. B **58**, 1343 (1998).
105. A. Schindlmayr, P. García-González, and R. W. Godby, Phys. Rev. B **64**, 235106 (2001).
106. P. García-González and R. W. Godby, Phys. Rev. Lett. **88**, 056406 (2002).
107. Y.-H. Kim, I.-H. Lee, S. Nagaraja, J.-P. Leburton, R. Q. Hood, and R. M. Martin, Phys. Rev. B **61**, 5202 (2000)
108. L. Pollack and J. P. Perdew, J. Phys.: Condens. Matter **12**, 1239 (2000)
109. P. García-González, Phys. Rev. B **62**, 2321 (2000).
110. W. Kohn, Y. Meir, and D. E. Makarov, Phys. Rev. Lett. **80**, 4153 (1998).
111. S. Kurth and J. P. Perdew, Phys. Rev. B **59**, 10146 (1999).
112. A. Seidl, A. Görling, P. Vogl, J. A. Majewski, and M. Levy, Phys. Rev. B **53**, 3764 (1996).
113. G. E. Engel, Phys. Rev. Lett. **78**, 3515 (1997).
114. P. Sánchez-Friera and R. W. Godby, Phys. Rev. Lett. **85**, 5611 (2000).
115. R. W. Godby, M. Schlüter, and L. J. Sham, Phys. Rev. B **37**, 10159 (1988).
116. S. Moroni, D. M. Ceperley, and G. Senatore, Phys. Rev. Lett. **75**, 689 (1995).
117. J. P. Perdew, K. Burke, and M. Ernzerhof, Phys. Rev. Lett. **77**, 3685 (1996).

6 A Tutorial on Density Functional Theory

Fernando Nogueira*, Alberto Castro[†],
and Miguel A.L. Marques[‡]

* Departamento de Física,
Universidade de Coimbra,
Rua Larga, 3004 – 516, Coimbra, Portugal
fnog@teor.fis.uc.pt

[†] Departamento de Física Teórica,
Universidad de Valladolid,
E–47011 Valladolid, Spain
alberto.castro@tddft.org

[‡] Donostia International Physics Center (DIPC),
P. Manuel Lardizábal 4,
20080 San Sebastián, Spain
marques@tddft.org

Fernando Nogueira

6.1 Introduction

The success of density functional theory (DFT) is clearly demonstrated by the overwhelming amount of research articles describing results obtained within DFT that were published in the last decades. There is also a fair number of books reviewing the basics of the theory and its extensions (e.g., the present volume, [1] and [2]). These works fall mainly into three classes: those dealing with the theory (proposing extensions, new functionals, etc.), those concerned with the technical aspects of the numerical implementations, and others – the vast majority – presenting results. In our opinion, any scientist working in the field should have a sound knowledge of the three classes. For example, a theorist developing a new functional should be aware of the difficulties in implementing it. Or the applied scientist, performing calculations on specific systems, should know the limitations of the theory and of the numerical implementation she/he is using. The goal of this chapter is to supply the beginner with a brief pedagogical overview of DFT, combining the above-mentioned aspects. However, we will not review its foundations – we redirect the reader to the chapter of J. Perdew and S. Kurth that opens this book. Obviously, we will not be able to provide many details, but we hope that the beginner obtains a general impression of the capabilities and limitations of DFT.

This chapter is written in the form of a tutorial, combining basic theoretical and numerical aspects with specific examples, running from the simplest hydrogen atom to more complex molecules and solids. For the examples we used only freely available codes [3], so that the reader may easily reproduce

the calculations. All input and output files can be found in the web site
http://www.tddft.org/DFT2001/. The chapter follows closely the outline
of the practical sessions held at Caramulo, during the DFT2001 summer
school. Some theoretical or numerical aspects that were required in the practical sessions were, however, not covered by any of the lectures in Caramulo
(e.g., pseudo-potentials). To fill this gap we provide in this chapter a brief
account of some of them. We do not intend to discuss every possible numeric
implementation of DFT. In particular, we do not include any explicit example of a localized basis set DFT calculation. Neither do we intend to present
an extensive survey of the numerical aspects of each technique. We expect,
however, that the technical details given are sufficient to enable the reader
to perform himself the simulations presented herein.

The outline of the chapter is the following: We start, in Sect. 6.2, by
giving a technical overview on how to solve the Kohn-Sham equations. The
next section is devoted to pseudo-potentials, an essential ingredient of many
DFT calculations. In Sect. 6.4 we present our first test case, namely atoms,
before we proceed to some plane-wave calculations in Sect. 6.5. The final
example, methane calculated using a real-space implementation, is presented
in Sect. 6.6. We will use atomic units throughout this chapter, except when
explicitly stated otherwise.

6.2 Solving the Kohn–Sham Equations

6.2.1 Generalities

It is usually stated that the Kohn-Sham equations are "simple" to solve. By
"simple" it is meant that for a given system, e.g., an atom, a molecule, or a
solid, the computational effort to solve the Kohn-Sham equations is smaller
than the one required by the traditional quantum chemistry methods, like
Hartree-Fock (HF) or configuration interaction (CI)[1]. But it does not mean
that it is easy or quick to write, or even to use, a DFT based computer
program. Typically, such codes have several thousand lines (for example, the
ABINIT [4] package – a plane-wave DFT code – recently reached 200,000
lines) and hundreds of input options. Even writing a suitable input file is
often a matter of patience and experience.

In spite of their differences, all codes try to solve the Kohn-Sham equations

$$\left[-\frac{\nabla^2}{2} + v_{KS}[n](\boldsymbol{r}) \right] \varphi_i(\boldsymbol{r}) = \varepsilon_i \varphi_i(\boldsymbol{r}) \ . \tag{6.1}$$

[1] This statement has to be taken with care, for it certainly depends on the approximation for the exchange-correlation potential. For example, it holds when
using the local-density approximation or any of the generalized gradient approximations. However, if we use the exact exchange functional, the calculations are
at least as computationally demanding as in Hartree-Fock.

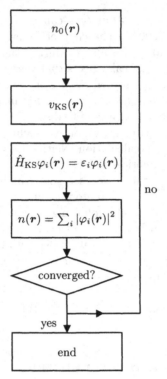

Fig. 6.1. Flow-chart depicting a generic Kohn-Sham calculation

The notation $v_{KS}[n]$ means that the Kohn-Sham potential, v_{KS}, has a functional dependence on n, the electronic density, which is defined in terms of the Kohn-Sham wave-functions by

$$n(\boldsymbol{r}) = \sum_i^{\text{occ}} |\varphi_i(\boldsymbol{r})|^2 \; . \tag{6.2}$$

The potential v_{xc} is defined as the sum of the external potential (normally the potential generated by the nuclei), the Hartree term and the exchange and correlation (xc) potential

$$v_{KS}[n](\boldsymbol{r}) = v_{\text{ext}}(\boldsymbol{r}) + v_{\text{Hartree}}[n](\boldsymbol{r}) + v_{xc}[n](\boldsymbol{r}) \; . \tag{6.3}$$

Due to the functional dependence on the density, these equations form a set of nonlinear coupled equations. The standard procedure to solve it is iterating until self-consistency is achieved. A schematic flow chart of the scheme is depicted in Fig. 6.1. Usually one supplies some model density, $n_0(\boldsymbol{r})$, to start the iterative procedure. In principle, any positive function normalized to the total number of electrons would work, but using an educated guess for $n_0(\boldsymbol{r})$ can speed-up convergence dramatically. For example, in a molecular or a solid-state system one could construct $n_0(\mathbf{r})$ from a sum of atomic densities

$$n_0(\mathbf{r}) = \sum_\alpha n_\alpha(\mathbf{r} - \mathbf{R}_\alpha) , \qquad (6.4)$$

where $\mathbf{R}_\alpha$ and n_α represent the position and atomic density of the nucleus α. For an atom, a convenient choice is the Thomas-Fermi density.

We then evaluate the Kohn-Sham potential (see 6.3) with this density. Each of the components of v_{KS} is calculated separately and each of them poses a different numerical problem. The external potential is typically a sum of nuclear potentials centered at the atomic positions,

$$v_{ext}(\mathbf{r}) = \sum_\alpha v_\alpha(\mathbf{r} - \mathbf{R}_\alpha) . \qquad (6.5)$$

In some applications, v_α is simply the Coulomb attraction between the bare nucleus and the electrons, $v_\alpha(\mathbf{r}) = -Z_\alpha/r$, where Z_α is the nuclear charge. In other cases the use of the Coulomb potential renders the calculation unfeasible, and one has to resort to pseudo-potentials (see Sect. 6.3.1).

The next term in v_{KS} is the Hartree potential,

$$v_{Hartree}(\mathbf{r}) = \int d^3 r' \frac{n(\mathbf{r}')}{|\mathbf{r} - \mathbf{r}'|} . \qquad (6.6)$$

There are several different techniques to evaluate this integral, either by direct integration (as it is done when solving the atomic Kohn-Sham equations), or by solving the equivalent differential (Poisson's) equation,

$$\nabla^2 v_{Hartree}(\mathbf{r}) = -4\pi n(\mathbf{r}) . \qquad (6.7)$$

As the choice of the best technique depends on the specific problem, we defer further discussion on the Hartree term to Sects. 6.2.2–6.2.4.

Finally, we have the xc potential, which is formally defined through the functional derivative of the xc energy,

$$v_{xc}(\mathbf{r}) = \frac{\delta E_{xc}}{\delta n(\mathbf{r})} . \qquad (6.8)$$

Perhaps more than a hundred approximate xc functionals have appeared in the literature over the past 30 years. The first to be proposed and, in fact, the simplest of all, is the local-density approximation (LDA). It is written as

$$E_{xc}^{LDA} = \int d^3 r \; \varepsilon^{HEG}(n)\big|_{n=n(\mathbf{r})} \quad ; \quad v_{xc}^{LDA}(\mathbf{r}) = \frac{d}{dn} \varepsilon^{HEG}(n)\big|_{n=n(\mathbf{r})} , \qquad (6.9)$$

where $\varepsilon^{HEG}(n)$ stands for the xc energy per unit volume of the homogeneous electron gas (HEG) of (constant) density n. Note that $\varepsilon^{HEG}(n)$ is a simple *function* of n, which was tabulated for several densities using Monte Carlo methods by Ceperley and Alder [5]. A number of different parameterizations exist for this function, like the PZ81 [6] and PW92 [7]. It is clear from these considerations that evaluating the LDA xc potential is as simple (and fast) as evaluating any rational or transcendental function. In the case of the generalized gradient approximations (GGA) the functional has a similar form, but now ε does not depend solely on the density n, but also on its gradient ∇n. The evaluation of the GGA xc potential is also fairly straightforward. Finally, we mention the third generation of density functionals, the orbital-dependent functionals (see the chapter by E. Engel in this book) like the exact exchange (EXX). In order to obtain the xc potential in this case, one is required to solve an integral equation[2]. This equation is quite complex, and its solution can easily become the most time-consuming part of the Kohn-Sham calculation. We should also notice that functionals like the EXX involve the evaluation of the so-called Coulomb integrals. These two-center integrals, that also appear in Hartree-Fock theory, pose another difficult problem to the computational physicist or chemist.

Now that we have the Kohn-Sham potential, we can solve the Kohn-Sham equation (6.1). The goal is to obtain the p lowest eigenstates of the Hamiltonian H_{KS}, where p is half the number of electrons (for a spin-unpolarized calculation). For an atom, or for any other case where the Kohn-Sham equations can be reduced to a one-dimensional differential equation, a very efficient integration method is commonly employed (see below). In other cases, when using basis sets, plane-waves, or real-space methods, one has to diagonalize the Hamiltonian matrix, $\hat{H}_{KS}$. We have to keep in mind that fully diagonalizing a matrix is a q^3 problem, where q is the dimension of the matrix (which is roughly proportional to the number of atoms in the calculation). Moreover, the dimension of the Hamiltonian is sometimes of the order of $10^6 \times 10^6 = 10^{12}$ elements[3]. It is clearly impossible to store such a matrix in any modern computer. To circumvent these problems, one usually resorts to iterative methods. In these methods it is never necessary to write the full Hamiltonian – the knowledge of how $\hat{H}_{KS}$ applies to a test wave-function is sufficient. These methods also scale much better with the dimension of the matrix. Nonetheless, diagonalizing the Kohn-Sham Hamiltonian is usually the most time-consuming part of an ordinary Kohn-Sham calculation.

We have now all the ingredients to obtain the electronic density from (6.2). The self-consistency cycle is stopped when some convergence criterion is reached. The two most common criteria are based on the difference of total energies or densities from iteration i and $i - 1$, i.e., the cycle is stopped when $\left|E^{(i)} - E^{(i-1)}\right| < \eta_E$ or $\int d^3r \left|n^{(i)} - n^{(i-1)}\right| < \eta_n$, where $E^{(i)}$ and $n^{(i)}$ are the total energy and density at iteration i, and η_E and η_n are user defined tolerances. If, on the contrary, the criteria have not been fulfilled, one restarts

---

[2] Or choose to apply the Krieger, Lee and Iafrate approximation [8].

[3] However, $\hat{H}_{KS}$ is usually a very sparse matrix. For example, in a typical real-space calculation only less than .1 % of the elements of $\hat{H}$ are different from 0.

the self-consistency cycle with a new density. It could simply be the output density of the previous cycle – unfortunately this would almost certainly lead to instabilities. To avoid them, one usually mixes this output density with densities from previous iterations. In the simplest scheme, linear mixing, the density supplied to start the new iteration, $n^{(i+1)}$ is a linear combination of the density obtained from (6.2), n', and the density of the previous iteration, $n^{(i)}$,

$$n^{(i+1)} = \beta n' + (1 - \beta)n^{(i)} , \tag{6.10}$$

where the parameter β is typically chosen to be around 0.3. More sophisticated mixing schemes have been proposed (e.g., Anderson or Broyden mixing [9–13]), in which $n^{(i+1)}$ is an educated extrapolation of the densities of several previous iterations.

At the end of the calculation, we can evaluate several observables, the most important of which is undoubtedly the total energy. From this quantity, one can obtain, e.g., equilibrium geometries, phonon dispersion curves, or ionization potentials. In Kohn-Sham theory, the total energy is written as

$$E = -\sum_i^{\text{occ}} \int d^3r \, \varphi_i^*(\mathbf{r}) \frac{\nabla^2}{2} \varphi_i(\mathbf{r}) + \int d^3r \, v_{\text{ext}}(\mathbf{r}) n(\mathbf{r}) + $$
$$+ \frac{1}{2} \int d^3r \int d^3r' \frac{n(\mathbf{r})n(\mathbf{r}')}{|\mathbf{r} - \mathbf{r}'|} + E_{\text{xc}} , \tag{6.11}$$

where the terms are respectively the non-interacting (Kohn-Sham) kinetic energy, the external potential, the Hartree and the xc energies. This formula can be further simplified by using the Kohn-Sham equation, (see 6.1), to yield

$$E = \sum_i^{\text{occ}} \varepsilon_i - \int d^3r \left[\frac{1}{2} v_{\text{Hartree}}(\mathbf{r}) + v_{\text{xc}}(\mathbf{r}) \right] n(\mathbf{r}) + E_{\text{xc}} . \tag{6.12}$$

This is the formula implemented in most DFT codes. Note that, when performing geometry optimization or nuclear dynamics, one needs to add to the total energy a repulsive Coulomb term that accounts for the interactions between the ions

$$E_{\text{nn}} = \sum_{\alpha,\beta} \frac{Z_\alpha Z_\beta}{|\mathbf{R}_\alpha - \mathbf{R}_\beta|} . \tag{6.13}$$

Calculating the sum over all atoms is fairly straightforward for finite systems, but non-trivial for extended systems: As the Coulomb interaction is very long ranged, the (infinite) sum in (6.13) is very slowly convergent. There is, however, a technique due to Ewald that allows us to circumvent this problem and evaluate (6.13) (see Sect. 6.2.3).

6.2.2   Atoms

In order to solve the Kohn-Sham equations (6.1) for atoms, one normally performs a spherical averaging of the density[4]. This averaging leads to a spherically symmetric Kohn-Sham potential. The Hartree potential is then trivially evaluated as

$$v_{\text{Hartree}}(r) = \frac{4\pi}{r} \int_0^r \mathrm{d}r' \, r'^2 n(r') + 4\pi \int_r^\infty \mathrm{d}r' \, r' n(r') \, . \tag{6.14}$$

and the Kohn-Sham wave-functions can be written as the product of a radial wave-function, $R_{nl}(r)$, and a spherical harmonic, $Y_{lm}(\theta, \phi)$:

$$\varphi_i(\boldsymbol{r}) = R_{nl}(r) Y_{lm}(\theta, \phi) \, . \tag{6.15}$$

The wave-functions are labeled using the traditional atomic quantum numbers: n for the principal quantum number and l, m for the angular momentum. The Kohn-Sham equation then becomes a "simple" one-dimensional second-order differential equation

$$\left[-\frac{1}{2} \frac{\mathrm{d}^2}{\mathrm{d}r^2} - \frac{1}{r} \frac{\mathrm{d}}{\mathrm{d}r} + \frac{l(l+1)}{2r^2} + v_{\text{KS}}(r) \right] R_{nl}(r) = \varepsilon_{nl} R_{nl}(r) \, , \tag{6.16}$$

that can be transformed into two coupled first-order differential equations

$$\frac{\mathrm{d}f_{nl}(r)}{\mathrm{d}r} = g_{nl}(r)$$

$$\frac{\mathrm{d}g_{nl}(r)}{\mathrm{d}r} + \frac{2}{r} g_{nl}(r) - \frac{l(l+1)}{r^2} f_{nl}(r) + 2 \left\{ \varepsilon_{nl} - v_{\text{KS}}(r) \right\} f_{nl}(r) = 0 \, , \tag{6.17}$$

where $f_{nl}(r) \equiv R_{nl}(r)$.

When $r \to \infty$, the coupled equations become

$$\frac{\mathrm{d}f_{nl}(r)}{\mathrm{d}r} = g_{nl}(r)$$

$$\frac{\mathrm{d}g_{nl}(r)}{\mathrm{d}r} + 2\varepsilon_{nl} f_{nl}(r) \simeq 0 \, , \tag{6.18}$$

provided that the Kohn-Sham potential goes to zero at large distances from the atom (which it does, see Fig. 6.6). This indicates that the solutions of (6.17) should behave asymptotically as

$$f_{nl}(r) \overset{r \to \infty}{\Longrightarrow} e^{-\sqrt{-2\varepsilon_{nl}}r} \tag{6.19}$$

$$g_{nl}(r) \overset{r \to \infty}{\Longrightarrow} -\sqrt{-2\varepsilon_{nl}} f_{nl}(r) \, .$$

---

[4] Although the assumption of a spherically symmetric potential (density) is only strictly valid in a closed shell system, the true many-body potential is indeed spherically symmetric. For open shell systems this assumption implies an identical filling of all degenerate atomic orbitals.

At the origin ($r \to 0$) the solutions are of the form

$$f_{nl}(r) \xrightarrow{r \to 0} A r^\alpha \tag{6.20}$$

$$g_{nl}(r) \xrightarrow{r \to 0} B r^\beta .$$

Substituting (6.20) into (6.17) gives $B = lA$, $\alpha = l$, and $\beta = l - 1$.

For a fixed ε_{nl} and A it is a simple task to integrate (6.17) from $r = 0$ to ∞ using (6.20) to provide the initial values. However, if ε_{nl} is not an eigenvalue of (6.16), the solution will diverge (i.e., it will not obey boundary conditions at infinity (6.19)). Fortunately, there is a simple procedure to obtain the ε_{nl} that yield solutions with the correct asymptotic behavior. The technique involves integrating (6.20) from $r = 0$ to a conveniently chosen point r_m (e.g., the classical turning point), and at the same time integrating (6.20) starting from a point very far away ("practical infinity", r_∞) to r_m. From the two values of $f_{nl}(r_m)$ and $g_{nl}(r_m)$ obtained in this way, it is then possible to improve our estimate of ε_{nl}.

The technique for simultaneously finding the eigenvalues ε_{nl} and the wave-functions proceeds as follows:

i) Choose an arbitrary value for ε_{nl} and $f_{nl}(r_\infty)$;
ii) Calculate $g_{nl}(r_\infty)$ using the boundary conditions (6.19);
iii) Integrate (6.17) from r_∞ to r_m (to get $f_{nl}^{in}(r)$ and $g_{nl}^{in}(r)$);
iv) Choose an arbitrary value for A, calculate $B = lA$, and use the boundary conditions (6.20) to get $f_{nl}(0)$ and $g_{nl}(0)$;
v) Integrate (6.17) from 0 to r_m (to get $f_{nl}^{out}(r)$ and $g_{nl}^{out}(r)$);
vi) Calculate $\gamma = g_{nl}^{in}(r_m)/g_{nl}^{out}(r_m)$ and scale $f_{nl}^{out}(r)$ and $g_{nl}^{out}(r)$ by this factor – now $g_{nl}(r)$ is continuous at the matching point ($\tilde{g}_{nl}^{out}(r_m) \equiv \gamma g_{nl}^{out}(r_m) = g_{nl}^{in}(r_m)$) but $f_{nl}(r)$ is not;
vii) Compute $\delta(\varepsilon_{nl}) = f_{nl}^{out}(r_m) - f_{nl}^{in}(r_m)$: The zeros of this function are the eigenvalues, so one can find them using, e.g., the bisection method (one has to provide an educated guess for the minimum and maximum value of the eigenvalues).

6.2.3   Plane-Waves

To calculate the total energy of solids, a plane-wave expansion of the Kohn-Sham wave-functions is very useful, as it takes advantage of the periodicity of the crystal [14–16]. For finite systems, such as atoms, molecules and clusters, plane-waves can also be used in a super-cell approach[5]. In this method, the

---

[5] The super-cell technique is restricted in its usual form to neutral systems due to the long-range interaction between a charged cluster and its periodic images: the Coulomb energy for charged periodic systems diverges and must be removed. Some common methods used to circumvent this difficulty are: i) To introduce a compensating jellium background that neutralizes the super-cell [17]; ii) To use a cutoff in the Coulomb interaction [18]; iii) To shield each charged cluster with a spherical shell having a symmetric charge which neutralizes the super-cell and cancels the electric dipole of the charged cluster [19].

finite system is placed in a unit cell of a fictitious crystal, and this cell is made large enough to avoid interactions between neighboring cells. The Kohn-Sham equations can then be solved, for any system, in momentum space. However, for finite systems a very large number of plane-waves is needed as the electronic density is concentrated on a small fraction of the total volume of the super-cell.

The valence wave-functions of the large Z atoms oscillate strongly in the vicinity of the atomic core due to the orthogonalization to the inner electronic wave-functions. To describe these oscillations a large number of plane-waves is required, difficulting the calculation of the total energy. However, the inner electrons are almost inert and are not significantly involved in bonding. This suggests the description of an atom based solely on its valence electrons, which feel an effective potential including both the nuclear attraction and the repulsion of the inner electrons. This approximation, the pseudo-potential approximation, will be presented in more detail in Sect. 6.3.1.

When using the pseudo-potential approximation, the external potential, v_{ext}, is simply the sum of the pseudo-potentials of all the atoms in the system. If atom α is located in the unit cell at $\boldsymbol{\tau}_\alpha$ and its pseudo-potential is $w_\alpha(\boldsymbol{r}, \boldsymbol{r}')$, the external potential is

$$w(\boldsymbol{r}, \boldsymbol{r}') = \sum_{j,\alpha} w_\alpha(\boldsymbol{r} - \boldsymbol{R}_j - \boldsymbol{\tau}_\alpha, \boldsymbol{r}' - \boldsymbol{R}_j - \boldsymbol{\tau}_\alpha) , \qquad (6.21)$$

where $\boldsymbol{R}_j$ are the lattice vectors. The pseudo-potential is considered in its more general non-local form, which implies that the second term of the right-hand side of (6.11) is rewritten as

$$\int \mathrm{d}^3r \, v_{\text{ext}}(\boldsymbol{r}) n(\boldsymbol{r}) \longrightarrow \sum_{i=1}^{N} \int \mathrm{d}^3r \int \mathrm{d}^3r' \, \varphi_i(\boldsymbol{r}) w(\boldsymbol{r}, \boldsymbol{r}') \varphi_i^*(\boldsymbol{r}') . \qquad (6.22)$$

According to Bloch's theorem, the Kohn-Sham wave-functions, $\varphi_{\boldsymbol{k},n}(\boldsymbol{r})$, can be written as

$$\varphi_{\boldsymbol{k},n}(\boldsymbol{r}) = \mathrm{e}^{\mathrm{i}\boldsymbol{k}\cdot\boldsymbol{r}} \sum_{\boldsymbol{G}} c_{\boldsymbol{k},n}(\boldsymbol{G}) \mathrm{e}^{\mathrm{i}\boldsymbol{G}\cdot\boldsymbol{r}} , \qquad (6.23)$$

where $\boldsymbol{k}$ is the wave vector, n the band index, and $\boldsymbol{G}$ are the reciprocal lattice vectors. The Kohn-Sham energies are $\varepsilon_{\boldsymbol{k},n}$, and the electronic density is

$$n(\boldsymbol{r}) = \sum_{\boldsymbol{k},n} \sum_{\boldsymbol{G},\boldsymbol{G}'} f(\varepsilon_{\boldsymbol{k},n}) c_{\boldsymbol{k},n}^*(\boldsymbol{G}') c_{\boldsymbol{k},n}(\boldsymbol{G}) \mathrm{e}^{\mathrm{i}(\boldsymbol{G}-\boldsymbol{G}')\cdot\boldsymbol{r}} , \qquad (6.24)$$

where the $f(\varepsilon_{\boldsymbol{k},n})$ denote the occupation numbers. The Fourier transform of the density is

$$n(\boldsymbol{G}) = \sum_{\boldsymbol{k},n} \sum_{\boldsymbol{G}'} f(\varepsilon_{\boldsymbol{k},n}) c_{\boldsymbol{k},n}^*(\boldsymbol{G}' - \boldsymbol{G}) c_{\boldsymbol{k},n}(\boldsymbol{G}') . \qquad (6.25)$$

The sums over k are performed over all Brillouin zone vectors, but can be reduced to sums on the irreducible Brillouin zone by taking advantage of the space group of the lattice[6].

There are thus two convergence parameters that need to be fine-tuned for every calculation: the Brillouin zone sampling and a cutoff radius in reciprocal space to truncate the sums over reciprocal lattice vectors (we cannot perform infinite summations!)

The kinetic energy is rewritten as

$$T = \frac{1}{2} \sum_{k,n} \sum_{G} f(\varepsilon_{k,n}) \, |c_{k,n}(G)|^2 \, |k + G|^2 \, , \qquad (6.26)$$

and the Hartree energy is given by

$$E_{\text{Hartree}} = \frac{\Omega}{2} \sum_{G} v_{\text{Hartree}}(G) n(G) \, , \qquad (6.27)$$

where Ω is the unit cell volume and the Hartree potential, $v_{\text{Hartree}}(G)$, is obtained using Poisson's equation

$$v_{\text{Hartree}}(G) = 4\pi \frac{n(G)}{G^2} \, . \qquad (6.28)$$

The electron-ion interaction energy, (6.22), is given by

$$E_{\text{ei}} = \sum_{k,n} \sum_{G,G'} f(\varepsilon_{k,n}) c_{k,n}^*(G) c_{k,n}(G') w(k + G, k + G') \, , \qquad (6.29)$$

and the Fourier transform of the total pseudo-potential is

$$w(k + G, k + G') = \sum_{\alpha} w_{\alpha}(k + G, k + G') e^{i(G-G')\cdot \tau_{\alpha}} \, . \qquad (6.30)$$

The Fourier transform of the individual pseudo-potentials, $v_{\alpha}(k, k')$, can be written in a simple form if the separable Kleinmnan and Bylander form is used (see Sect. 6.3.8).

Both E_{ei} (due to the local part of the pseudo-potential) and the Hartree potential diverge at $G = 0$. The ion-ion interaction energy, E_{nn}, also diverges. However, the sum of these three divergent terms is a constant, if the system is electrically neutral. This constant is [14–16]

$$\lim_{G,G'\to 0} \left[\sum_{k,n} f(\varepsilon_{k,n}) c_{k,n}^*(G) c_{k,n}(G') w(k + G, k + G') + \right.$$

$$\left. + \frac{\Omega}{2} v_{\text{Hartree}}(G) n(G) \right] + E_{\text{nn}} = E_{\text{rep}} + E_{\text{Ewald}} \, , \qquad (6.31)$$

---

[6] To further simplify these sums, it is possible to do a smart sampling of the irreducible Brillouin zone, including in the sums only some special k vectors [20–23].

where

$$E_{\text{rep}} = Z_{\text{total}} \frac{1}{\Omega} \sum_{\alpha} \Lambda_{\alpha} \tag{6.32}$$

and

$$\Lambda_{\alpha} = \frac{1}{\Omega} \int d^3r \left[v_{\alpha,\text{local}}(r) + \frac{Z_{\alpha}}{r} \right] . \tag{6.33}$$

In these expressions Z_{α} is the electric charge of ion α, and $v_{\alpha,\text{local}}(r)$ is the local part of the pseudo-potential of atom α (equations (6.74) and (6.75)). The non-divergent part of the ion-ion interaction energy, E_{Ewald}, is calculated using a trick due to Ewald [24]. One separates it in two parts, one short-ranged that is summed in real space, and a long-range part that is treated in Fourier space. By performing this splitting, one transforms a slowly convergent sum into two rapidly convergent sums

$$E_{\text{Ewald}} = \frac{1}{2} \sum_{\alpha,\alpha'} Z_{\alpha} \Gamma_{\alpha,\alpha'} Z_{\alpha'} , \tag{6.34}$$

with the definition

$$\Gamma_{\alpha,\alpha'} = \frac{4\pi}{\Omega} \sum_{G \neq 0} \frac{\cos\left[G \cdot (\tau_{\alpha} - \tau_{\alpha'})\right]}{G^2} e^{-\frac{G^2}{4\eta^2}} +$$

$$+ \sum_{j} \frac{\text{erfc}\left(\eta \left| R_j + \tau_{\alpha} - \tau_{\alpha'} \right|\right)}{\left| R_j + \tau_{\alpha} - \tau_{\alpha'} \right|} - \frac{\pi}{\eta^2 \Omega} - \frac{2\eta}{\sqrt{\pi}} \delta_{\alpha\alpha'} . \tag{6.35}$$

($\text{erfc}(x)$ is the complimentary error function.) Note that this term has only to be evaluated once at the beginning of the self-consistency cycle, for it does not depend on the density. The parameter η is arbitrary, and is chosen such that the two sums converge quickly.

In momentum space, the total energy is then

$$E_{\text{tot}} = T + E'_{\text{Hartree}} + E'_{\text{ei}} + E_{\text{xc}} + E_{\text{Ewald}} + E_{\text{rep}} , \tag{6.36}$$

with the terms $G, G' = 0$ excluded from the Hartree and pseudo-potential contributions. Finally, the Kohn-Sham equations become

$$\sum_{G'} \hat{H}_{G,G'}(k) c_{k,n}(G') = \varepsilon_{k,n} c_{k,n}(G) , \tag{6.37}$$

where

$$\hat{H}_{G,G'}(k) = \frac{1}{2} |k + G|^2 \delta_{G,G'} +$$

$$+ w(k + G, k + G') + v_{\text{Hartree}}(G - G') + v_{\text{xc}}(G - G') , \tag{6.38}$$

and are solved by diagonalizing the Hamiltonian.

6.2.4   Real-Space

In this scheme, functions are not expanded in a basis set, but sampled in a real-space mesh [25]. This mesh is commonly chosen to be uniform (the points are equally spaced in a cubic lattice), although other options are possible. Convergence of the results has obviously to be checked against the grid spacing. One big advantage of this approach is that the potential operator is diagonal. The Laplacian operator entering the kinetic energy is discretized at the grid points r_i using a finite order rule,

$$\nabla^2 \varphi(r_i) = \sum_j c_j \varphi(r_j) \ . \tag{6.39}$$

For example, the lowest order rule in one dimension, the three point rule reads

$$\left. \frac{d^2}{dr^2} \varphi(r) \right|_{r_i} = \frac{1}{4} \left[\varphi(r_{i-1}) - 2\varphi(r_i) + \varphi(r_{i+1}) \right] \ . \tag{6.40}$$

Normally, one uses a 7 or 9-point rule.

Another important detail is the evaluation of the Hartree potential. It cannot be efficiently obtained by direct integration of (6.6). There are however several other options: (i) solving Poisson's equation, (6.7), in Fourier space – as in the plane-wave method; (ii) recasting (6.7) into a minimization problem and applying, e.g., a conjugate gradients technique; (iii) using multi-grid methods [25–27]. The last of the three is considered to be the most efficient technique.

In our opinion, the main advantage of real-space methods is the simplicity and intuitiveness of the whole procedure. First of all, quantities like the density or the wave-functions are very simple to visualize in real space. Furthermore, the method is fairly simple to implement numerically for 1-, 2-, or 3-dimensional systems, and for a variety of different boundary conditions. For example, one can study a finite system, a molecule, or a cluster without the need of a super-cell, simply by imposing that the wave-functions are zero at a surface far enough from the system. In the same way, an infinite system, a polymer, a surface, or bulk material can be studied by imposing the appropriate cyclic boundary conditions. Note also that in the real-space method there is only one convergence parameter, namely the grid-spacing.

Unfortunately, real-space methods suffer from a few drawbacks. For example, most of the real-space implementations are not variational, i.e., we may find a total energy lower than the true energy, and if we reduce the grid-spacing the energy can actually increase. Moreover, the grid breaks translational symmetry, and can also break other symmetries that the system may possess. This can lead to the artificial lifting of some degeneracies, to the appearance of spurious peaks in spectra, etc. Of course all these problems can be minimized by reducing the grid-spacing.

6.3   Pseudo-potentials

6.3.1   The Pseudo-potential Concept

The many-electron Schrödinger equation can be very much simplified if electrons are divided in two groups: valence electrons and inner core electrons. The electrons in the inner shells are strongly bound and do not play a significant role in the chemical binding of atoms, thus forming with the nucleus an (almost) inert core. Binding properties are almost completely due to the valence electrons, especially in metals and semiconductors.

This separation suggests that inner electrons can be ignored in a large number of cases, thereby reducing the atom to a ionic core that interacts with the valence electrons. The use of an effective interaction, a pseudo-potential, that approximates the potential felt by the valence electrons, was first proposed by Fermi in 1934 [28]. Hellmann in 1935 [29] suggested that the form

$$w(r) = -\frac{1}{r} + \frac{2.74}{r}e^{-1.16r} \qquad (6.41)$$

could represent the potential felt by the valence electron of potassium. In spite of the simplification pseudo-potentials introduce in calculations, they remained forgotten until the late 50's. It was only in 1959, with Phillips and Kleinman [30–32], that pseudo-potentials began to be extensively used.

Let the exact solutions of the Schrödinger equation for the inner electrons be denoted by $|\psi_c\rangle$, and $|\psi_v\rangle$ those for the valence electrons. Then

$$\hat{H}|\psi_n\rangle = E_n|\psi_n\rangle , \qquad (6.42)$$

with $n = c, v$. The valence orbitals can be written as the sum of a smooth function (called the pseudo wave-function), $|\varphi_v\rangle$, with an oscillating function that results from the orthogonalization of the valence to the inner core orbitals

$$|\psi_v\rangle = |\varphi_v\rangle + \sum_c \alpha_{cv}|\psi_c\rangle , \qquad (6.43)$$

where

$$\alpha_{cv} = -\langle\psi_c|\varphi_v\rangle . \qquad (6.44)$$

The Schrödinger equation for the smooth orbital $|\varphi_v\rangle$ leads to

$$\hat{H}|\varphi_v\rangle = E_v|\varphi_v\rangle + \sum_c (E_c - E_v)|\psi_c\rangle\langle\psi_c|\varphi_v\rangle . \qquad (6.45)$$

This equation indicates that states $|\varphi_v\rangle$ satisfy a Schrödinger-like equation with an energy-dependent pseudo-Hamiltonian

$$\hat{H}^{\mathrm{PK}}(E) = \hat{H} - \sum_c (E_c - E)|\psi_c\rangle\langle\psi_c| . \qquad (6.46)$$

It is then possible to identify

$$\hat{w}^{\mathrm{PK}}(E) = \hat{v} - \sum_{c}(E_c - E)|\psi_c\rangle\langle\psi_c|\,, \qquad (6.47)$$

where $\hat{v}$ is the true potential, as the effective potential in which valence electrons move. However, this pseudo-potential is non-local and depends on the eigen-energy of the electronic state one wishes to find.

At a certain distance from the ionic core $\hat{w}^{\mathrm{PK}}$ becomes $\hat{v}$ due to the decay of the core orbitals. In the region near the core, the orthogonalization of the valence orbitals to the strongly oscillating core orbitals forces valence electrons to have a high kinetic energy (The kinetic energy density is essentially a measure of the curvature of the wave-function.) The valence electrons feel an effective potential which is the result of the screening of the nuclear potential by the core electrons, the Pauli repulsion and xc effects between the valence and core electrons. The second term of (6.47) represents then a repulsive potential, making the pseudo-potential much weaker than the true potential in the vicinity of the core. All this implies that the pseudo wave-functions will be smooth and will not oscillate in the core region, as desired.

A consequence of the cancellation between the two terms of (6.47) is the surprisingly good description of the electronic structure of solids given by the nearly-free electron approximation. The fact that many metal and semiconductor band structures are a small distortion of the free electron gas band structure suggests that the valence electrons do indeed feel a weak potential. The Phillips and Kleinman potential explains the reason for this cancellation.

The original pseudo-potential from Hellmann (6.41) can be seen as an approximation to the Phillips and Kleinman form, as in the limit $r \to \infty$ the last term can be approximated as $Ae^{-r/R}$, where R is a parameter measuring the core orbitals decay length.

The Phillips and Kleinman potential was later generalized [33,34] to

$$\hat{w} = \hat{v} + \sum_{c}|\psi_c\rangle\langle\xi_c|\,, \qquad (6.48)$$

where ξ_c is some set of functions.

The pseudo-potential can be cast into the form

$$w(\boldsymbol{r},\boldsymbol{r}') = \sum_{l}\sum_{m=-l}^{l} Y_{lm}^*(\hat{r})w_l(r,r')Y_{lm}(\hat{r}')\,, \qquad (6.49)$$

where Y_{lm} are the spherical harmonics. This expression emphasizes the fact that w as a function of r and r' depends on the angular momentum. The most usual forms for $w_l(r,r')$ are the separable Kleinman and Bylander form [35]

$$w_l(r,r') = v_l(r)v_l(r')\,, \qquad (6.50)$$

and the semi-local form

$$w_l(r, r') = w_l(r)\delta(r - r') \,. \tag{6.51}$$

6.3.2   Empirical Pseudo-potentials

Until the late 70's the method employed to construct a pseudo-potential was based on the Phillips and Kleinman cancellation idea. A model analytic potential was constructed and its parameters were fitted to experimental data. However, these models did not obey condition (6.43).

One of the most popular model potentials was introduced by Heine and Abarenkov in 1964 [36–38]. The Heine-Abarenkov potential is

$$w^{\mathrm{HA}}(r) = \begin{cases} -z/r & , \text{if } r > R \\ -A_l \hat{P}_l & , \text{if } r \le R \,, \end{cases} \tag{6.52}$$

with $\hat{P}_l$ an angular momentum projection operator. The parameters A_l were adjusted to the excitation energies of valence electrons and the parameter R is chosen, for example, to make A_0 and A_1 similar (leading to a local pseudo-potential for the simple metals).

A simplification of the Heine-Abarenkov potential was proposed in 1966 by Ashcroft [39,40]

$$w^{\mathrm{A}}(r) = \begin{cases} -z/r & , \text{if } r > R \\ 0 & , \text{if } r \le R \,. \end{cases} \tag{6.53}$$

In this model potential it is assumed that the cancellation inside the core is perfect, i.e., that the kinetic term cancels exactly the Coulomb potential for $r < R$. To adjust R, Ashcroft used data on the Fermi surface and on liquid phase transport properties.

The above mentioned and many other model potentials are discontinuous at the core radius. This discontinuity leads to long-range oscillations of their Fourier transforms, hindering their use in plane-wave calculations. A recently proposed model pseudo-potential overcomes this difficulty: the evanescent core potential of Fiolhais et al. [41]

$$w^{\mathrm{EC}}(r) = -\frac{z}{R} \left\{ \frac{1}{x} \left[1 - (1 + \beta x) \exp^{-\alpha x}\right] - A \exp^{-x} \right\} \,, \tag{6.54}$$

with $x = r/R$, where R is a decay length and $\alpha > 0$. Smoothness of the potential and the rapid decay of its Fourier transform are guaranteed by imposing that the first and third derivatives are zero at $r = 0$, leaving only two parameters to be fitted (α and R). These are chosen by imposing one of several conditions [41–46]: total energy of the solid is minimized at the observed electron density; the average interstitial electron density matches the all-electron result; the bulk moduli match the experimental results; etc.

Although not always bringing great advances, several other model potentials were proposed [47–49]. Also, many different methods for adjusting the parameters were suggested [50]. The main application of these model potentials was to the theory of metallic cohesion [51–55].

6.3.3   *Ab-initio* Pseudo-potentials

A crucial step toward more realistic pseudo-potentials was given by Topp and Hopfield [49,56], who suggested that the pseudo-potential should be adjusted such that they describe the valence charge density accurately. Based on that idea, modern pseudo-potentials are obtained inverting the free atom Schrödinger equation for a given reference electronic configuration [57], and forcing the pseudo wave-functions to coincide with the true valence wave-functions beyond a certain distance r_l. The pseudo wave-functions are also forced to have the same norm as the true valence wave-functions.

These conditions can be written as

$$R_l^{\mathrm{PP}}(r) = R_{nl}^{\mathrm{AE}}(r) \qquad\qquad , \text{if } r > r_l$$

$$\int_0^{r_l} dr \left| R_l^{\mathrm{PP}}(r) \right|^2 r^2 = \int_0^{r_l} dr \left| R_{nl}^{\mathrm{AE}}(r) \right|^2 r^2 , \text{if } r < r_l ,$$

(6.55)

where $R_l(r)$ is the radial part of the wave-function with angular momentum l, and PP and AE denote, respectively, the pseudo wave-function and the true (all-electron) wave-function. The index n in the true wave-functions denotes the valence level. The distance beyond which the true and the pseudo wave-functions are equal, r_l, is also l-dependent.

Besides (6.55), there are still two other conditions imposed on the pseudo-potential: the pseudo wave-functions should not have nodal surfaces and the pseudo energy-eigenvalues should match the true valence eigenvalues, i.e.,

$$\varepsilon_l^{\mathrm{PP}} = \varepsilon_{nl}^{\mathrm{AE}} .$$

(6.56)

The potentials thus constructed are called norm-conserving pseudo-potentials, and are semi-local potentials that depend on the energies of the reference electronic levels, $\varepsilon_l^{\mathrm{AE}}$.

In summary, to obtain the pseudo-potential the procedure is: i) The free atom Kohn-Sham radial equations are solved taking into account all the electrons, in some given reference configuration

$$\left[-\frac{1}{2} \frac{d^2}{dr^2} + \frac{l(l+1)}{2r^2} + v_{\mathrm{KS}}^{\mathrm{AE}} \left[n^{\mathrm{AE}} \right](r) \right] r R_{nl}^{\mathrm{AE}}(r) = \varepsilon_{nl}^{\mathrm{AE}} r R_{nl}^{\mathrm{AE}}(r) , \qquad (6.57)$$

where a spherical approximation to Hartree and exchange and correlation potentials is assumed and relativistic effects are not considered. The Kohn-Sham potential, $v_{\mathrm{KS}}^{\mathrm{AE}}$, is given by

$$v_{\mathrm{KS}}^{\mathrm{AE}} \left[n^{\mathrm{AE}} \right](r) = -\frac{Z}{r} + v_{\mathrm{Hartree}} \left[n^{\mathrm{AE}} \right](r) + v_{\mathrm{xc}} \left[n^{\mathrm{AE}} \right](r) . \qquad (6.58)$$

ii) Using norm-conservation (6.55), the pseudo wave-functions are determined. Their shape in the region $r < r_l$ needs to be previously defined, and it is here that many modern potentials differ from one another. iii) Knowing the pseudo wave-function, the pseudo-potential results from the inversion of the radial Kohn-Sham equation for the pseudo wave-function and the valence electronic density

$$w_{l,\mathrm{scr}}(r) = \varepsilon_l^{\mathrm{PP}} - \frac{l(l+1)}{2r^2} + \frac{1}{2r R_l^{\mathrm{PP}}(r)} \frac{\mathrm{d}^2}{\mathrm{d}r^2} \left[r R_l^{\mathrm{PP}}(r) \right] . \tag{6.59}$$

The resulting pseudo-potential, $w_{l,\mathrm{scr}}$, still includes screening effects due to the valence electrons that have to be subtracted to yield

$$w_l(r) = w_{l,\mathrm{scr}}(r) - v_{\mathrm{Hartree}} \left[n^{\mathrm{PP}} \right] (r) - v_{\mathrm{xc}} \left[n^{\mathrm{PP}} \right] (r) . \tag{6.60}$$

The cutoff radii, r_l, are not adjustable pseudo-potential parameters. The choice of a given set of cutoff radii establishes only the region where the pseudo and true wave-functions coincide. Therefore, the cutoff radii can be considered as a measure of the quality of the pseudo-potential. Their smallest possible value is determined by the location of the outermost nodal surface of the true wave-functions. For cutoff radii close to this minimum, the pseudo-potential is very realistic, but also very strong. If very large cutoff radii are chosen, the pseudo-potentials will be smooth and almost angular momentum independent, but also very unrealistic. A smooth potential leads to a fast convergence of plane-wave basis calculations [58]. The choice of the ideal cutoff radii is then the result of a balance between basis-set size and pseudo-potential accuracy.

6.3.4 Hamann Potential

One of the most used parameterizations for the pseudo wave-functions is the one proposed in 1979 by Hamann, Schlüter, and Chiang [59] and later improved by Bachelet, Hamann and Schlüter [60] and Hamann [61].

The method proposed consists of using an intermediate pseudo-potential, $\bar{w}_l(r)$, given by

$$\bar{w}_l(r) + v_{\mathrm{Hartree}} \left[n^{\mathrm{PP}} \right] (r) + v_{\mathrm{xc}} \left[n^{\mathrm{PP}} \right] (r) =$$
$$= v_{\mathrm{KS}}^{\mathrm{AE}} \left[n^{\mathrm{AE}} \right] (r) \left[1 - f \left(\frac{r}{r_l} \right) \right] + c_l f \left(\frac{r}{r_l} \right) , \tag{6.61}$$

where $f(x) = \mathrm{e}^{-x^\lambda}$, and $\lambda = 4.0$ [59] or $\lambda = 3.5$ [60,61]. The Kohn-Sham equations are solved using this pseudo-potential, and the constants c_l are adjusted in order to obey (6.56). Notice that the form of the wave-functions implies that (6.55) is verified for some $\tilde{r}_l > r_l$. As the two effective potentials are identical for $r > \tilde{r}_l$, and given the fast decay of $f(x)$, the intermediate

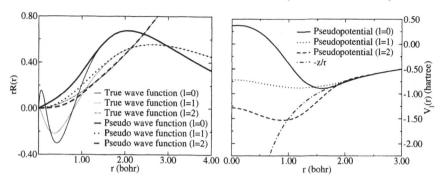

Fig. 6.2. Hamann pseudo-potential for Al, with $r_0 = 1.24$, $r_1 = 1.54$, and $r_2 = 1.40$ bohr: pseudo wave-functions vs. true wave-functions (left) and pseudo-potentials (right)

pseudo wave-functions, $\bar{R}_l(r)$, coincide, up to a constant, with the true wave-functions in that region.

In the method proposed by Hamann [61], the parameters c_l are adjusted so that

$$\left.\frac{\mathrm{d}}{\mathrm{d}r} \ln \left[r R_{nl}^{\mathrm{AE}}(r) \right]\right|_{r=\tilde{r}_l} = \left.\frac{\mathrm{d}}{\mathrm{d}r} \ln \left[r \bar{R}_l(r) \right]\right|_{r=\tilde{r}_l} . \tag{6.62}$$

This way, the method is not restricted to bound states.

To impose norm-conservation, the final pseudo wave-functions, $R_l^{\mathrm{PP}}(r)$, are defined as a correction to the intermediate wave-functions

$$R_l^{\mathrm{PP}}(r) = \gamma_l \left[\bar{R}_l(r) + \delta_l g_l(r) \right] , \tag{6.63}$$

where γ_l is the ratio $R_{nl}^{\mathrm{AE}}(r)/\bar{R}_l(r)$ in the region where $r > \tilde{r}_l$ and $g_l(r) = r^{l+1} f(r/r_l)$. The constants δ_l are adjusted to conserve the norm.

Figure 6.2 shows the Hamann pseudo-potential for Al, with $r_0 = 1.24$, $r_1 = 1.54$ and $r_2 = 1.40$ bohr. Note that the true and the pseudo wave-functions do not coincide at r_l – this only happens at $r > \tilde{r}_l$.

6.3.5   Troullier–Martins Potential

A different method to construct the pseudo wave-functions was proposed by Troullier and Martins [58,62], based on earlier work by Kerker [63]. This method is much simpler than Hamann's and emphasizes the desired smoothness of the pseudo-potential (although it introduces additional constraints to obtain it). It achieves softer pseudo-potentials for the $2p$ valence states of the first row elements and for the d valence states of the transition metals. For other elements both methods produce equivalent potentials.

The pseudo wave-functions are defined as

$$R_l^{\mathrm{PP}}(r) = \begin{cases} R_{nl}^{\mathrm{AE}}(r) & , \text{if } r > r_l \\ r^l e^{p(r)} & , \text{if } r < r_l , \end{cases} \tag{6.64}$$

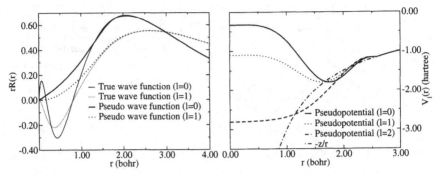

Fig. 6.3. Troullier-Martins pseudo-potential for Al, with $r_0 = r_1 = r_2 = 2.60$ bohr: pseudo wave-functions vs. true wave-functions (left) and pseudo-potentials (right)

with

$$p(r) = c_0 + c_2 r^2 + c_4 r^4 + c_6 r^6 + c_8 r^8 + c_{10} r^{10} + c_{12} r^{12} . \tag{6.65}$$

The coefficients of $p(r)$ are adjusted by imposing norm-conservation, the continuity of the pseudo wave-functions and their first four derivatives at $r = r_l$, and that the screened pseudo-potential has zero curvature at the origin. This last condition implies that

$$c_2^2 + c_4(2l + 5) = 0 , \tag{6.66}$$

and is the origin of the enhanced smoothness of the Troullier and Martins pseudo-potentials.

Figure 6.3 shows the Troullier and Martins pseudo-potential for Al, with $r_0 = r_1 = r_2 = 2.60$ bohr. The $3d$ wave-functions are not shown since the state is unbound for this potential.

There are many other not so widely used norm-conserving pseudo-potentials [64–68]. Note that, in some cases, norm-conservation was abandoned in favor of increased pseudo-potential smoothness [69].

6.3.6 Non-local Core Corrections

It is tempting to assume that the Kohn-Sham potential depends linearly on the density, so that the unscreening of the pseudo-potential can be performed as in (6.60). Unfortunately, even though the Hartree contribution is indeed linearly dependent on the density, the xc term is not

$$v_{xc} \left[n^{AE} \right] (r) \equiv v_{xc} \left[n^{core} + n^{PP} \right] (r) \tag{6.67}$$
$$\neq v_{xc} \left[n^{core} \right] (r) + v_{xc} \left[n^{PP} \right] (r) .$$

In some cases, like the alkali metals, the use of a nonlinear core-valence xc scheme may be necessary to obtain a transferable pseudo-potential. In these cases, the unscreened potential is redefined as

$$w_l(r) = w_{l,\text{scr}}(r) - v_{\text{Hartree}}\left[n^{\text{PP}}\right](r) - v_{\text{xc}}\left[\tilde{n}^{\text{core}} + n^{\text{PP}}\right](r) , \qquad (6.68)$$

and the core density is supplied together with the pseudo-potential. In a code that uses pseudo-potentials, one has simply to add the valence density to the given atomic core density to obtain the xc potential. To avoid spoiling the smoothness of the potential with a rugged core density, usually a partial core density [70,71], $\tilde{n}^{\text{core}}$, is built and supplied instead of the true core density

$$\tilde{n}^{\text{core}}(r) = \begin{cases} n^{\text{core}}(r) & \text{for } r \geq r_{\text{nlc}} \\ P(r) & \text{for } r < r_{\text{nlc}} \end{cases} . \qquad (6.69)$$

The polynomial $P(r)$ decays monotonically and has vanishing first and second derivatives at the origin. At r_{nlc} it joins smoothly the true core density (it is continuous up to the third derivative). The core cutoff radius, r_{nlc}, is typically chosen to be the point where the true atomic core density becomes smaller that the atomic valence density. It can be chosen to be larger than this value but if it is too large the description of the non-linearities may suffer. Note that, as the word partial suggests,

$$\int_0^{r_{\text{nlc}}} dr\, \tilde{n}^{\text{core}}(r)\, r^2 < \int_0^{r_{\text{nlc}}} dr\, n^{\text{core}}(r)\, r^2 . \qquad (6.70)$$

These corrections are more important for the alkali metals and other elements with few valence electrons and core orbitals extending into the tail of the valence density (e.g., Zn and Cd).

In some cases, the use of the generalized gradient approximation (GGA) for exchange and correlation leads to the appearance of very short-ranged oscillations in the pseudo-potentials (see Fig. 6.4). These oscillations are artifacts of the GGA that usually disappear when non-local core corrections are considered. Nevertheless, they do not pose a real threat for plane-wave calculations, since they are mostly filtered out by the energy cutoff.

6.3.7 Pseudo-potential Transferability

A useful pseudo-potential needs to be transferable, i.e., it needs to describe accurately the behavior of the valence electrons in several different chemical environments. The logarithmic derivative of the pseudo wave-function determines the scattering properties of the pseudo-potential. Norm-conservation forces these logarithmic derivatives to coincide with those of the true wave-functions for $r > r_l$. In order for the pseudo-potential to be transferable, this equality should hold at all relevant energies, and not only at the energy, ε_l, for which the pseudo-potential was adjusted. Norm-conservation assures that this is fulfilled for the nearby energies, as [49,72]

$$\left. \frac{d}{d\varepsilon_l} \frac{d}{dr} \ln R_l(r) \right|_{r=R} = -\frac{2}{r^2 R_l^2(r)} \int_0^R dr\, |R_l(r)|^2\, r^2 . \qquad (6.71)$$

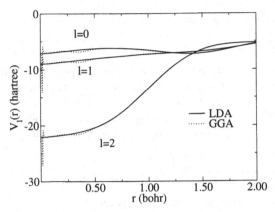

Fig. 6.4. Troullier-Martins pseudo-potential for Cu, with $r_0 = r_2 = 2.2$ and $r_1 = 2.4$ bohr. Notice that the LDA and GGA pseudo-potential are essentially identical, the main difference being the GGA potential oscillations near the origin

It is however necessary to take into account that the environment surrounding the electrons can be different from the one in the reference situation. Thus, although the pseudo-potential remains the same, the effective potential changes (the Hartree and xc potentials depend on the density). Therefore, the logarithmic derivative is not an absolute test of the transferability of a pseudo-potential [73]. The ideal method to assess the transferability of a potential consists in testing it in diverse chemical environments. The most usual way of doing this is to test its transferability to other atomic configurations and even to the ionized configurations. The variation of the total energy of the free atom with the occupancy of the valence orbitals is another test of transferability [74]. As the potential is generated for a given reference electronic configuration, it can be useful to choose the configuration that best resembles the system of interest [61]. However, the potential does not (should not) depend too much on the reference configuration.

6.3.8   Kleinman and Bylander Form of the Pseudo-potential

The semi-local form of the pseudo-potentials described above leads to a complicated evaluation of their action on a wave-function

$$\langle \boldsymbol{r} \,|\hat{w}|\, \Psi \rangle = \int \mathrm{d}^3 r'\, w(\boldsymbol{r}, \boldsymbol{r}') \Psi(\boldsymbol{r}') =$$

$$= \sum_l \sum_{m=-l}^{l} Y_{lm}(\hat{r}) w_l(r) \int \mathrm{d}^3 r'\, \delta(r - r') Y_{lm}^*(\hat{r}') \Psi(\boldsymbol{r}') \,. \quad (6.72)$$

Unfortunately, the last integral must be calculated for each r. In a plane-wave expansion, this involves the product of an $N_{\mathrm{PW}} \times N_{\mathrm{PW}}$ matrix with the vector representing the wave-function. This operation is of order $N_{PW} \times N_{PW}$, and N_{PW}, the number of plane-waves in the basis set, can be very large.

The semi-local potential can be rewritten in a form that separates long and short range components. The long range component is local, and corresponds to the Coulomb tail. Choosing an arbitrary angular momentum component (usually the most repulsive one) and defining

$$\Delta w_l(r) = w_l(r) - w_{\text{local}}(r) \,. \tag{6.73}$$

the pseudo-potential can be written as

$$w(\boldsymbol{r}, \boldsymbol{r}') = w_{\text{local}}(r) + \sum_l \Delta w_l(r) \sum_{m=-l}^{l} Y_{lm}^*(\hat{r}')Y_{lm}(\hat{r})\delta(r - r') \,. \tag{6.74}$$

Kleinman and Bylander [35] suggested that the non-local part of (6.74) are written as a separable potential, thus transforming the semi-local potential into a truly non-local pseudo-potential. If $\varphi_{lm}(\boldsymbol{r}) = R_l^{PP}(r)Y_{lm}(\hat{r})$ denotes the pseudo wave-functions obtained with the semi-local pseudo-potential, the Kleinman and Bylander (KB) form is given by

$$w^{KB}(\boldsymbol{r}, \boldsymbol{r}') = w_{\text{local}}(r) + \sum_l \Delta w_l^{KB}(\boldsymbol{r}, \boldsymbol{r}') =$$

$$= w_{\text{local}}(r) + \sum_l \sum_{m=-l}^{l} \frac{\varphi_{lm}(\boldsymbol{r})\Delta w_l(r)\Delta w_l(r')\varphi_{lm}(\boldsymbol{r}')}{\int d^3r \, \Delta w_l(r) \left|\varphi_{lm}(\boldsymbol{r})\right|^2} \,, \tag{6.75}$$

which is, in fact, easier to apply than the semi-local expression.

The KB separable form has, however, some disadvantages, leading sometimes to solutions with nodal surfaces that are lower in energy than solutions with no nodes [75,76]. These (ghost) states are an artifact of the KB procedure. To eliminate them one can use a different component of the pseudo-potential as the local part of the KB form or choose a different set of core radii for the pseudo-potential generation. As a rule of thumb, the local component of the KB form should be the most repulsive pseudo-potential component. For example, for the Cu potential of Fig. 6.4, the choice of $l = 2$ as the local component leads to a ghost state, but choosing instead $l = 0$ remedies the problem.

6.4 Atomic Calculations

As our first example we will present several atomic calculations. These simple systems will allow us to gain a fist impression of the capabilities and limitations of DFT. To solve the Kohn-Sham equations we used the code of J. L. Martins [77]. The results are then compared to Hartree-Fock calculations performed with GAMESS [78]. As an approximation to the xc potential, we

Table 6.1. Ionization potentials calculated either by taking the difference of total energies between the neutral and the singly ionized atom (diff.), or from the eigenvalue of the highest occupied orbital (HOMO). We note that in the case of Hartree-Fock, $-\varepsilon_{\mathrm{HOMO}}$ is only an approximation to the ionization potential

atom	LDA		GGA		Hartree-Fock		expt.
	diff.	$-\varepsilon_{\mathrm{HOMO}}$	diff.	$-\varepsilon_{\mathrm{HOMO}}$	diff.	$-\varepsilon_{\mathrm{HOMO}}$	
H	0.479	0.269	0.500	0.279	0.500	0.500	0.500
Ar	0.586	0.382	0.581	0.378	0.543	0.590	0.579
Hg	0.325	0.205	0.311	0.194	0.306	0.320	0.384
Hg (rel)	0.405	0.261	0.391	0.249		0.320	0.384

took the LDA, in the parameterization of Perdew and Zunger [6], and one GGA, flavor Perdew, Becke and Ernzerhof [79]. Furthermore, all calculations were done within the spin-polarized version of DFT.

The simplest atom one can study is hydrogen. As hydrogen has only one electron, its ground-state can be obtained analytically. One could expect that DFT yields precise results for such a trivial case. Surprisingly this is not true for several of the functionals currently in use, such as the LDA or most of the GGAs. In Table 6.1 we present calculations of the ionization potential (IP) for hydrogen. We note that in Kohn-Sham theory there are at least two ways to determine this quantity: (i) The eigenvalue of the highest occupied Kohn-Sham state is equal to minus the ionization potential, $\mathrm{IP} = -\varepsilon_{\mathrm{HOMO}}$; (ii) By using the definition of the IP as the difference of total energies, $\mathrm{IP} = E(\mathrm{X}^+) - E(\mathrm{X})$, where X is the atomic species. Even though the IPs calculated from (ii) come out fairly well for both LDA and GGA (the GGA are, in fact, slightly better), the $-\varepsilon_{\mathrm{HOMO}}$ are far too small, almost by a factor of two. On the other hand, Hartree-Fock is exact for this one-electron problem. To explain this discrepancy we have to take a closer look at the xc potential. As hydrogen has only one electron, the Kohn-Sham potential has to reduce to the external potential, $-1/r$. This implies that the xc for hydrogen is simply $v_{\mathrm{xc}}(\boldsymbol{r}) = -v_{\mathrm{Hartree}}(\boldsymbol{r})$. More precisely, it is the exchange potential that cancels the Hartree potential, while the correlation is zero. In the LDA and the GGA, neither of these conditions is satisfied. It is, however, possible to solve the hydrogen problem exactly within DFT by using some more sophisticated xc potentials, like the exact exchange [80], or the self-interaction corrected LDA [6] functionals.

Our first many-electron example is argon. Argon is a noble gas with the closed shell configuration $1s^2 2s^2 2p^6 3s^2 3p^6$, so its ground-state is spherical. In Fig. 6.5 we plot the electron density for this atom as a function of the distance to the nucleus. The function $n(r)$ decays monotonically, with very little structure, and is therefore not a very elucidative quantity to behold. However, if we choose to represent $r^2 n(r)$, we can clearly identify the shell structure of the atom: Three maxima, corresponding to the center of the three shells,

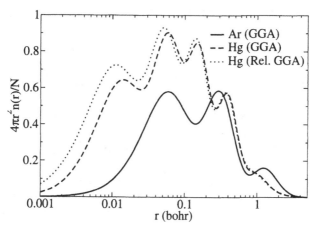

Fig. 6.5. Radial electronic density of the argon and mercury atoms versus the distance to the nucleus. Both the solid and dashed curves were obtained using the GGA to approximate the xc potential. For comparison the density resulting from a relativistic GGA calculation for mercury is also shown. The density is normalized so that the area under each curve is 1

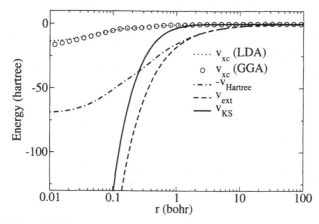

Fig. 6.6. LDA and GGA xc potentials for the argon atom. The dashed-dotted line corresponds to minus the Hartree potential evaluated with the GGA density. The LDA Hartree potential is however indistinguishable from this curve. Furthermore, the dashed line represents the argon nuclear potential, $-18/r$, and the solid line the total Kohn-Sham potential

and two minima separating these regions. The xc correlation potential used in the calculation was the GGA, but the LDA density looks almost indistinguishable from the GGA density. This is a fairly general statement – the LDA and most of the GGAs (as well as other more complicated functionals) yield very similar densities in most cases. The potentials and the energies can nevertheless be quite different.

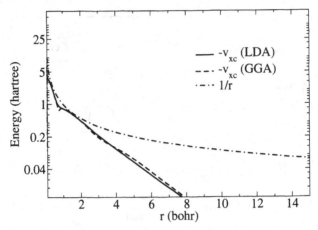

Fig. 6.7. LDA and GGA xc potentials for the argon atom in a logarithmic scale. For the sake of comparison we also plot the function $1/r$

Having the density it is a simple task to compute the Hartree and xc potentials. These, together with the nuclear potential $v_{\text{ext}}(r) = -Z/r$, are depicted in Fig. 6.6. The Hartree potential is always positive and of the same order as the external potential. On the other hand, the xc potential is always negative and around 5 times smaller. Let us now suppose that an electron is far away from the nucleus. This electron feels a potential which is the sum of the nuclear potential and the potential generated by the remaining $N - 1$ electrons. The further away from the nucleus, the smaller will be the dipole and higher-moment contributions to the electric field. It is evident from these considerations that the Kohn-Sham potential has to decay asymptotically as $-(Z - N + 1)/r$. As the external potential decays as $-Z/r$, and the Hartree potential as N/r, one readily concludes that the xc potential has to behave asymptotically as $-1/r$. In fact it is the exchange part of the potential that has to account for this behavior, whilst the correlation potential decays with a higher power of $1/r$. To better investigate this feature, we have plotted, in logarithmic scale, $-v_{\text{xc}}$, in the LDA and GGA approximations, together with the function $1/r$ (see Fig. 6.7). Clearly both the LDA and the GGA curve have a wrong (exponential) asymptotic behavior. From the definition of the LDA, (see 6.9), it is quite simple to derive this fact. The electronic density for a finite system decays exponentially for large distances from the nucleus. The quantity ε^{HEG} entering the definition is, as mentioned before, a simple function, not much more complicated than a polynomial. By simple inspection, it is then clear that inserting an exponentially decaying density in (6.9) yields an exponentially decaying xc potential.

The problem of the exponential decay can yet be seen from a different perspective. For a many-electron atom the Hartree energy can be written, in terms of the Kohn-Sham orbitals, as

$$E_{\text{Hartree}} = \frac{1}{2} \int d^3r \int d^3r' \sum_{ij}^{\text{occ}} \frac{|\varphi_i(\boldsymbol{r})|^2 |\varphi_j(\boldsymbol{r}')|^2}{|\boldsymbol{r} - \boldsymbol{r}'|} . \tag{6.76}$$

Note that in the sum the term with $i = j$ is *not* excluded. This diagonal represents the interaction of one electron with itself, and is therefore called the self-interaction term. It is clearly a spurious term, and is exactly canceled by the diagonal part of the exchange energy. It is easy to see that neither the LDA nor the GGA exchange energy cancel exactly the self-interaction. This is, however, not the case in more sophisticated functionals like the exact exchange or the self-interaction-corrected LDA.

The self-interaction problem is responsible for some of the failures of the LDA and the GGA, namely (i) the too small ionization potentials when calculated from $\varepsilon_{\text{HOMO}}$; (ii) the non-existence of Rydberg series; (iii) the incapacity to bind extra electrons, thus rendering almost impossible the calculation of electron-affinities (EA).

In Table 6.1 we show the IPs calculated for the argon atom. It is again evident that $-\varepsilon_{\text{HOMO}}$ is too small [failure (i)], while the IPs obtained through total energy differences are indeed quite close to the experimental values, and in fact better than the Hartree-Fock results. Note that the LDA result is too large, but is corrected by the gradient corrections. This is again a fairly universal feature of the LDA and the GGA: The LDA tends to overestimate energy barriers, which are then corrected by the GGA to values closer to the experimental results.

Up to now we have disregarded relativistic corrections in our calculations. These, however, become important as the atomic number increases. To illustrate this fact, we show in Fig. 6.5 the radial electronic density of mercury ($Z = 80$) and in Table 6.1 its IP obtained from both a relativistic and a non-relativistic calculation. From the plot it is clear that the density changes considerably when introducing relativistic corrections, especially close to the nucleus, where these corrections are stronger. Furthermore, the relativistic IP is much closer to the experimental value. But, what do we mean by "relativistic corrections"? Even though a relativistic version of DFT (and relativistic functionals) have been proposed (see the chapter by R. Dreizler in this volume), very few calculations were performed within this formalism. In the context of standard DFT, "relativistic" calculation normally means the solution of a: (a) Dirac-like equation but adding a non-relativistic xc potential; (b) Pauli equation, i.e., including the mass polarization, Darwin and spin-orbit coupling terms; (c) Scalar-relativistic Pauli equation, i.e., including the mass polarization, Darwin and either ignoring the spin-orbit term, or averaging it; (d) ZORA equation (see [81,82]). Our calculations were performed with the recipe (a).

To complete this section on atomic calculations, we would like to take a step back and look at the difficulty in calculating electronic affinities (EA)

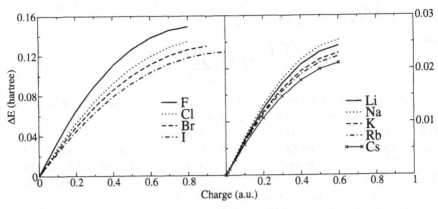

Fig. 6.8. $E(X) - E(X^{-\alpha})$ versus α for the halogen and alkali atoms

within the LDA and the GGA. For that purpose we performed GGA calculations for several atomic species, namely the halogen and alkali series, that we charged with a fraction, α, of an extra electron. The results are summarized in Fig. 6.8, where we depicted the difference of total energies between the charged and the neutral species, $E(X) - E(X^{-\alpha})$. Only Iodine was able to accept a full extra electron, while all other atoms bounded between 0.5 and 0.7 electrons. Even though a "proper" calculation of the EA is not possible in these cases, practical recipes do exist. We can, e.g., extrapolate $E(X) - E(X^{-\alpha})$ to $\alpha = 1$, and use this value as an estimation of the EA. In Table 6.2 we show the EAs obtained through a very simple polynomial extrapolation. The results compare fairly well for the halogens, while for the alkali atoms they exhibit errors of around 30%. However, we would like to stress that the situation is far from satisfactory from the theoretical point of view, and can only be solved by using better xc functionals.

Table 6.2. Electronic affinities for the halogen and alkali atoms. All values were obtained from extrapolation of $E(X) - E(X^{-\alpha})$ to $\alpha = 1$, except in the case of iodine (the only of this set of atoms able to bind an extra electron)

	F	Cl	Br	I	Li	Na	K	Rb	Cs
DFT	0.131	0.139	0.131	0.123	0.0250	0.0262	0.0240	0.0234	0.0222
expt.	0.125	0.133	0.124	0.112	0.0227	0.0201	0.0184	0.0179	0.0173

6.5   Plane-Wave Calculations

In this section we will present some simple calculations using a plane-wave expansion of the Kohn-Sham orbitals [4]. The plane-wave basis set is or-

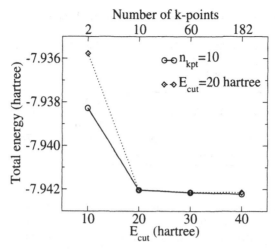

Fig. 6.9. Convergence of total energy of bulk Si with plane-wave energy cutoff and number of k-points used in the sampling of the irreducible wedge of the Brillouin zone (within the LDA). The analysis of convergence with energy cutoff was done at a fixed Monkhorst-Pack sampling [23] using 10 k-points, and the convergence with k-point sampling was studied at a fixed energy cutoff of 20 hartree

thonormal and the convergence of the calculations increases systematically with the number of plane-waves. Gaussian basis sets, on the contrary, do not provide a clear and systematic way to improve the convergence of the calculations and do not form an orthonormal set. As a result, the calculations often depend on the choice of basis set. Another advantage of plane-waves is that the evaluation of forces for molecular dynamics is straightforward (the Pulay forces [83,11] are identically zero). These advantages lead the combination of pseudo-potentials, plane-waves, and Kohn-Sham equations to be known as the "standard model of solid-state theory".

As the first example of the use of a plane-wave expansion of the Kohn-Sham equations we shall calculate some properties of bulk silicon and examine its band-structure. All the results for bulk Si (diamond lattice) were obtained with a Troullier-Martins pseudo-potential with $r_0 = r_1 = r_2 = 1.89$ bohr. The local component used in the Kleinman and Bylander form of the pseudo-potential was the d-component. The variation of the total energy with respect to energy cutoff was assessed and a cutoff of 20 hartree was shown to lead to energies converged up to 0.001 hartree (see Fig. 6.9). The irreducible wedge of the Brillouin zone was sampled with different Monkhorst-Pack schemes [23] and the scheme using 10 k-points was deemed sufficient to converge the total energy again up to 0.001 hartree.

The calculations for bulk silicon were done using both the LDA (Perdew-Wang 92 parameterization [7]) and the GGA (Perdew-Burke-Ernzerhof functional [79]). We note that we always used a pseudo-potential compatible with

Table 6.3. Comparison of some bulk properties of silicon obtained with the LDA and the GGA: equilibrium lattice constant (a), bulk modulus (B) and cohesive energy (E_c). Bulk moduli were obtained by fitting the Murnaghan equation of state [84] to the calculated total energy vs. volume curve. The experimental results (expt.) are those cited in [85]

	LDA	GGA	expt.
a (Å)	5.378	5.463	5.429
B (Mbar)	0.965	0.882	0.978
E_c (eV/atom)	6.00	5.42	4.63

the approximation for the xc potential, i.e., for the LDA calculations we used a pseudo-potential generated with the LDA, and the same for the GGA. Although sometimes there is no discernible difference between the results obtained with pseudo-potentials generated with different xc functionals (but using the same cutoffs), one should always use the same functional for the calculation as the one used in the generation of the pseudo-potential [71].

In Table 6.3 we summarize the results obtained for some bulk properties of silicon. It is immediately apparent that the LDA under-estimates the equilibrium lattice parameter, while the GGA over-estimates it. This is a typical result: the LDA, in general, over-binds by 1–2% and the GGA produces larger bond lengths, correcting the LDA, but sometimes over-corrects it. In the present case the GGA leads to a lattice parameter 0.5% larger than the experimental value. A similar statement can be made for the cohesive energy $(E_c = E_{\text{bulk}}/N_{\text{atom}} - E_{\text{atom}})$: the LDA predicts a cohesive energy larger than the experimental value, and the GGA corrects it.

The band-structure of silicon obtained in this calculation is shown in Fig. 6.10. It was calculated at the LDA equilibrium lattice constant, even in the GGA case. These band-structures exhibit the well-known "band-gap problem" of DFT: the predicted band-gap is too small roughly by a factor of two. This is true for the LDA and the GGA. In fact, the GGA does not show a great improvement, even when the band-structure is calculated at its predicted equilibrium lattice constant (Table 6.4). The failure of these two DFT schemes in predicting the band-gap of silicon is not a surprise. Even if the true xc potential was known, the difference between the conduction and valence bands in a KS calculation would differ from the true band-gap (E_g). The true band-gap may be defined as the ground-state energy difference between the N and $N \pm 1$ systems

$$E_g = E(N+1) + E(N-1) - 2E(N) . \tag{6.77}$$

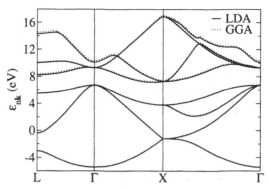

Fig. 6.10. Band structure of Si, obtained at the LDA equilibrium lattice constant

The difference between the highest occupied level and the lowest unoccupied level of the N-electron system is, on the other hand,

$$\varepsilon_{N+1}^{KS}(N) - \varepsilon_{N}^{KS}(N) = E_g - \left[\varepsilon_{N+1}^{KS}(N+1) - \varepsilon_{N+1}^{KS}(N)\right]$$
$$\equiv E_g - \Delta_{\mathrm{xc}} . \tag{6.78}$$

Δ_{xc} is then a measure of the shift in the Kohn-Sham potential due to an infinitesimal variation of the density (in an extended system, the densities of the N and $N + 1$ systems are almost identical). This shift is rigid (see the discussions in Chaps. 1 and 5), and is entirely due to a discontinuity in the derivative of the xc energy functional. It cannot therefore be accounted for by simple analytical, continuous approximations to exchange and correlation, like the LDA or the GGA. One could however argue that the error in the LDA band-gaps should come from two different sources: Δ_{xc} and the use of an approximate functional for exchange and correlation. If the latter were the most important, one could hope that better approximations would yield band-gaps in closer agreement with experiment. However, it appears that the "exact" Kohn-Sham band-gap does not differ much from the LDA band-gap, Δ_{xc} being the major culprit of the band-gap problem.

Usually, the LDA conduction bands are shifted from the correct bands by a quantity that is only weakly dependent on k. A common solution to the band-gap problem is then to rigidly shift upward the Kohn-Sham conduction bands. This is called the "scissors operator".

A system which is much more difficult to handle within a first-principles pseudo-potential, plane-wave, density functional method is copper (as all the other noble and transition metals). Metals require a very good sampling of the irreducible wedge of the Brillouin zone in order to properly describe the Fermi surface. This makes them computationally more demanding. But copper presents yet another difficulty: It is mandatory that the $3d$-electrons are taken into account, as they contribute significantly to bonding and to the valence band structure. Therefore, these electrons cannot be frozen into the

Table 6.4. Comparison of the band-gap (E_g) and of gaps at some special points in the Brillouin zone (Γ, X and L). The column labeled GGA* refers to values obtained with the GGA at the LDA equilibrium lattice constant, and GGA labels the results obtained with the GGA at the GGA equilibrium lattice constant. The experimental results (expt.) are those cited in [86]. All values are in eV

	LDA	GGA*	GGA	expt.
E_g	0.45	0.53	0.61	1.17
Γ	2.57	2.59	2.57	3.34
X	3.51	3.59	3.56	1.25
L	2.73	2.84	2.64	2.4

Table 6.5. Lattice parameter (a), bulk modulus (B), and cohesive energy (E_c) of Cu, calculated with the LDA (Perdew-Wang 92 functional [7]) and GGA (Perdew-Burke-Ernzerhof functional [79]). Bulk moduli were obtained by fitting the Murnaghan equation of state [84] to the calculated total energy versus volume curve. The experimental results are those cited in [87]

	LDA	GGA	expt.
a (Å)	3.571	3.682	3.61
B (Mbar)	0.902	0.672	1.420
E_c (eV/atom)	4.54	3.58	3.50

core. However, their inclusion in the set of valence electrons means that there will be at least 11 valence electrons (one could also include the $3s$ and $3p$ electrons) and that the pseudo-potential will be very hard. The combination of these two factors makes the calculations almost prohibitive.

The use of soft pseudo-potentials like the Troullier-Martins pseudo-potential alleviates the problem. Table 6.5 and Fig. 6.11 show some results for bulk Cu obtained with a Troullier-Martins pseudo-potential with $r_0 = r_2 = 2.05$ bohr and $r_1 = 2.30$ bohr. The local component used in the Kleinman and Bylander form of the pseudo-potential was the s-component and a partial core correction was included with $r_{\text{nlc}} = 0.8$ bohr. The pseudo-potential thus obtained is soft enough to allow for well converged plane-wave calculations with an energy cutoff of 60 hartree. The Brillouin zone was sampled with a Monkhorst-Pack scheme using 60 k-points and a Gaussian broadening of the levels with a 0.01 hartree width. The convergence of the calculations against energy cutoff, k-point sampling and width of the smearing gaussian was better than 0.001 hartree. The calculations were done using both the LDA and the GGA for exchange and correlation. The LDA used was the Perdew-Wang 92 [7] parameterization of the Ceperley-Alder results [5] and the GGA was the Perdew-Burke-Ernzerhof [79] functional. As in the case of silicon, and for the sake of consistency, the pseudo-potentials employed in both calculations were consistent with the xc functional.

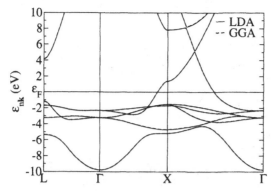

Fig. 6.11. Calculated band structure of Cu, obtained at the LDA equilibrium lattice constant

Table 6.6. Theoretical band-widths and energies at some high symmetry points in the Brillouin zone. The column labeled GGA* refers to values obtained with the GGA at the LDA equilibrium lattice constant, and GGA labels the results obtained with the GGA at the GGA equilibrium lattice constant. Results are compared to a GW calculation [88] and to averages over several experiments [89] (expt.). All values are in eV

		LDA	GGA*	GGA	GW	expt.
Positions	Γ_{12}	−2.31	−2.31	−2.12	−2.81	−2.78
of d bands	X_5	−1.53	−1.53	−1.44	−2.04	−2.01
	L_3	−1.68	−1.69	−1.58	−2.24	−2.25
	$\Gamma_{12} - \Gamma_{25'}$	0.91	0.90	0.78	0.60	0.81
Widths	$X_5 - X_3$	3.17	3.15	2.73	2.49	2.79
of d bands	$X_5 - X_1$	3.62	3.62	3.14	2.90	3.17
	$L_3 - L_3$	1.57	1.56	1.34	1.26	1.37
	$L_3 - L_1$	3.69	3.66	3.23	2.83	2.91
Positions	Γ_1	−9.77	−9.77	−9.02	−9.24	−8.60
of s, p bands	$L_{2'}$	−1.16	−1.19	−0.88	−0.57	−0.85
L gap	$L_1 - L_{2'}$	4.21	4.16	3.92	4.76	4.95

From Table 6.5 it is apparent that the LDA predicted, as usual, a lattice parameter smaller than the experimental one, while the GGA over-corrected this error. The over-binding of the LDA is also present in the cohesive energy, which is 30% larger than the experimental value. The GGA fared much better, producing an error of only 2%.

From Table 6.6 one can see that the LDA predicts d bands that are more delocalized than the experimental ones and are also 0.5 eV closer to the Fermi level. As the LDA is supposed to work well only for smoothly

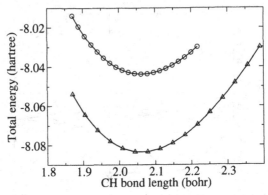

Fig. 6.12. Methane total energy vs. CH bond length: results obtained with the LDA (circles, Perdew-Wang 92 functional [7]) and the PBE [79] GGA (triangles). The Troullier-Martins pseudo-potential used for carbon had all the cutoff radii equal to 1.3 bohr. For hydrogen a pseudo-potential was also generated, with the same cutoff radii. Calculations were converged to better than 1 mhartree at an energy cutoff of 60 hartree and when using a 20 bohr cubic super-cell

varying densities, it comes at no surprise that highly localized states are not correctly described by it. The GGA does not improve on this result if the band-structure is calculated at the LDA lattice constant. If, however, one uses the predicted GGA lattice constant, then the width of the d-bands comes closer to the experimental values albeit getting even closer to the Fermi level. The GW results presented were calculated at the experimental lattice constant and show a very good agreement with experiment for the positions of the d-bands. Nevertheless, the widths of the bands are more precisely described by a much simpler GGA calculation.

As a last example of the use of plane-wave basis sets, we will look at methane. To deal with finite systems one has to resort to the super-cell technique. As we are using periodic boundary conditions, we will only be able to simulate a finite system if we place it inside a very large cell. If this cell is big enough, the system (molecule, cluster, etc.) will not interact with its periodic images. This means that, besides the usual convergence checks, one has also to check that the calculation converges with increasing cell size. Fortunately, in this case it is sufficient to use the Γ-point for sampling the irreducible wedge of the Brillouin zone.

The calculation of the equilibrium geometry is usually performed by minimizing the total energy using some conjugate-gradients (or more sophisticated) methods. However, for this simple example, we can just vary the CH bond length and plot the total energy. This is shown in Fig. 6.12.

From the energy curve it is also simple to extract the vibrational frequency of the CH bond. Close to the minimum, the energy depends quadratically on

Table 6.7. CH bond length and vibrational frequency (w) of the CH bond of CH_4, calculated with the LDA (Perdew-Wang 92 functional [7]) and GGA (Perdew-Burke-Ernzerhof functional [79])

	LDA	GGA	expt.
CH bond length (bohr)	2.06	2.06	2.04
w (cm^{-1})	3422	3435	2917

the bond length,

$$E \approx E_{eq} + \frac{1}{2} m \omega^2 (r - r_{eq})^2 , \qquad (6.79)$$

where E_{eq} is the total energy at the equilibrium CH bond length (r_{eq}), ω the vibrational frequency, and m is an "effective" mass of the system, which for this specific case reads

$$\frac{1}{m} = \frac{1}{m_C} + \frac{1}{4 m_H} , \qquad (6.80)$$

where m_C and m_H are the masses of the carbon and hydrogen atom, respectively. In Table 6.7 we summarize the results obtained for methane.

The results show that both the LDA and the GGA are over-estimating the CH bond length and the vibrational frequency. These calculations were repeated using a real-space method (see next section).

6.6   Real-Space Calculations

To illustrate the use of real-space methods, we again chose to study methane (CH_4). For all calculations, we used the program octopus [90] (see also http://www.tddft.org/programs/octopus), which was written by some of the authors, and is freely available under an open source license. Furthermore, we employed the Troullier-Martins pseudo-potentials which are distributed with the code, and the GGA in the parameterization of Perdew, Burke and Ernzerhof.

The first step of any calculation is the determination of the grid-spacing that is necessary to converge the energy to the required precision. This study is presented in Fig. 6.13. It is clear that the real-space technique is not variational, because the total energy does not decrease monotonically, but instead oscillates as we reduce the grid-spacing. To have the total energy and the Kohn-Sham eigenvalues converged to better than 0.005 hartree ($\approx 0.1\,eV$) a grid-spacing of at least 0.35 bohr is necessary. This was therefore the grid-spacing we used to obtain the following results. Note that the optimum grid-spacing depends on the strength of the pseudo-potential used: The deeper the pseudo-potential, the tighter the mesh has to be.

The variation of the total energy with the C-H bond length is shown in Fig. 6.14. Remarkably, the calculated equilibrium C-H bond length, r_{eq},

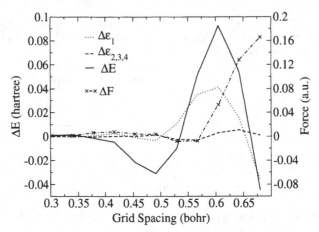

Fig. 6.13. Convergence of energies and forces versus grid-spacing. E is the total energy, F the absolute value of total force on a H atom, ε_1 the Kohn-Sham eigenvalue of the HOMO-1 state, and $\varepsilon_{2,3,4}$ the Kohn-Sham eigenvalue of the HOMO state (which is triply degenerate). For the sake of clarity, we plot the difference between these quantities and their converged values

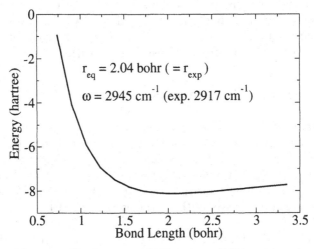

Fig. 6.14. Total energy of CH_4 versus C-H bond length

comes out on-top of the experimental value. The calculated value for the vibrational frequency of the CH bond ($\omega = 2945\,\mathrm{cm}^{-1}$) is slightly above the experimental result ($2917\,\mathrm{cm}^{-1}$), but the agreement is still very good.

For illustrative purposes we depict, in Fig. 6.15, the density and the Kohn-Sham orbitals of CH_4 in its equilibrium configuration. It is clear that very little information can be extracted by looking directly at the density, since it appears to be a very smooth function without any particular point of interest. It is therefore surprising that the density, by itself, is able to determine all observables of the system. The Kohn-Sham eigenfunctions do not have any physical interpretation – they are simply mathematical objects used to obtain the electronic density. However, they do resemble very much to the traditional "molecular orbitals" used in chemistry, and are widely used as such. Note that the last three orbitals, (c), (d) and (e) are degenerate, and that the sum of their partial densities retains the tetrahedral symmetry of CH_4.

To conclude our section on real-space methods we present, in Fig. 6.16, a plot of the so-called "egg-box" effect. As mentioned before, the numerical grid breaks translational symmetry. This implies that the result of the calculation is dependent on where we position the molecule relatively to the grid. As most of the times the grids are uniform, the error will be periodic, with a period equal to the grid spacing. Plotting the error in the total energy as a function of the position of the molecule leads to a curve that resembles an egg-box. This error is inherent to all real-space implementations, but can be systematically reduced by decreasing the grid-spacing. In this particular case, the maximum "egg-box" error is of the order of 2 mhartree, for a grid spacing of 0.35 bohr. Clearly, the magnitude of the error increases for larger grid-spacings and stronger pseudo-potentials. Note that this "egg-box" effect leads to a spurious force term when performing molecular dynamics or geometry minimizations, so special care has to be taken in these cases.

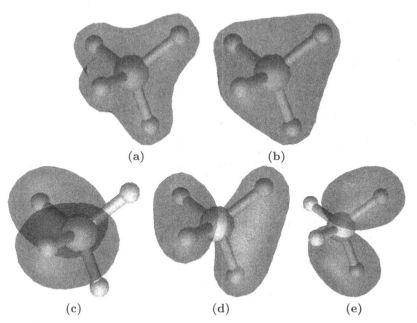

(a)   (b)

(c)   (d)   (e)

Fig. 6.15. Density (a), HOMO-1 (b) and the 3 degenerate HOMO (c, d and e) Kohn-Sham orbitals of CH_4

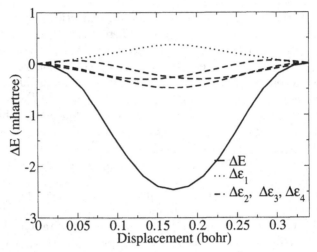

Fig. 6.16. Egg-box effect in CH_4. The x-axis represents the distance of the carbon atom to the central point of the grid. E is the total energy, ε_1 the Kohn-Sham eigenvalue of the HOMO-1 state, and $\varepsilon_{2,3,4}$ the Kohn-Sham eigenvalues of the HOMO state (which is triply degenerate). For the sake of clarity, we plot the difference between these quantities and their values when the carbon atom is located at the central grid-point

References

1. R. G. Parr and W. Yang, *Density-Functional Theory of Atoms and Molecules* (Oxford University Press, New York, 1989).
2. R. M. Dreizler and E. K. U. Gross, *Density Functional Theory: An Approach to the Quantum Many-Body Problem* (Springer-Verlag, Berlin, 1990).
3. The Free Software project for Atomic-scale Simulation aims at spreading the use of free software in atomic-scale simulations (URL http://www.fsatom.org).
4. The ABINIT code is a common project of the Université Catholique de Louvain, Corning Incorporated, the Université de Liège, the Commissariat à l'Energie Atomique, and other contributors (URL http://www.abinit.org/).
5. D. M. Ceperley and B. J. Alder, Phys. Rev. Lett. **45**, 566 (1980).
6. J. P. Perdew and A. Zunger, Phys. Rev. B **23**, 5048 (1981).
7. J. P. Perdew and Y. Wang, Phys. Rev. B **45**, 13244 (1992).
8. J. B. Krieger, Y. Li, and G. J. Iafrate, Phys. Rev. A **45**, 101 (1992).
9. D. G. Anderson, J. Assoc. Comput. Mach. **12**, 547 (1964).
10. D. R. Hamann, Phys. Rev. Lett. **42**, 662 (1979).
11. P. Pulay, Chem. Phys. Lett. **73**, 393 (1980).
12. C. G. Broyden, Math. Comp. **19**, 577 (1965).
13. D. D. Johnson, Phys. Rev. B **38**, 12807 (1988).
14. J. Ihm, A. Zunger, and M. L. Cohen, J. Phys. C: Solid State Phys. **12**, 4409 (1979), (E) *ibid.* **13** (1980) 3095.
15. J. Ihm, Rep. Prog. Phys. **51**, 105 (1988).
16. W. E. Pickett, Comp. Phys. Rep. **9**, 115 (1989).

17. N. Bingelli, J. L. Martins, and J. R. Chelikowsky, Phys. Rev. Lett. **68**, 2956 (1992).
18. M. R. Jarvis, I. D. White, R. W. Godby, and M. C. Payne, Phys. Rev. B **56**, 14972 (1997).
19. F. Nogueira, J. L. Martins, and C. Fiolhais, Eur. Phys. J. D **9**, 229 (2000).
20. A. Baldereschi, Phys. Rev. B **7**, 5212 (1973).
21. D. J. Chadi and M. L. Cohen, Phys. Rev. B **8**, 5747 (1973).
22. D. J. Chadi, Phys. Rev. B **16**, 1746 (1977).
23. H. J. Monkhorst and J. D. Pack, Phys. Rev. B **13**, 5188 (1976).
24. M. T. Yin and M. L. Cohen, Phys. Rev. B **26**, 3259 (1982).
25. T. L. Beck, Rev. Mod. Phys. **72**, 1041 (2000).
26. W. L. Briggs, *A Multigrid Tutorial* (SIAM, Philadelphia, 1987).
27. P. Wesseling, *An Introduction to Multigrid Methods* (Wiley, New York, 1992).
28. E. Fermi, Il Nuovo Cimento **11**, 157 (1934).
29. H. Hellmann, J. Chem. Phys. **3**, 61 (1935).
30. J. C. Phillips and L. Kleinman, Phys. Rev. **116**, 287 (1959).
31. L. Kleinman and J. C. Phillips, Phys. Rev. **118**, 1153 (1960).
32. E. Antončík, J. Phys. Chem. Solids **10**, 314 (1959).
33. M. H. Cohen and V. Heine, Phys. Rev. **122**, 1821 (1961).
34. B. J. Austin, V. Heine, and L. J. Sham, Phys. Rev. **127**, 276 (1962).
35. L. Kleinman and D. M. Bylander, Phys. Rev. Lett. **48**, 1425 (1982).
36. I. V. Abarenkov and V. Heine, Phil. Mag. **XII**, 529 (1965).
37. V. Heine and I. Abarenkov, Phil. Mag. **9**, 451 (1964).
38. A. O. E. Animalu and V. Heine, Phil. Mag. **12**, 1249 (1965).
39. N. W. Ashcroft, Phys. Lett. **23**, 48 (1966).
40. N. W. Ashcroft and D. C. Langreth, Phys. Rev. **155**, 682 (1967).
41. C. Fiolhais, J. P. Perdew, S. Q. Armster, J. M. MacLaren, and M. Brajczewska, Phys. Rev. B **51**, 14001 (1995), (E) *ibid.* **53** (1996) 13193.
42. F. Nogueira, C. Fiolhais, J. He, J. P. Perdew, and A. Rubio, J. Phys.: Condens. Matter **8**, 287 (1996).
43. C. Fiolhais, F. Nogueira, and C. Henriques, Prog. Surf. Sci. **53**, 315 (1996).
44. L. Pollack, J. P. Perdew, J. He, M. Marques, F. Nogueira, and C. Fiolhais, Phys. Rev. B **55**, 15544 (1997).
45. F. Nogueira, C. Fiolhais, and J. P. Perdew, Phys. Rev. B **59**, 2570 (1999).
46. J. P. Perdew, F. Nogueira, and C. Fiolhais, Theochem **9**, 229 (2000).
47. R. W. Shaw, Jr., Phys. Rev **174**, 769 (1968).
48. J. Callaway and P. S. Laghos, Phys. Rev. **187**, 192 (1969).
49. W. C. Topp and J. J. Hopfield, Phys. Rev. B **7**, 1295 (1973).
50. M. L. Cohen and V. Heine, Solid State Phys. **24**, 37 (1970).
51. W. A. Harrison, *Pseudopotentials in the Theory of Metals* (W. A. Benjamin, New York, 1966).
52. V. Heine and D. Weaire, Solid State Phys. **24**, 249 (1970).
53. J. Hafner and V. Heine, J. Phys. F.: Met. Phys. **13**, 2479 (1983).
54. J. Hafner and V. Heine, J. Phys. F: Met. Phys. **16**, 1429 (1986).
55. J. Hafner, *From Hamiltonians to Phase Diagrams* (Springer Verlag, Berlin, 1987).
56. T. Starkloff and J. D. Joannopoulos, Phys. Rev. B **16**, 5212 (1977).
57. A. Zunger and M. L. Cohen, Phys. Rev. B **18**, 5449 (1978).
58. N. Troullier and J. L. Martins, Phys. Rev. B **43**, 1993 (1991).

59. D. R. Hamann, M. Schlüter, and C. Chiang, Phys. Rev. Lett. **43**, 1494 (1979).
60. G. B. Bachelet, D. R. Hamann, and M. Schlüter, Phys. Rev. B **26**, 4199 (1982).
61. D. R. Hamann, Phys. Rev. B **40**, 2980 (1989).
62. N. Troullier and J. L. Martins, Solid State Commun. **74**, 613 (1990).
63. G. P. Kerker, J. Phys. C: Solid State Phys. **13**, L189 (1980).
64. E. L. Shirley, D. C. Allan, R. M. Martin, and J. D. Joannopoulos, Phys. Rev. B **40**, 3652 (1989).
65. A. M. Rappe, K. M. Rabe, E. Kaxiras, and J. D. Joannopoulos, Phys. Rev. B **41**, 1227 (1990).
66. G. Kresse, J. Hafner, and R. J. Needs, J. Phys.: Condens. Matter **4**, 7451 (1992).
67. A. M. Rappe and J. D. Joannopoulos, in *Computer Simulation in Materials Science*, edited by M. Meyer and V. Pontikis (Kluwer Academic Publishers, Dordrecht, 1991), pp. 409–422.
68. C. Hartwigsen, S. Goedecker, and J. Hutter, Phys. Rev. B **58**, 3641 (1998).
69. D. Vanderbilt, Phys. Rev. B **41**, 7892 (1990).
70. S. G. Louie, S. Froyen, and M. L. Cohen, Phys. Rev. B **26**, 1738 (1982).
71. M. Fuchs, M. Bockstedte, E. Pehlke, and M. Scheffler, Phys. Rev. B **57**, 2134 (1998).
72. R. W. Shaw, Jr. and W. A. Harrison, Phys. Rev. **163**, 604 (1967).
73. S. Goedecker and K. Maschke, Phys. Rev. A **45**, 88 (1992).
74. M. Teter, Phys. Rev. B **48**, 5031 (1993).
75. X. Gonze, P. Käckell, and M. Scheffler, Phys. Rev. B **41**, 12264 (1990).
76. X. Gonze, R. Stumpf, and M. Scheffler, Phys. Rev. B **44**, 8503 (1991).
77. URL http://bohr.inesc.pt/~jlm/pseudo.html/.
78. M. W. Schmidt, K. K. Baldridge, J. A. Boatz, S. T. Elbert, M. S. Gordon, J. J. Jensen, S. Koseki, N. Matsunaga, K. A. Nguyen, S. Su, T. L. Windus, M. Dupuis, and J. A. Montgomery, J. Comput. Chem. **14**, 1347 (1993).
79. J. P. Perdew, K. Burke, and M. Ernzerhof, Phys. Rev. Lett. **77**, 3865 (1996).
80. M. A. L. Marques, A. Castro, and A. Rubio, J. Chem. Phys. **115**, 3006 (2001).
81. C. Chang, M. Pélissier, and P. Durand, Phys. Scr. **34**, 394 (1986).
82. J. L. Heully, I. Lindgren, E. Lindroth, S. Lundqvist, and A. M. Mårtensson-Pendrill, J. Phys. B **19**, 2799 (1986).
83. P. Pulay, Mol. Phys. **17**, 197 (1969).
84. A. B. Alchagirov, J. P. Perdew, J. C. Boettger, R. C. Albers, and C. Fiolhais, Phys. Rev. B **63**, 224115 (2001).
85. T. J. Lenosky, J. D. Kress, I. Kwon, A. F. Voter, B. Edwards, D. F. Richards, S. Yang, and J. B. Adams, Phys. Rev. B **55**, 1528 (1997).
86. M. Städele, M. Moukara, J. A. Majewski, P. Vogl, and A. Görling, Phys. Rev. B **59**, 10031 (1999).
87. J. H. Rose, J. R. Smith, F. Guinea, and J. Ferrante, Phys. Rev. B **29**, 2963 (1984).
88. A. Marini, G. Onida, and R. D. Sole, Phys. Rev. Lett. **88**, 016403 (2002).
89. R. Courths and S. Hüfner, Phys. Rep. **112**, 53 (1984).
90. M. A. L. Marques, A. Castro, G. F. Bertsch, and A. Rubio, Comput. Phys. Commun. **151**, 60 (2003).

Lecture Notes in Physics

For information about Vols. 1–577
please contact your bookseller or Springer-Verlag
LNP Online archive: http://www.springerlink.com/series/lnp/